£12·30

DEVICES:
Discrete and Integrated

DEVICES:

Louis Nashelsky
Queensborough Community College

Robert Boylestad
Thayer School, Dartmouth College

Discrete
and Integrated

Prentice-Hall, Inc., Englewood Cliffs, New Jersey 07632

Library of Congress Cataloging in Publication Data

Nashelsky, Louis.
   Devices: discrete and integrated.

   Includes index.
   1. Semiconductors. 2. Integrated circuits.
I. Boylestad, Robert L., joint author. II. Title.
TK7871.85.N363   621.3815'2   80-19403
ISBN 0-13-208165-2

DEVICES: DISCRETE AND INTEGRATED
Louis Nashelsky/Robert Boylestad

Editorial/production supervision by
   Gary Samartino and Ros Herion
Interior design by Mark Binn and
   Gary Samartino
Page layout by J. E. Markus
Cover design by Mark Binn
Manufacturing buyer: Joyce Levatino

The cover art, Cosmograph #196, is used
with the permission of Photo Lettering, Inc.

© 1981 by PRENTICE-HALL, INC., Englewood Cliffs, New Jersey 07632

All rights reserved. No part of this book may be
reproduced in any form, or by any means,
without permission in writing from the publisher.

10 9 8 7 6 5 4 3 2 1

Printed in the United States of America

PRENTICE-HALL INTERNATIONAL, INC., *London*
PRENTICE-HALL OF AUSTRALIA, PTY. LTD., *Syndey*
PRENTICE-HALL OF CANADA, LTD., *Toronto*
PRENTICE-HALL OF INDIA PRIVATE LIMITED, *New Delhi*
PRENTICE-HALL OF JAPAN, INC., *Tokyo*
PRENTICE-HALL OF SOUTHEAST ASIA PTE. LTD., *Singapore*
WHITEHALL BOOKS LIMITED, *Wellington, New Zealand*

D
621.3815'2
NAS

Dedicated to
Walter and Renee Barth
and
Harald and Johanna Olufsen

Contents

PREFACE *xi*

Part I: DISCRETE

1. SEMICONDUCTOR THEORY

    1.1 Introduction *3*
    1.2 General Characteristics *4*
    1.3 Energy Levels *7*
    1.4 Extrinsic Materials—n- and p-Type *9*
    1.5 Drift and Diffusion Currents *12*
    1.6 Manufacturing Techniques *13*

2. DIODES

    2.1 Introduction *20*
    2.2 Ideal Diode *20*
    2.3 Basic Construction and Characteristics *23*
    2.4 Transition and Diffusion Capacitance *30*
    2.5 Reverse Recovery Time *31*
    2.6 Temperature Effects *32*
    2.7 Diode Specification Sheets *37*
    2.8 Semiconductor Diode Notation *41*
    2.9 Diode Ohmmeter Check *42*

- 2.10 Semiconductor Diode Fabrication *42*
- 2.11 Diode Arrays—Integrated Circuits *46*
- 2.12 Dc Conditions *49*
- 2.13 Static Resistance *51*
- 2.14 Dynamic Resistance *52*
- 2.15 Average Ac Resistance *55*
- 2.16 Equivalent Circuits *56*
- 2.17 Clippers and Clampers *62*

3. ZENER AND OTHER DEVICES

- 3.1 Introduction *76*
- 3.2 Zener Diodes *76*
- 3.3 Schottky Barrier (Hot-Carrier) Diode *84*
- 3.4 Varicap (Varactor) Diodes *88*
- 3.5 Power Diodes *93*
- 3.6 Tunnel Diodes *94*
- 3.7 Photodiode *97*
- 3.8 Photoconductive Cell *101*
- 3.9 IR Emitters *103*
- 3.10 Light-Emitting Diodes *105*
- 3.11 Liquid Crystal Displays *110*
- 3.12 Solar Cells *114*
- 3.13 Thermistors *119*

4. BIPOLAR TRANSISTORS

- 4.1 Introduction *128*
- 4.2 Transistor Construction *129*
- 4.3 Transistor Operation *130*
- 4.4 Transistor Amplifying Action *132*
- 4.5 Common-Base Configuration *133*
- 4.6 Common-Emitter Configuration *136*
- 4.7 Common-Collector Configuration *142*
- 4.8 Transistor Maximum Ratings *144*
- 4.9 Transistor Specification Sheet *145*
- 4.10 Transistor Fabrication *151*
- 4.11 Transistor Casing and Terminal Identification *153*
- 4.12 Transistor Testing *155*

5. DC AND AC BJT ANALYSIS

- 5.1 Introduction *161*
- 5.2 Operating Point *161*
- 5.3 Common-Base Bias Circuit *163*

- 5.4 Common-Emitter Circuit Connection *166*
- 5.5 Dc Bias Circuit With Emitter Resistor *169*
- 5.6 Dc Bias Circuit Independent of Beta *171*
- 5.7 Common-Collector (Emitter-Follower) Dc Bias Circuit *173*
- 5.8 Common-Emitter Configuration—Graphical Analysis *176*
- 5.9 Proper Biasing Insurance *178*
- 5.10 Small-Signal (Ac) Analysis *179*
- 5.11 Transistor Hybrid Equivalent Circuit *182*
- 5.12 Variations of Transistor Parameters *187*
- 5.13 Small-Signal Analysis of the Basic Transistor Amplifier Using the Hybrid Equivalent Circuit *189*
- 5.14 Approximate Base, Collector, and Emitter Equivalent Circuits *198*
- 5.15 An Alternative Approach *204*
- 5.16 Summary Table *216*
- 5.17 Darlington Compound Configuration *217*

6. FIELD-EFFECT TRANSISTORS

- 6.1 Introduction *237*
- 6.2 JFET Construction *237*
- 6.3 JFET Characteristics *239*
- 6.4 JFET Operation *245*
- 6.5 MOSFET Construction and Characteristics *252*
- 6.6 MOSFET Operation *256*

7. *PNPN* AND MISCELLANEOUS DEVICES

- 7.1 Introduction *264*
- 7.2 Silicon-Controlled Rectifier (SCR) *264*
- 7.3 Basic Silicon-Controlled Rectifier (SCR) Operation *265*
- 7.4 SCR Characteristics and Ratings *268*
- 7.5 SCR Construction and Terminal Identification *271*
- 7.6 SCR Applications *272*
- 7.7 Silicon-Controlled Switch *275*
- 7.8 Gate Turn-Off Switch *279*
- 7.9 Light-Activated SCR *280*
- 7.10 Shockley Diode *284*
- 7.11 Diac *284*
- 7.12 Triac *287*
- 7.13 Unijunction Transistor *291*
- 7.14 V-FET *296*
- 7.15 Phototransistors *297*
- 7.16 Opto-Isolators *300*
- 7.17 Programmable Unijunction Transistor *303*

Part II: INTEGRATED

8. INTEGRATED CIRCUIT FABRICATION

8.1 Introduction *315*
8.2 Planar Process *316*
8.3 Monolithic Circuit Elements *318*
8.4 Masks *322*
8.5 Monolithic Integrated Circuit—The NAND Gate *323*
8.6 Thin and Thick Film Integrated Circuits *332*
8.7 Hybrid Integrated Circuits *333*

9. LINEAR ICS

9.1 Differential Amplifier Basics *336*
9.2 Op-amp IC Basics *341*
9.3 Op-amp Operation and Applications *346*
9.4 Voltage Regulators *351*

10. DIGITAL IC UNITS

10.1 Digital Fundamentals *358*
10.2 Integrated-Circuit (IC) Logic Devices *363*
10.3 Bistable Multivibrator Circuits *368*
10.4 Digital IC Units *376*
10.5 Digital Memory Units *381*

11. LINEAR/DIGITAL IC UNITS

11.1 Introduction *387*
11.2 Comparators *387*
11.3 Digital/Analog Converters *395*
11.4 Interfacing *399*
11.5 Timers *404*

**APPENDIX A: Hybrid Parameters—
Conversion Equations (Exact and Approximate)** *410*

APPENDIX B: Charts and Tables *413*

ANSWERS TO SELECTED ODD-NUMBERED PROBLEMS *417*

INDEX *420*

Preface

The text provides an introduction to electronic devices—those used in discrete or single component circuits and integrated circuits. The presentation is at a level suitable for an engineering technology or electrical engineering curriculum.

It is expected that the student will have a fundamental background in algebraic manipulations and dc and ac circuit analysis. The text introduces the basic electronic devices and their operation with numerous examples of applications. The introductory chapters deal with the majority of the basic electronic devices available today, their theory of operation, typical device characteristics, device specifications, and a number of typical applications. The later chapters cover basic integrated electronic circuits—considered the basic device in many electronic applications. An introduction to these basic IC units and typical applications is provided.

Chapter 1 provides coverage of fundamental semiconductor theory with a description of p- and n-type materials and the p-n junction. A clear understanding of the theory of a p-n junction device is basic to the understanding of discrete and integrated devices in the text. Chapter 2 introduces the basic p-n junction diode device—its construction, characteristics, and operation. After covering the essential features of the diode, its manufacture, specifications, and important operating features are introduced. These include dc operation, static resistance, ac operation, dynamic resistance, and average ac resistance. The diode equivalent circuit is described, and basic applications of diodes to clipper and clamper circuits are presented. The text continues with coverage of the wide variety of two terminal solid-state devices in Chapter 3. This material is suitable for selected coverage by the instructor.

The basic three-terminal bipolar junction transistor (BJT) device is introduced in Chapter 4; its construction, operation, ratings, and connection in basic circuit configurations are covered. This most popular discrete amplifying device should be covered in some detail; hence, Chapter 5 provides a comprehensive coverage of the operation and analysis of BJTs and their use in the various basic circuit configurations—common-base, common-collector, and common-emitter. Analysis of these BJT circuits is provided using both the more classic hybrid parameters and the alternate practical approach using the pi-model with the BJT device defined by its current gain and emitter resistance. Although analysis using either device model provides the same solution (given the proper parameter values) and direct conversion of device parameters from one form to the other can be made (as described in the chapter), various circuits are more easily understood and explained using one or the other device equivalent representation. In addition, instructors often prefer one or the other device representation for analyzing various amplifier circuits. Sections 5.1 through 5.10 would be used in either case. Sections 5.11 through 5.14 concentrate on the hybrid representation, and Section 5.15 demonstrates the alternate device model in analyzing BJT circuits. Chapter 6 then covers the equally important solid-state device, the field-effect transistor (FET). The construction and characteristics of both junction FET (JFET) and metal-oxide semiconductor FET (MOSFET) are described. Discrete circuit applications of each transistor type are also provided for common amplifier connections. A large variety of two-, three-, and four-terminal semiconductor devices are covered in Chapter 7. The instructor may select various topics for class coverage and may also assign many of the chapter sections as home study assignments.

The first seven chapters cover circuit applications associated mainly with discrete circuits—those built using individual transistors, resistors, diode and capacitor components. Chapters 8 through 11 extend these devices to their use in integrated circuits—those built with all components formed at the same time. Chapter 8 describes the basic process used to manufacture semiconductor devices and integrated circuits.

The basic linear IC units are covered in Chapter 9, starting with the difference amplifier configuration. Operational amplifier (op-amp) circuits using difference amplifier stages are then introduced—the op-amp being the basic linear IC device for a large variety of applications. A number of these op-amp applications are presented using the virtual ground technique of circuit analysis. The chapter concludes with an introduction to voltage regulator IC units—a widely used linear IC device.

Chapter 10 covers a range of digital IC units, including logic gates, multivibrator circuits, and their application in larger digital units. The chapter covers those units that combine linear and digital IC applications such as comparator units, digital-analog converter units, devices to interface various signals used commonly in the digital area, and timer circuits that contain both linear and digital circuits in a single IC package.

The text contains numerous problems to provide the instructor and student with a methodology for studying and emphasizing the main points of each chapter. In addition, a glossary is provided at the end of each chapter to allow the student to refresh his mind regarding the knowledge acquired in that chapter and provide a reference to the main areas and topics covered.

We wish to express our gratitude to Dave Boelio of Prentice-Hall for his interest, guidance, and support in writing this text. He has remained a driving force since its inception that is deeply appreciated by us both. Thanks also are due for the typing efforts of Mrs. Alice Donnelly, Mrs. Marie Kielly, and Mrs. Kathryn Perry and for the many hours spent by Mr. David Caswell and Mr. Vincent Derrick in reviewing the problem selection.

LOUIS NASHELSKY, *Great Neck, N.Y.*
ROBERT BOYLESTAD, *Hanover, N.H.*

DEVICES:
Discrete and Integrated

PART I

Discrete

CHAPTER 1

Semiconductor Theory

1.1 INTRODUCTION

The few decades following the introduction of the semiconductor transistor in the late 1940s have seen a very dramatic change in the electronics industry. The miniaturization that has resulted leaves us to wonder about its limits. Complete systems now appear on a wafer thousands of times smaller than a single element in earlier networks.

The advantages associated with semiconductor systems as compared to the tube networks of prior years are, for the most part, immediately obvious: smaller and lightweight, no heater requirement or heater loss (as required for tubes), more rugged construction, more efficient, and not requiring a warm-up period. The tube still has a few isolated areas of application (very high power and high frequencies) but the use of tubes is becoming increasingly smaller in scope.

The miniaturization of recent years has resulted in semiconductor systems so small that the primary purpose of the container is simply to provide some means of handling the device and ensuring that the leads remain properly fixed to the semiconductor wafer. The limits of miniaturization appear to be limited by three factors: the quality of the semiconductor material itself, the network design technique, and the limits of the manufacturing and processing equipment.

In this chapter we examine the first of the criteria mentioned above: the semiconductor materials. We look at their general characteristics and the manufacturing techniques. The design aspect and limits of current technology are not to be considered in depth in this book. Our purpose here is simply to introduce the characteristics, purpose, operating mode, and areas of application of those devices presently receiving the greatest attention in the electronics industry.

1.2 GENERAL CHARACTERISTICS

The label *semiconductor* itself provides a hint as to its characteristics. The prefix *semi* is normally applied to anything midway between two limits. The term *conductor* is applied to any material that will permit a generous flow of charge due to the application of a limited amount of external pressure. A semiconductor, therefore, is a material that has a conductivity level somewhere between the extremes of an insulator (very low conductivity) and a conductor, such as copper, which has a high level of conductivity. Inversely related to the conductivity of a material is its resistance to the flow of charge or current. That is, the higher the conductivity level, the lower the resistance level. In tables, the term *resistivity* (ρ, Greek letter rho) is often used when comparing the resistance levels of materials. The resistivity of a material can be examined by noting the resistance of a sample having a length of 1 cm and a cross-sectional area of 1 cm², as shown in Fig. 1.1. Recall that the equation for the resistance of a material (at a particular temperature) is determined by $R = \rho l/A$, where R is measured in ohms, l is the length of the sample, A is its incident surface area, and ρ is the resistivity. If $l = 1$ cm and $A = 1$ cm², then $R = \rho$, as indicated above. The units for ρ as defined by the equation are as follows:

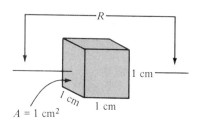

Figure 1.1

$$\rho = \frac{RA}{l} \Rightarrow \frac{\Omega \text{ cm}^2}{\text{cm}} = \boxed{\Omega \text{ cm}} \tag{1.1}$$

Please be assured that this book is not heavily bent toward mathematical developments and complex algebraic techniques. The authors feel that a clear understanding of dimensions is an absolute necessity for proper development of an engineering background. We would assume that the somewhat abstract unit of measure for resistivity now has a measure of understanding and clarity. Incidentally, the actual resistance of a semiconductor material as measured above is called its *bulk* resistance. The resistance introduced by connecting the leads to the bulk material is called the *ohmic contact* resistance. These terms will appear in the description of devices to be introduced throughout.

In Table 1.1, typical resistivity values are provided for three broad categories of materials.

TABLE 1.1 Typical Resistivity Values (At 300°K—Room Temperature)

Conductor	Semiconductor	Insulator
$\rho \cong 10^{-6}$ Ω cm (copper)	$\rho \cong 50$ Ω cm (germanium) $\rho \cong 50 \times 10^3$ Ω cm (silicon)	$\rho \cong 10^{12}$ Ω cm (mica)

Although you may be familiar with the electrical properties of copper and mica from your past studies, the characteristics of the semiconductor materials of germanium (Ge) and silicon (Si) may be relatively new. As you will find in the chapters to follow, they are certainly not the only two semiconductor materials. They are, however, the two materials that have received the broadest range of interest in the development of semiconductor devices. In recent years the shift has been steadily toward silicon and away from germanium, but germanium is still in modest production.

Note in Table 1.1 the extreme range between the conductor and insulating materials for the 1-cm length of the material. Eighteen places separate the placement of the decimal point for one number from the other. Ge and Si have received the attention they have for a number of reasons. One very important consideration is the fact that they can be manufactured to a very high purity level. In fact, recent advances have reduced impurity levels in the pure material to 1 part in 10 billion (1:10,000,000,000). One might ask if these low impurity levels are really necessary. They certainly are if you consider that the addition of one part impurity (of the proper type) per million in a wafer of silicon material can change that material from a relatively poor conductor to a good conductor of electricity. We are obviously dealing with a whole new spectrum of comparison levels when we deal with the semiconductor medium. The ability to change the characteristics of the material significantly through this process, which is known as "doping," is yet another reason why Ge and Si have received such wide attention. Further reasons include the fact that their characteristics can be altered significantly through the application of heat or light—an important consideration in the development of heat- and light-sensitive devices.

Some of the unique qualities of Ge and Si noted above are due to their atomic structure. The atoms of both materials form a very definite pattern that is periodic in nature (i.e., continually repeats itself). One complete pattern is called a *crystal* and the periodic arrangement of the atoms a *lattice*. For Ge and Si the crystal has the three-dimensional diamond structure of Fig. 1.2. Any material composed solely of repeating crystal structures of the same kind is called a *single-crystal* structure. For semiconductor materials of practical application in the electronics field, this single-crystal feature exists, and, in addition, the periodicity of the structure does not change significantly with the addition of impurities in the doping process.

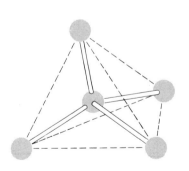

Figure 1.2 Ge and Si single crystal structure.

Let us now examine the structure of the atom itself and note how it might affect the electrical characteristics of the material. As you are aware, the atom is composed of three basic particles: the *electron*, the *proton*, and the *neutron*. In the atomic lattice, the neutrons and protons form the *nucleus,* while the electrons revolve around the nucleus in a fixed *orbit*. The Bohr model of two most commonly used semiconductors, *germanium* and *silicon,* are shown in Fig. 1.3.

As indicated by Fig. 1.3a, the germanium atom has 32 orbiting electrons, while silicon has 14 oribiting electrons. In each case, there are 4 electrons in the outermost

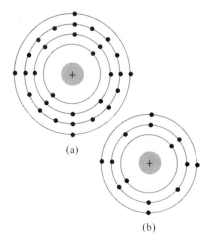
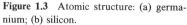
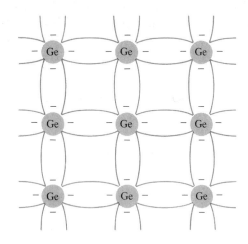

Figure 1.3 Atomic structure: (a) germanium; (b) silicon.

Figure 1.4 Covalent bonding of the germanium atom.

(valence) shell. The potential *(ionization potential)* required to remove any one of these 4 valence electrons is lower than that required for any other electron in the structure. In a pure germanium or silicon crystal these 4 valence electrons are bonded to 4 adjoining atoms, as shown in Fig. 1.4 for germanium. Both Ge and Si are referred to as *tetravalent* atoms because they each have four valence electrons.

This type of bonding, formed by *sharing* electrons, is called *covalent bonding*. Although the covalent bond will result in a stronger bond between the valence electrons and their parent atom, it is still possible for the valence electrons to absorb sufficient kinetic energy from natural causes to break the covalent bond and assume the "free" state. These natural causes include effects such as light energy in the form of photons and thermal energy from the surrounding medium. At room temperature there are approximately 1.5×10^{10} free carriers in a cubic centimeter of intrinsic silicon material. *Intrinsic* materials are those semiconductors that have been carefully refined to reduce the impurities to a very low level—essentially as pure as can be made available through modern technology. The free electrons in the material due only to natural causes are referred to as *intrinsic carriers*. At the same temperature, intrinsic germanium material will have approximately 2.5×10^{13} free carriers per cubic centimeter. The ratio of the number of carriers in germanium to that of silicon is greater than 10^3 and would indicate that germanium is a much better conductor at room temperature. This may be true, but both are still considered poor conductors in the intrinsic state. Note in Table 1.1 that the resistivity also differs by a ratio of about 1000:1, with silicon having the larger value. This should be the case of course, since resistivity and conductivity are inversely related.

A change in the temperature of a semiconductor material can increase the number of free electrons quite substantially. As the temperature rises from absolute zero (0°K), an increasing number of valence electrons absorb sufficient thermal energy to break the covalent bond and contribute to the number of free carriers as described above. This increased number of carriers will increase the conductivity index and result in a lower resistance level. Semiconductor materials such as Ge and Si that show a reduction in resistance with increase in temperature are said to have a *negative temperature coefficient*. You will probably recall that the resistance of most conductors will increase with temperature. This is due to the fact that the numbers of carriers in a conductor will not increase significantly with temperature, but their vibration pattern above a relatively fixed location will make it increasingly difficult for electrons to pass through. An increase in temperature therefore results in an increased resistance level and a *positive temperature coefficient*.

1.3 ENERGY LEVELS

In the isolated atomic structure there are discrete (individual) energy levels associated with each orbiting electron as shown in Fig. 1.5a. Each material will, in fact, have its own set of permissible energy levels for the electrons in its atomic structure. The more distant the electron from the nucleus, the higher the energy

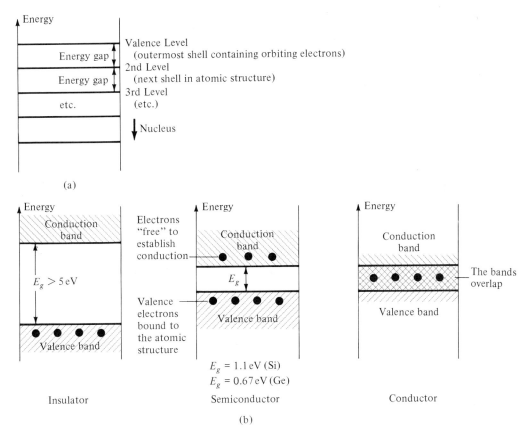

Figure 1.5 Energy levels: (a) discrete levels in isolated atomic structures; (b) conduction and valence bands of an insulator, semiconductor, and conductor.

SEC. 1.3 ENERGY LEVELS

state, and any electron that has left its parent atom has a higher energy state than any electron in the atomic structure. Between the discrete energy levels are gaps in which no electrons in the isolated atomic structure can appear. As the atoms of a material are brought closer together to form the crystal lattice structure, there is an interaction between atoms that will result in the electrons in a particular orbit of one atom having slightly different energy levels from electrons in the same orbit of an adjoining atom. The net result is an expansion of the discrete levels of possible energy states for the valence electrons to that of bands as shown in Fig. 1.5b. Note that there are still boundary levels and maximum energy states in which any electron in the atomic lattice can find itself, and there remains a *forbidden* region between the valence band and the ionization level. Recall that ionization is the mechanism whereby an electron can absorb sufficient energy to break away from the atomic structure and join the "free" carriers in the conduction band. You will note that energy is measured in *electron volts* (eV). The unit of measure is appropriate, since

$$W(\text{energy}) = P(\text{power}) \cdot t(\text{time})$$

but
$$P = VI$$

resulting in
$$W = VIt$$

but
$$I = \frac{Q}{t} \quad \text{or} \quad Q = It$$

and
$$\boxed{W = QV} \quad \text{joules} \quad (1.2)$$

Substituting the charge of an electron and a potential difference of 1 volt into Eq. (1.2) will result in an energy level referred to as one *electron volt*. Since energy is also measured in joules and the charge of one electron = 1.6×10^{-19} coulomb,

$$W = QV = (1.6 \times 10^{-19} \text{ C})(1 \text{ V})$$

and
$$\boxed{1 \text{ eV} = 1.6 \times 10^{-19} \text{ J}} \quad (1.3)$$

The small unit of measure is required to avoid reference to very small numbers in the discussion to follow.

At 0°K or absolute zero, all the valence electrons of the semiconductor materials are in the valence bands. However, at room temperature (300°K) a large number of electrons have acquired sufficient energy to enter the conduction band, that is, to bridge the 1.1-eV gap for silicon and 0.67-eV for germanium. The obviously lower E_g for germanium accounts for the increased number of carriers in that material as compared to silicon at room temperature. Note for the insulator that the energy gap is typically 5 eV or more. Very few electrons can acquire the required energy at room temperature, with the result that the material remains an insulator. The conductor has electrons in the conduction band even at 0°K. Quite obviously, therefore, at room temperature there are more than enough free carriers to sustain a heavy flow of charge or current.

We see in Section 1.4 that if certain impurities are added to the intrinsic semiconductor materials, the result will be permissible energy states in the forbidden band and a net reduction in E_g for both semiconductor materials—consequently increased carrier density in the conduction band at room temperature!

1.4 EXTRINSIC MATERIALS—*n*- AND *p*-TYPE

The characteristics of semiconductor materials can be altered significantly by the addition of certain impurity atoms into the relatively pure semiconductor material. These impurities, although only added to perhaps 1 part in 10 million, can alter the band structure sufficiently to totally change the electrical properties of the material. A semiconductor material that has been subjected to this *doping* process is called an *extrinsic* material. There are two extrinsic materials of immeasurable importance to semiconductor device fabrication; *n*-type and *p*-type. Each will be described in some detail in the following paragraphs.

n-Type Material

Both the *n*- and *p*-type materials are formed by adding a predetermined number of impurity atoms into a germanium or silicon base. The *n*-type is created by adding those impurity elements that have *five* valence electrons *(pentavalent)*, such as *antimony, arsenic,* and *phosphorus*. The effect of such impurity elements is indicated in Fig. 1.6 (using antimony as the impurity in a germanium base). Note that the four covalent bonds are still present. There is, however, an additional fifth electron due to the impurity atom, which is *unassociated* with any particular covalent bond. This remaining electron, loosely bound to its parent (antimony) atom, is relatively free to move within the newly formed *n*-type material. Since the inserted impurity atom has donated a relatively "free" electron to the structure, impurities with five valence electrons are called *donor* atoms. It is important to realize that even though a large number of "free" carriers have been established in the *n*-type material, it is still electrically *neutral* since ideally the number of positively charged protons in the nuclei is still equal to the number of "free" and orbiting negatively charged electrons in the structure.

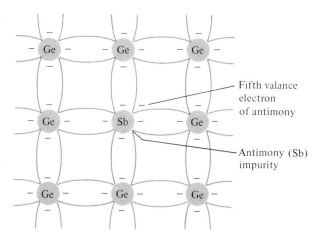

Figure 1.6 Antimony impurity in *n*-type material.

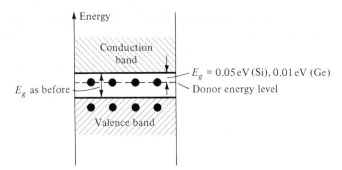

Figure 1.7 Effect of donor impurities on the energy band structure.

The effect of this doping process on the relative conductivity can best be described through the use of the energy-band diagram of Fig. 1.7. Note that a discrete energy level (called the *donor* level) appears in the forbidden band with an E_g significantly less than that of the intrinsic material. Those "free" electrons due to the added impurity sit at this energy level and have absolutely no difficulty absorbing a sufficient measure of thermal energy to move into the conduction band at room temperature. The result is that at room temperature, there are a large number of carriers (electrons) in the conduction level and the conductivity of the material increases significantly. At room temperature in an intrinsic Si material there is about one free electron for every 10^{12} atoms (1 to 10^9 for Ge). If our dosage level were 1 in 10 million (10^7), the ratio ($10^{12}/10^7 = 10^5$) would indicate that the carrier concentration has increased by a ratio of 100,000:1.

p-Type Material

The *p*-type material is formed by doping a pure germanium or silicon crystal with impurity atoms having *three* valence electrons. The elements most frequently used for this purpose are *boron, gallium,* and *indium.* The effect of one of these elements, boron, on a base of germanium is indicated in Fig. 1.8.

Note that there is now an insufficient number of electrons to complete the covalent bonds of the newly formed lattice. The resulting vacancy is called a *hole* and is represented by a small circle or positive sign due to the absence of a

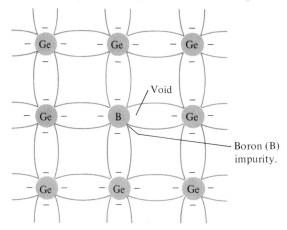

Figure 1.8 Boron impurity in *p*-type material.

negative charge. Since the resulting vacancy will readily *accept* a "free" electron, the impurities added are called *acceptor* atoms. The resulting *p*-type material is electrically neutral, for the same reasons as for the *n*-type material.

The effect of the hole on conduction is shown in Fig. 1.9. If a valence electron acquires sufficient kinetic energy to break its covalent bond and fills the void created by a hole, then a vacancy, or hole, will be created in the covalent bond that released the electron. There is therefore a transfer of holes to the left and electrons to the right, as shown in Fig. 1.9. The direction to be used in this text is that of *conventional* flow, which is indicated by the direction of hole flow.

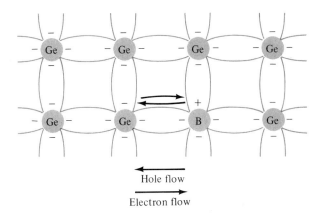

Figure 1.9 Electron vs. hole flow.

In the intrinsic state, the number of free electrons in Ge or Si is due only to those few electrons in the valence band that have acquired sufficient energy from thermal or light sources to break the covalent bond or to the few impurities that could not be removed. The vacancies left behind in the covalent bonding structure represent our very limited supply of holes. In an *n*-type material, the number of holes has not changed significantly from this intrinsic level. The net result, therefore, is that the number of electrons far outweigh the number of holes. For this reason the electron is called the *majority carrier* and the hole the *minority carrier*, as shown in Fig. 1.10a. Note in Fig. 1.10b that the reverse is true for the *p*-type material. When the fifth electron of a donor atom leaves the parent atom, the atom remaining acquires a net positive charge: hence the positive sign in the donor-

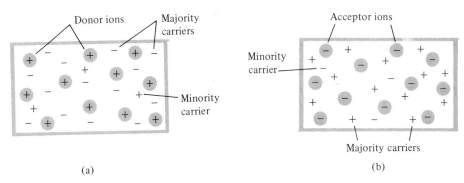

Figure 1.10 (a) *n*-type material; (b) *p*-type material.

SEC. 1.4 EXTRINSIC MATERIALS—n- AND p-TYPE

ion representation. For similar reasons, the negative sign appears in the acceptor ion.

The *n*- and *p*-type materials represent the basic building blocks of semiconductor devices. We will find in Chapter 2 that the joining of a single *n*-type material with a *p*-type material will result in a semiconductor element of considerable importance in electronic systems.

1.5 DRIFT AND DIFFUSION CURRENTS

The flow of charge or current through a semiconductor material is normally referred to as one of two types: drift and diffusion. *Drift current* relates directly to the mechanism encountered in the flow of charge in a conductor. When an emf is applied across the material as shown in Fig. 1.11, the electrons are naturally drawn to the positive end of the sample. However, collisions with the other atoms, ions, and carriers encountered in their movement may result in an erratic path, as shown in the figure. The net result, however, is a drift of carriers to the positive end.

The concept of *diffusion current* is best described by considering the effect of placing a drop of dye into a clear pool of water. The heavy concentration of dye will eventually diffuse through the clear water. The darker color of the heavy concentration of dye will give way to a much lighter shade as it spreads out through the liquid. The same effect will take place in a semiconductor material

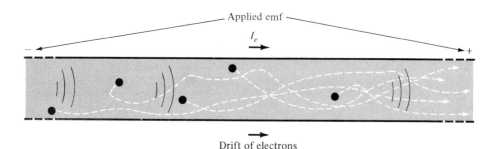

Figure 1.11 Drift current.

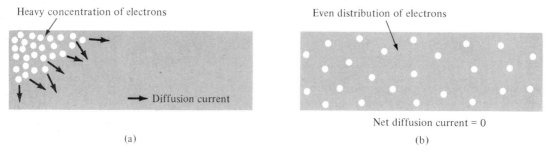

Figure 1.12 Diffusion current: (a) heavy introduction of carriers in a region of the semiconductor material; (b) the steady-state condition.

if a heavy concentration of carriers is introduced to a region as shown in Fig. 1.12a. In time they will distribute themselves evenly through the material, as shown in Fig. 1.12b. This movement is due only to interaction between neighboring atoms; there is no applied source of energy as is required for drift current. Diffusion current is an important consideration in the examination of minority-carrier flow in *n*- and *p*-type materials. The diffusion phenomenon is also important as a technology for the doping process and must be carefully investigated when models are constructed for semiconductor devices (diffusion capacitance, etc.).

1.6 MANUFACTURING TECHNIQUES

The first step in the manufacture of any semiconductor device is to obtain semiconductor materials, such as germanium or silicon, of the desired purity level. Impurity levels of *less* than *one* part in *one billion* (1 in 1,000,000,000) are required for most semiconductor fabrications today. The basic processes involved in the production of semiconductor materials with this low level of impurities are indicated in Fig. 1.13.

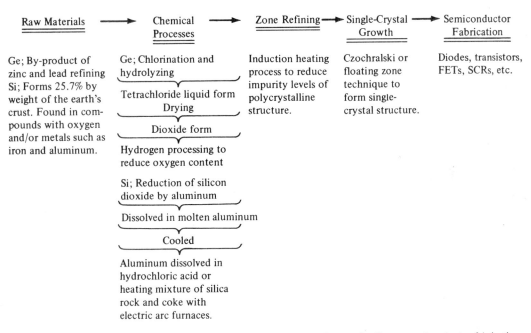

Figure 1.13 Sequence of events leading to semiconductor fabrication.

As indicated in Fig. 1.13, the raw materials are first subjected to a series of chemical reactions and a zone refining process to form a polycrystalline crystal of the desired purity level. The atoms of a polycrystalline crystal are haphazardly arranged, while in the single crystal desired the atoms are arranged in a symmetrical, uniform, geometrical lattice structure.

Zone refining apparatus is shown in Fig. 1.14. It consists of a graphite or quartz boat for minimum contamination, a quartz container, and a set of RF (radio-frequency) induction coils. Either the coils or boat must be movable along the length of the quartz container. The same result will be obtained in either case, although moving coils are discussed here since it appears to be the more popular method. The interior of the quartz container is filled with either an inert (little or no chemical reaction) gas, or vacuum, to reduce further the chance of contamination. In the zone refining process, a bar of germanium is placed in the boat with the coils at one end of the bar as shown in Fig. 1.14. The radio-frequency signal is then applied to the coil, which will induce a flow of charge (eddy currents) in the germanium ingot. The magnitude of these currents is increased until sufficient heat is developed to melt that region of the semiconductor material. The impurities in the ingot will enter a more liquid state than the surrounding semiconductor material. If the induction coils of Fig. 1.14 are now slowly moved to the right to induce melting in the neighboring region, the "more fluidic" impurities will "follow" the molten region. The net result is that a large percentage of the impurities will appear at the right end of the ingot when the induction coils have reached this end. This end piece of impurities can then be cut off and the entire process repeated until the desired purity level is reached.

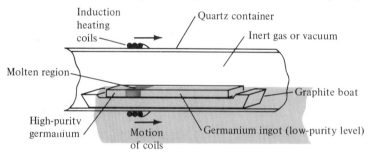

Figure 1.14 Zone refining process.

The final operation before semiconductor fabrication can take place is the formation of a single crystal of germanium or silicon. This can be accomplished using either the *Czochralski* or the *floating zone* technique, the latter being the more recently devised. The apparatus employed in the Czochralski technique is shown in Fig. 1.15a. The polycrystalline material is first transformed to the molten state by the RF induction coils. A single-crystal "seed" of the desired impurity level is then immersed in the molten germanium and gradually withdrawn while the shaft holding the seed is slowly turning. As the "seed" is withdrawn, a single-crystal germanium lattice structure will grow on the "seed" as shown in Fig. 1.15a. The resulting single-crystal ingot can be as large as 7 to 10 in. in length and 1 to 3 in. in diameter (Fig. 1.15b).

The floating zone technique eliminates the need for having both a zone refining and single-crystal forming process. Both can be accomplished at the same time using this technique. A second advantage of this method is the absence of the graphite or quartz boat, which often introduces impurities into the germanium or silicon ingot. Two clamps hold the bar of germanium or silicon in the vertical position within a set of movable RF induction coils as shown in Fig. 1.16. A small single-crystal "seed" of the desired impurity level is deposited at the lower

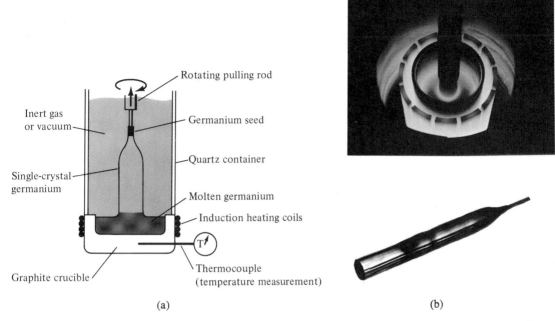

Figure 1.15 Czochralski technique. (b: *top*, courtesy Texas Instruments Incorporated; *bottom*, courtesy Motorola Incorporated.)

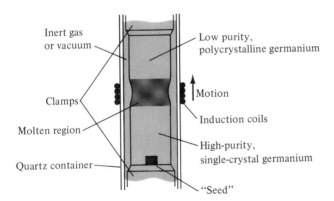

Figure 1.16 Floating zone technique.

end of the bar and heated with the germanium bar until the molten state is reached. The induction coils are then slowly moved up the germanium or silicon ingot while the bar is slowly rotating. As before, the impurities follow the molten state, resulting in an improved impurity level single-crystal germanium lattice below the molten zone. Through proper control of the process, there will always be sufficient surface tension present in the semiconductor material to ensure that the ingot does not rupture in the molten zone.

The single-crystal structure produced can then be cut into wafers sometimes as thin as $\frac{1}{1000}$ (or 0.001) of an inch ($\cong \frac{1}{5}$ the thickness of this paper). This cutting process can be accomplished using the setup of Fig. 1.17a or b. In Fig. 1.17a,

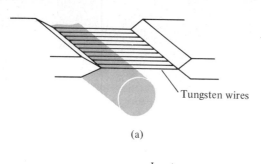

(a)

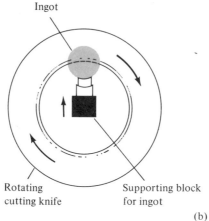

(b)

Figure 1.17 Slicing the single-crystal ingot into wafers. (Courtesy Texas Instruments Incorporated.)

tungsten wires (0.001 in. in diameter) with abrasive deposited surfaces are connected to supporting blocks at the proper spacing and then the entire system is moved back and forth as a saw. The system of Fig. 1.17b is self-explanatory.

In the next few chapters, as we introduce the various devices, we will pick up with the semiconductor wafer produced above and describe the full construction technique.

Other semiconductor materials will be introduced as their area of application is considered.

PROBLEMS

§ 1.2

1. In your own words, define semiconductor, resistivity, bulk resistance, and ohmic contact resistance.

2. (a) Using Table 1.1, determine the resistance of a silicon sample having an area of 1 cm² and a length of 3 cm.

(b) Repeat part (a) if the length is 1 cm and the area 4 cm².
(c) Repeat part (a) if the length is 8 cm and the area 0.5 cm².
(d) Repeat part (a) for copper and compare the results.

3. Sketch the atomic structure of copper and discuss why it is a good conductor and how its structure is different from germanium and silicon.

4. Define, in your own words, an intrinsic material, a negative temperature coefficient, and covalent bonding.

5. Consult your reference library and list three materials that have a negative temperature coefficient and three that have a positive temperature coefficient.

§ 1.3

6. How much energy in joules is required to move a charge of 6 C through a difference in potential of 3 V?

7. If 48 eV of energy is required to move a charge through a potential difference of 12 V, determine the charge involved.

8. Consult your reference library and determine the level of E_g for GaP and ZnS, two semiconductor materials of practical value. In addition, determine the written name for each material. What is the level of ρ for each material?

§ 1.4

9. Describe the difference between *n*-type and *p*-type semiconductor materials.

10. Describe the difference between donor and acceptor impurities.

11. Describe the difference between majority and minority carriers.

12. Sketch the atomic structure of silicon and insert an impurity of arsenic as demonstrated for germanium in Fig. 1.6.

13. Repeat Problem 12 but insert an impurity of indium.

14. Consult your reference library and find another explanation of hole versus electron flow. Using both descriptions, describe in your own words the process of hole conduction.

§ 1.5

15. Describe another example of the diffusion process.

16. How is drift current different from diffusion current?

§ 1.6

17. Describe the Czochralski method for fabricating a single crystal of germanium or silicon.

18. How is the floating zone technique different from the Czochralski method?

19. What is induction heating? Describe in your own words.

GLOSSARY

Semiconductor A material whose electrical properties lie somewhere between that of a conductor and an insulator.

Resistivity A constant of proportionality that relates the resistance of a sample to its geometric shape. Denoted by the symbol ρ (rho).

Bulk Resistance The actual resistance of an isolated sample of the semiconductor material.

Ohmic Contact Resistance The resistance introduced through the connection of the leads to the bulk material.

Germanium A material with four valence electrons commonly used as a base in the manufacture of semiconductor devices.

Silicon The material with four valence electrons most commonly used as a base in the manufacture of semiconductor devices.

Crystal One complete pattern of an atomic array.

Lattice The periodic arrangement of the atoms in a material.

Single-Crystal Structure Any material composed solely of repeating crystal structures of the same kind.

Electron The particle in the atomic structure of negative charge that orbits the nucleus of the atom.

Proton The element of the atomic structure with a positive charge located within the nucleus of the atom.

Neutron The uncharged particle within the nucleus of the atom.

Valence Electrons The electrons in the outermost shell of the atomic structure.

Ionization Potential The potential required to remove a valence electron from the atomic structure.

Tetravalent Atoms Atoms that have four valence electrons.

Covalent Bonding The bonding between atoms established by the sharing of atoms.

Intrinsic Material Material that has been carefully refined to reduce the impurities to a very low level.

Negative Temperature Coefficient A property of materials that reveals that the resistance will decrease with increase in temperature.

Positive Temperature Coefficient A property of materials that reveals that the resistance will increase with increase in temperature.

Forbidden Region The region between the valence band and the ionization level.

Electron Volt (eV) A measure of the resulting energy transfer when an electron passes through a potential difference of 1 V.

Extrinsic Material A semiconductor material in which impurities have been placed through a process referred to as doping.

n-Type Material A material in which electrons have been introduced in excess of that required to complete the covalent bonding.

p-Type Material A material that has been doped in such a way as to ensure that there is an insufficient number of electrons to complete the covalent bonding between the atoms.

Conventional Flow The flow of charge associated with the direction of hole flow in an electrical system.

Majority Carrier The carrier in a material that has the greatest density.

Minority Carrier The carrier in a material that has the least density.

Drift Current Current initiated by an applied emf across a conductor.

Diffusion Current The flow of charge that results through the motion of charge away from a heavy concentration of like charges to a more even distribution.

Czochralski Technique A method of semiconductor fabrication involving the pulling of a single-crystal seed immersed in a tub of molten germanium or silicon.

Floating Zone Technique A method of semiconductor fabrication involving the use of induction coils.

CHAPTER 2
Diodes

2.1 INTRODUCTION

The semiconductor diode is one of the basic building blocks of the wide variety of electronic systems in use today. It will appear in a range of applications, extending from the simple to the very complex. In addition to the details of its construction and characteristics, we examine a few practical applications of the device. The very important data and graphs to be found on specification sheets will also be covered to ensure an understanding of the terminology employed and demonstrate the wealth of information typically available from manufacturers.

Before examining the construction and characteristics of an actual device, we first consider the ideal device, to provide a basis for comparison.

2.2 IDEAL DIODE

The *ideal diode* is a *two-terminal* device having the symbol and characteristics shown in Fig. 2.1a and b, respectively.

In the description of the elements to follow, it is critical that the various *letter symbols, voltage polarities,* and *current directions* be defined. If the polarity of the applied voltage is consistent with that shown in Fig. 2.1a, the portion of the characteristics to be considered in Fig. 2.1b is to the right of the vertical axis. If a reverse voltage is applied, the characteristics to the left are pertinent. If the current through the diode has the direction indicated in Fig. 2.1a, the portion of the characteristics to be considered is above the horizontal axis, while a reversal

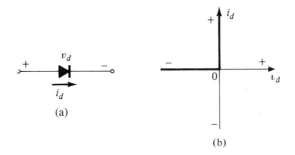

Figure 2.1 Ideal diode: (a) symbol; (b) characteristics.

in direction would require the use of the characteristics below the axis. For the majority of the device characteristics to appear in this text the *ordinate* will be the *current axis*, while the *abscissa* will be the *voltage axis*.

One of the important parameters for the diode is the resistance at the point or region of operation. If we consider the region defined by the direction of i_d and polarity of v_d in Fig. 2.1a (upper-right quadrant of Fig. 2.1b), we shall find that the value of the forward resistance, R_f, as defined by Ohm's law is

$$R_f = \frac{V_f}{I_f} = \frac{0}{2, 3 \text{ mA}, \ldots, \text{ or any positive value}} = 0 \, \Omega$$

where V_f is the forward voltage across the diode and I_f is the forward current through the diode. *The ideal diode, therefore, is a short circuit for the forward region of conduction* ($i_d \neq 0$).

If we now consider the region of negatively applied potential (third quadrant) of Fig. 2.1b,

$$R_r = \frac{V_r}{I_r} = \frac{-5, -20, \text{ or any reverse-bias potential}}{0}$$

$$= \text{very large number, which for our purposes we shall consider to be infinite } (\infty)$$

where V_r is the reverse voltage across the diode and I_r is the reverse current in the diode. *The ideal diode, therefore, is an open circuit in the region of nonconduction* ($i_d = 0$).

In review, the conditions depicted in Fig. 2.2 are true.

In general, it is relatively simple to determine whether a diode is in the region of conduction or nonconduction by simply noting the direction of the current i_d to be established by an applied emf. For conventional flow (opposite to that of

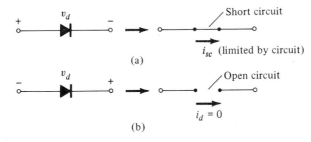

Figure 2.2 (a) Conduction and (b) nonconduction states of the ideal diode as determined by the applied bias.

SEC. 2.2 IDEAL DIODE

electron flow), if the resultant diode current has the same direction as the arrowhead of the diode symbol, the diode is operating in the conducting region. This is depicted in Fig. 2.3.

As an introductory example of one practical application of the diode let us consider the process of rectification, by which an alternating voltage having zero average value is converted to one having a dc or average value greater than zero. The circuit required is shown in Fig. 2.4 with an ideal diode.

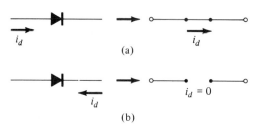

Figure 2.3 (a) Conduction and (b) nonconduction states of the ideal diode as determined by current direction of applied network.

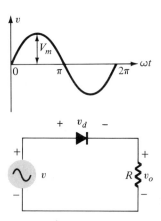

Figure 2.4 Basic rectifying circuit.

For the region defined by $0 \to \pi$ of the sinusoidal input voltage v, the polarity of the voltage drop across the diode would be such that the short-circuit representation would result and the circuit would appear as shown in Fig. 2.5a. For the region $\pi \to 2\pi$, the open-circuit representation would be applicable and the circuit would appear as shown in Fig. 2.5b.

For future reference, note the polarities of the input v for each circuit. For sinusoidal inputs the polarity indicated will be for the positive portion of the sinusoidal waveform, as shown in Fig. 2.4. For the situation shown in Fig. 2.5a, the output voltage v_o, will appear exactly the same as the input voltage, v, as long as the diode is forward-biased. In Fig. 2.5b, because of the open-circuit representation of the ideal diode, the output voltage v_o equals zero from π to 2π of the impressed voltage v. The complete resultant output waveform is shown in Fig. 2.5c for the entire sinusoidal input. For each cycle of the input voltage v, the waveform of v_o will repeat itself so that each waveform has the same frequency. A closer examination of the various figures will also reveal that the impressed emf v and v_o are *in phase;* that is, the positive pulse of each appears during the same time period. Phase relationships will become increasingly important when we consider semiconductor amplifiers.

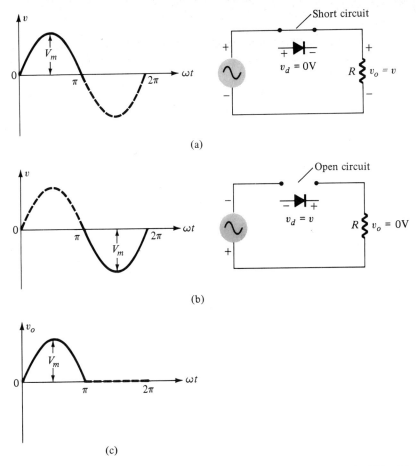

Figure 2.5 Rectifying action of the circuit of Fig. 2.4.

2.3 BASIC CONSTRUCTION AND CHARACTERISTICS

In Chapter 1 n- and p-type materials were introduced. The semiconductor diode is formed by simply bringing these materials together (constructed from the same base—Ge or Si), as shown in Fig. 2.6, using techniques to be described later in the chapter. At the instant the two materials are "joined" the electrons and holes in the region of the junction will combine resulting in a lack of carriers in the region near the junction. This region of uncovered positive and negative ions is called the *depletion* region due to the depletion of carriers in this region.

No Applied Bias

The minority carriers in the n-type material that find themselves within the depletion region will pass directly into the p-type material. The closer the minority carrier is to the junction, the greater the attraction for the layer of negative ions and the less the opposition of the positive ions in the depletion region of the *n*-type material. For the purposes of future discussions we shall assume that all the minority carriers of the *n*-type material that find themselves in the depletion region due to their random motion will pass directly into the *p*-type material. Similar discussion can be applied to the minority carriers (electrons) of the *p*-type material. This carrier flow has been indicated in Fig. 2.6 for the minority carriers of each material.

The majority carriers in the *n*-type material must overcome the attractive forces of the layer of positive ions in the *n*-type material and the shield of negative ions in the *p*-type material in order to migrate into the neutral region of the *p*-type material. The number of majority carriers is so large in the *n*-type material, however, that there will be invariably a small number of majority carriers with sufficient kinetic energy to pass through the depletion region into the *p*-type material. Again, the same type of discussion can be applied to the majority carriers of the *p*-type material. The resulting flow due to the majority carriers is also shown in Fig. 2.6.

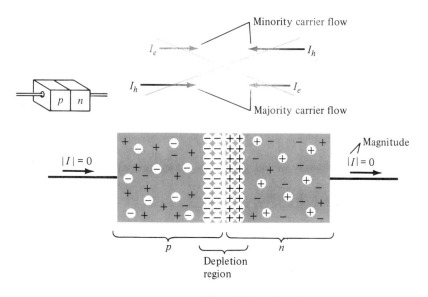

Figure 2.6 *p-n* junction with no external bias.

A close examination of Fig. 2.6 will reveal that the relative magnitudes of the flow vectors are such that the net flow in either direction is zero. This cancellation of vectors has been indicated by crossed lines. The length of the vector representing hole flow has been drawn longer than that for electron flow to demonstrate that the magnitude of each need not be the same for cancellation and that the doping levels for each material may result in an unequal carrier flow of holes and electrons. In summary, *the net flow of charge in any one direction with no applied emf is zero.*

Reverse-Bias Condition

If an external potential of V volts is applied across the p-n junction such that the positive terminal is connected to the n-type material and the negative terminal is connected to the p-type material as shown in Fig. 2.7, the number of uncovered negative ions in the depletion region of the n-type material will increase due to the large number of "free" electrons drawn to the positive potential of the applied *emf*. For similar reasons, the number of uncovered negative ions will increase in the p-type material. The net effect, therefore, is a widening of the depletion region. This widening of the depletion region will establish too great a barrier for the majority carriers to overcome, effectively reducing the majority carrier flow to zero (Fig. 2.7).

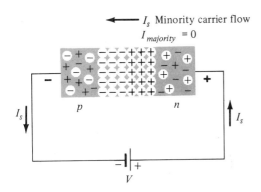

Figure 2.7 Reverse-biased p-n junction.

The number of minority carriers, however, that find themselves entering the depletion region will not change, resulting in minority-carrier flow vectors of the same magnitude indicated in Fig. 2.7 with no applied *emf*. The current that exists under these conditions is called the *reverse saturation current* and is represented by the subscript *s*. It is seldom more than a few microamperes in magnitude except for high-power devices. The term "saturation" comes from the fact that it reaches its maximum value quickly and does not change significantly with increase in the reverse-bias potential. The situation depicted in Fig. 2.6 is referred to as a *reverse-bias* condition.

Forward-Bias Condition

A *forward-bias* condition is established by applying the positive potential to the p-type material and the negative potential to the n-type material (for future reference, note that the forward-bias condition is defined by the corresponding first letter in *p*-type and *p*ositive or in *n*-type and *n*egative), as shown in Fig. 2.8. Note that the minority-carrier flow has not changed in magnitude, but the reduction in the width of the depletion region has resulted in a heavy majority

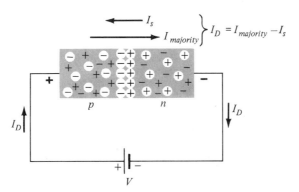

Figure 2.8 Forward-biased p-n junction.

SEC. 2.3 BASIC CONSTRUCTION AND CHARACTERISTICS

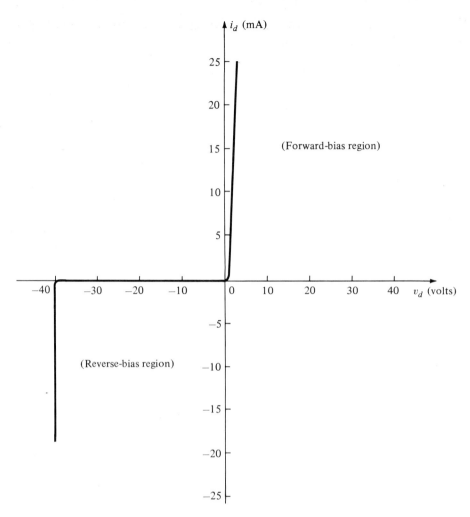

Figure 2.9 Semiconductor diode characteristics—continuous scale on the vertical and horizontal axes.

flow across the junction. The magnitude of the majority-carrier flow will increase exponentially with increasing forward bias, as indicated in Figs. 2.9 and 2.10. Note in Fig. 2.9 the similarities with the ideal diode except at a very negative voltage. The offset in the first quadrant and the sharp drop in the third quadrant will be examined in this chapter. To reiterate, the first quadrant represents the forward-bias region, and the third quadrant the reverse-bias region. Note the extreme change in scales for both the voltage and current in Fig. 2.10. For the current it is approximately a 5000:1 change. The vertical scale for the majority of smaller units is the milliampere, as shown in Fig. 2.10. However, semiconductor diodes are available today with a full ampere in the vertical scale, even though the diameter of the casing may not be greater than $\frac{1}{4}$ in.

It can be demonstrated through the use of solid-state physics that this diode current can be mathematically related to temperature (T_K) and applied bias (V) in the following manner:

$$I = I_s(e^{kV/T_K} - 1) \qquad (2.1)$$

where
I_s = reverse saturation current

$k = 11{,}600/\eta$ with $\eta = 1$ for Ge and 2 for Si

$T_K = T_C + 273°$

Note the exponential factor that will result in a very sharp increase in I with increasing levels of V. The characteristics of a commercially available germanium (Ge) diode will differ slightly from the characteristics of Fig. 2.10 because of the *body* or *bulk* resistance of the semiconductor material and the *contact* resistance between the semiconductor material and the external metallic conductor. They will cause the curve to shift slightly in the forward-bias region, as indicated by

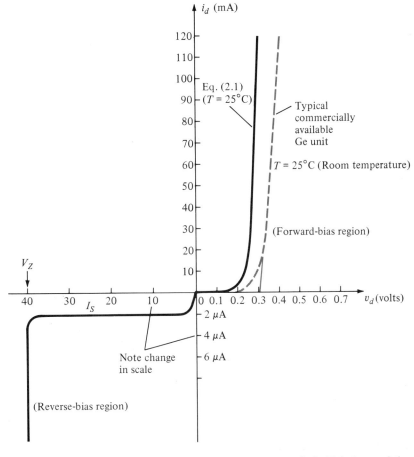

Figure 2.10 Semiconductor diode (Ge) characteristics.

SEC. 2.3 BASIC CONSTRUCTION AND CHARACTERISTICS

the dashed line in Fig. 2.10. As construction techniques improve and these undesired resistance levels are reduced, the commercially available unit will approach the characteristic defined by Eq. (2.1).

In an effort to demonstrate that Eq. (2.1) does in fact represent the curves of Fig. 2.10, let us determine the current I for the forward-bias voltage of 0.22 V at room temperature (25°C).

$$T_K = T_C + 273° = 25° + 273° = 298°$$

$$k(\text{Ge}) = \frac{11{,}600}{1} = 11{,}600$$

$$\frac{kV}{T_K} = \frac{(11{,}600)(0.22)}{298} = 8.56$$

and $\quad I = I_s(e^{8.56} - 1) = (1 \times 10^{-6})(5238 - 1) = 5237 \times 10^{-6}$

so that $\quad\quad\quad\quad\quad\quad I \cong 5.237 \text{ mA}$

as verified by Fig. 2.10.

Temperature can have a marked effect on the diode current. This is clearly demonstrated by the factor T_K in Eq. (2.1). The effect of varying T_K will be determined for the forward-bias condition in the exercises appearing at the end of the chapter. In the reverse-bias region it has been found experimentally that the *reverse saturation current I_s will almost double in magnitude for every 10°C change in temperature*. It is not uncommon for a germanium diode with an I_s in the order of 1 or 2 μA at 25°C to have a leakage current of 100 μA = 0.1 mA at a temperature of 100°C. Current levels of this magnitude in the reverse-bias region would certainly question our desired open-circuit condition in the reverse-bias region. Fortunately, typical values of I_s for silicon at room temperature range from 1/100 to 1/1000 that of a similar application germanium diode so that even at higher temperature levels I_s does not usually reach levels of serious concern. In the example above, with $I_s = 1$ μA for germanium, if I_s were, at the most, 1/100 of 1 μA = 0.01 μA for a silicon diode, then at 100°C it would be only (1/100)(100 μA) = 1 μA. The stability of an electronic system is highly dependent on the temperature sensitivity of its components. The fact that I_s will only approach 1 μA at 100°C for a silicon diode while it approaches 0.1 mA (100 μA) for a germanium diode is one very important reason that silicon devices enjoy a significantly higher level of attention. Fundamentally, the open-circuit equivalent in the reverse-bias region is better realized at any temperature with silicon than with germanium.

Zener Region

Note the sharp change in the characteristics of Fig. 2.11 at the reverse-bias potential V_Z (the subscript Z refers to the name Zener). This constant-voltage effect is induced by a high reverse-bias voltage across the diode. When the applied reverse potential becomes more and more negative, a point is eventually reached where the few free minority carriers have developed sufficient velocity to liberate additional carriers through ionization. That is, they collide with the valence elec-

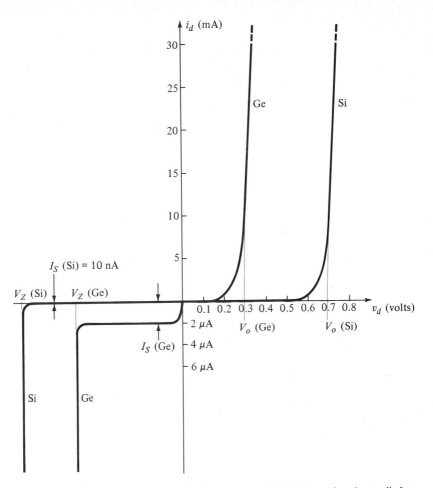

Figure 2.11 Comparison of Si and Ge semiconductor diodes.

trons and impart sufficient energy to them to permit them to leave the parent atom. These additional carriers can then aid the ionization process to the point where a high *avalanche* current is established and the *avalanche breakdown* region determined.

The avalanche region (V_Z) can be brought closer to the vertical axis by increasing the doping levels in the *p*- and *n*-type materials. However, as V_Z decreases to very low levels, such as -5 V, another mechanism, called *Zener breakdown*, will contribute to the sharp change in the characteristic. It occurs because there is a strong electric field in the region of the junction that can disrupt the bonding forces within the atom and "generate" carriers. Although the Zener breakdown mechanism is only a significant contributor at lower levels of V_Z, this sharp change in the characteristic at any level is called the *Zener region* and diodes employing this unique portion of the characteristic of a *p-n* junction are called *Zener diodes*. They will be described in detail in Chapter 3.

The Zener region of the semiconductor diode described must be avoided if the response of a system is not to be completely altered by the sharp change in characteristics in this reverse-voltage region. The maximum reverse-bias potential that can be applied before entering this region is called the *peak inverse voltage* (referred to simply as the PIV rating), or the peak reverse voltage (denoted by PRV rating). If an application requires a PIV rating greater than that of a single unit, a number of diodes of the same characteristics can be connected in series. Diodes are also connected in parallel to increase the current-carrying capacity.

Silicon Versus Germanium

Silicon diodes have, in general, higher PIV and current ratings and wider temperature ranges than germanium diodes. PIV ratings for silicon can be in the neighborhood of 1000 V, whereas the maximum value for germanium is closer to 400 V. Silicon can be used for applications in which the temperature may rise to about 200°C (400°F), whereas germanium has a much lower maximum rating (100°C). The disadvantage of silicon, however, as compared to germanium, as indicated in Fig. 2.10, is the higher forward-bias voltage required to reach the region of upward swing. It is typically of the order of magnitude of 0.7 V for *commercially* available silicon diodes and 0.3 V for germanium diodes. The increased offshoot for silicon is due primarily to the factor η in Eq. (2.1). This factor only plays a part in determining the shape of the curve at very low current levels. Once the curve starts its vertical rise, the factor η drops to 1 (the continuous value for germanium). This is evidenced by the similarities in the curves once the offshoot potential is reached. The potential at which this rise occurs is very important in the circuit analysis to follow and therefore requires the specific notation V_o, as indicated on the figure. In review:

$$V_o = 0.7 \text{ (Si)}$$
$$V_o = 0.3 \text{ (Ge)}$$

Obviously, the closer the upward swing is to the vertical axis, the more "ideal" the device. However, the other characteristics of silicon as compared to germanium still make it the choice in the majority of commercially available units.

2.4 TRANSITION AND DIFFUSION CAPACITANCE

Electronic devices are inherently sensitive to very high frequencies. Most shunt capacitive effects that can be ignored at lower frequencies because the reactance $X_C = 1/2\pi f C$ is very large (open-circuit equivalent) cannot be ignored at very high frequencies. X_C will become sufficiently small due to the high value of f to introduce a low-reactance "shorting" path. In the *p-n* semiconductor diode, there are two capacitive effects to be considered. Both types of capacitance are present in the forward- and reverse-bias regions, but one so outweighs the other in each region that we consider the effects of only one in each region. In the reverse-bias region we have the *transition-* or *depletion*-region capacitance (C_T), while in the forward-bias region we have the *diffusion* (C_D) or *storage* capacitance.

Recall that the basic equation for the capacitance of a parallel plate capacitor is defined by $C = \epsilon A/d$, where ϵ is the permittivity of the dielectric (insulator) between the plates of area A separated by a distance d. In the reverse-bias region there is a depletion region (free of carriers) that behaves essentially like an insulator between the layers of opposite charge. Since the depletion region will increase with increased reverse-bias potential, the resulting transition capacitance will decrease, as shown in Fig. 2.12. The fact that the capacitance is dependent on the applied reverse-bias potential has application in a number of electronic systems. In fact, in Chapter 3 a diode will be introduced whose existence is wholly dependent on this phenomena.

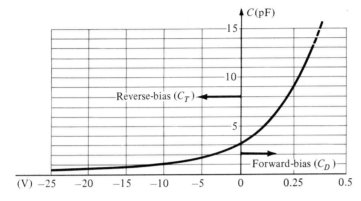

Figure 2.12 Transition and diffusion capacitance vs. applied bias for a silicon diode.

Although the effect described above will also be present in the forward-bias region, it is overshadowed by a capacitance effect directly dependent on the rate at which charge is injected into the regions just outside the depletion region. In other words, directly dependent on the resulting current of the diode. Increased levels of current will result in increased levels of diffusion capacitance. However, increased levels of current result in reduced levels of associated resistance (to be demonstrated shortly), and the resulting time constant $(\tau = RC)$, which is very important in high-speed applications, does not become excessive.

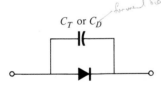

Figure 2.13 Including the effect of the transition or diffusion capacitance on the semiconductor diode.

The capacitive effects described above are represented by a capacitor in parallel with the ideal diode, as shown in Fig. 2.13. For low- or midfrequency applications (except in the power area), however, the capacitor is normally not included in the diode symbol.

2.5 REVERSE RECOVERY TIME

There are certain pieces of data that are normally provided on diode specification sheets provided by manufacturers. One such quantity that has not been considered yet is the reverse recovery time denoted by t_{rr}. In the forward-bias state it has been shown in an earlier section that there are a large number of electrons from the n-type material progressing through the p-type material and a large number of holes in the n-type—a requirement for conduction. The electrons in the p-type and of holes progressing through the n-type material establish a large number of minority carriers in each material. If the applied voltage should be reversed to establish a reverse-bias situation, we would ideally like to see the diode change instantaneously from the conduction state to the nonconduction state. However, because of the large number of minority carriers in each material, the diode will simply reverse as shown in Fig. 2.14 and stay at this measurable level for the period of time t_d (storage time) required for the minority carriers to return to their majority-carrier state in the opposite material. Eventually, when this storage phase has passed, the current will reduce in level to that associated with the nonconduction state. This second period of time is denoted by t_t (transition interval). The reverse recovery time is the sum of these two intervals: $t_{rr} = t_d + t_t$. Naturally, it is an important consideration in high-speed switching applications. Most commercially available switching diodes have a t_{rr} in the range of a few nanoseconds to 1 μs. Units are available, however, with a t_{rr} of only a few hundred picoseconds (10^{-12}).

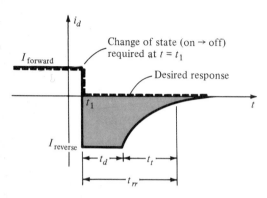

Figure 2.14 Defining the reverse recovery time.

2.6 TEMPERATURE EFFECTS

Temperature is a very important consideration in the design or analysis of electronic systems. It will affect virtually all of the characteristics of any semiconductor device. The change in characteristics of a semiconductor diode due to temperature variations above and below room temperature (25°C) is shown in Fig. 2.15. Note the reduced levels of forward voltage drop but increased levels of saturation current at 100°C. The Zener potential is also experiencing a pronounced change in level.

Heat Sinks

Increased levels of current through any semiconductor device will result in increased junction temperatures. Although silicon materials can handle currents approaching 1000 A/in.², the changes in characteristics with temperature may result in an unstable system. Germanium materials have maximum allowable operating junction temperatures ranging from 85 to 100°C, while the range for silicon is 150 to 200°C (just one more case for the increased level of interest in silicon devices). Quite frequently, the maximum allowable junction temperature is exceeded before the maximum power dissipation level is established. In fact, we will find in this section that the maximum dissipation level is very dependent on temperature.

With no applied bias and consequently zero current through the device, the temperature of the junction of the semiconductor diode is essentially the same as that of the surrounding air or medium. This temperature, called the *ambient* temperature, has the symbol T_A and is measured (as are all temperature levels in this discussion) in degrees Celsius.

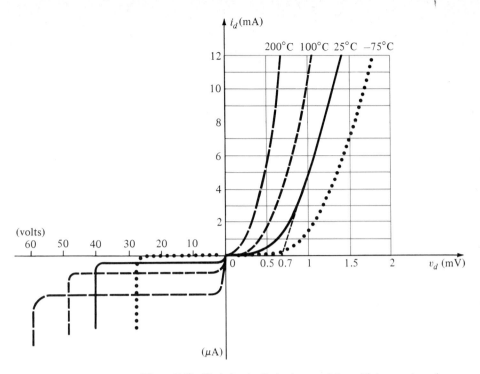

Figure 2.15 Variation in diode characteristics with temperature change.

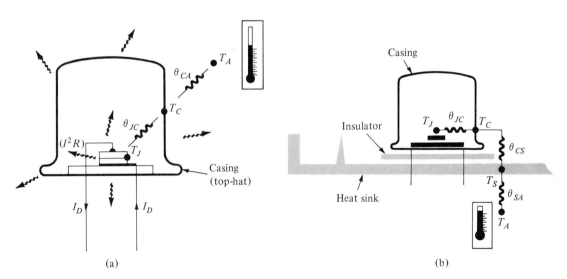

Figure 2.16 Path of thermal conductivity: (a) without a heat sink; (b) with a heat sink.

As current begins to flow through the device, there is an I^2R loss at the junction that will increase the temperature of the junction. The resulting heat is then transferred to the case and eventually to the surrounding air, as shown in Fig. 2.16a. If a *heat sink* is applied as shown in Fig. 2.16b, the resulting heat will pass from the case to the sink before being emitted to the surrounding medium.

SEC. 2.6 TEMPERATURE EFFECTS

The sole purpose of the heat sink is to provide a large surface area through which the heat can be quickly removed from the device—thus permitting the device to work at higher dissipation levels. The insulator appearing in Fig. 2.16b is often necessary to insulate the semiconductor device from the heat sink. Its effect on the flow of heat will be considered shortly.

The ease with which heat can be transferred from one element to another is a measure of the *thermal conductivity* between the two. Inversely related to the conductivity is the *thermal resistance*—a measure of how much the medium will oppose the flow of heat. In our present design considerations, therefore, we are trying to establish the lowest possible thermal resistance path from the junction to the surrounding air. Thermal resistance has the symbol Θ, is measured in °C/W, and has the resistance symbol as appearing in Fig. 2.16. In both parts of Fig. 2.16 the path of heat flow is indicated and the temperature of importance noted. A schematic representation for both systems appears in Fig. 2.17. For the system of Fig. 2.16a, Θ_{CA} is simply the thermal resistance path from case to the surrounding medium—without consideration of a heat sink.

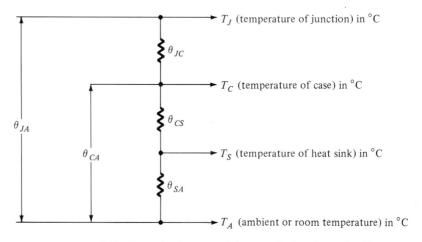

Figure 2.17 Thermal resistance path between the junction and ambient temperatures.

The series path between elements will result in a total thermal resistance from junction to surrounding air of

$$\Theta_{JA} = \Theta_{JC} + \Theta_{CS} + \Theta_{SA} \quad \text{with heat sink} \quad (°C/W) \qquad (2.2a)$$

$$\Theta_{JA} = \Theta_{JC} + \Theta_{CA} \quad \text{without heat sink} \quad (°C/W) \qquad (2.2b)$$

The temperature of the junction is related to Θ_{JA}, the power dissipated, and the ambient temperature T_A by

$$T_J = P_D \Theta_{JA} + T_A \quad (°C) \qquad (2.3)$$

In words, it simply states that the junction temperature is equal to the temperature of the surrounding air plus the increased temperature due to the heat conversion at the junction. The better we are at removing the heat from the junction through lower values of Θ_{JA}, the higher the power rating can be for the same junction temperature.

In total,

$$T_J = P_D(\Theta_{JC} + \underbrace{\Theta_{CS} + \Theta_{SA}}_{\Theta_{CA}}) + T_A \qquad (°C) \qquad (2.4)$$

In Eq. (2.4), the maximum T_J is usually specified on a data sheet. It is always less than the maximum permissible value for that material. As indicated, T_A is simply the temperature of the surrounding medium measured in degrees Celsius. Θ_{JA} is established by the device chosen—its power level, case construction, and so on. P_D is the level of power dissipation determined by $P = V_D I_D$ for a diode. Θ_{CA} or Θ_{JA} is normally provided on the data sheet for a specific semiconductor device. Θ_{CS} is the interface thermal resistance that depends on the casing design and how it is mounted to the heat sink. The use of an insulator will increase the thermal resistance above that obtained with direct mounting. Keep in mind that the term "insulator" here is referring to current flow and not heat flow. Our goal is to design an insulator that will block the flow of charge but not severely affect heat-flow levels. For a TO-3 case (top-hat appearance), an average value of Θ_{CS} is 0.1°C/W for direct mounting and 0.5°C/W with an insulator. Θ_{SA} is provided by the heat-sink design, as shown in Fig. 2.18.

(a)

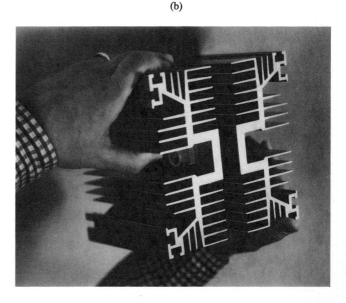

(b)

Figure 2.18 Tran-tec semiconductor coolers: (a) Model 19, $\theta_{SA} = 2.5°C/W$; (b) Model 1128, $\theta_{SA} = 0.8°C/W$. (Courtesy Tran-Tec Corporation.)

SEC. 2.6 TEMPERATURE EFFECTS

EXAMPLE 2.1

(a) Determine whether the junction temperature of a semiconductor diode is being exceeded under the following operating conditions:

$$P_D = 20 \text{ W}$$
$$T_A = 25°C$$
$$T_J \text{ (max) (specified)} = 150°C$$
$$\Theta_{JC} = 2.0°C/W \quad \Theta_{CS} = 0.5°C/W \quad \Theta_{SA} = 2.5°C/W$$

(b) If $\Theta_{JA} = 20°C/W$ for the device without a heat sink, has the junction temperature been exceeded?

Solution:

(a) From Eq. (2.4):

$$T_J = P_D(\Theta_{JC} + \Theta_{CS} + \Theta_{SA}) + T_A$$
$$= 20(2.0 + 0.5 + 2.5) + 25$$
$$= 100 + 25 = \mathbf{125°C}$$
$$T_J \text{ (specified)} = 150°C > 125°C \quad \text{(safe operation)}$$

(b) From Eq. (2.3):

$$T_J = P_D \Theta_{JA} + T_A$$
$$= 20(20) + 25$$
$$= 400 + 25 = \mathbf{425°C} \gg 150°C \quad \text{(permanent damage to the device can be expected)}$$

In Example 2.1 the maximum permissible dissipation without a heat sink could be determined from Eq. (2.5):

$$\boxed{P_D = \frac{T_J - T_A}{\Theta_{JA}}} \quad \text{(watts)} \quad (2.5)$$

That is,

$$P_{D_{\max}} = \frac{150 - 25}{20} = \frac{125}{20} = \mathbf{6.25 \text{ W}}$$

The heat-sink requirement could be determined by

$$\boxed{\Theta_{SA} = \frac{T_J - T_A}{P_D} - (\Theta_{JC} + \Theta_{CS})} \quad (°C/W) \quad (2.6)$$

Power Derating Curve

It should be clear from the discussion above that the maximum power rating is closely related to the case temperature of the device. Once a certain case temperature is reached, the maximum power rating as provided in the data sheet will start to drop off linearly (straight line), as shown in Fig. 2.19. The resulting plot is called a *power derating curve* for the device. On many specification sheets for semiconductor diodes, this will show a plot of forward current versus case temperature, but the effect on the quantity of interval is the same. The *derating factor* or measure of how quickly the curve will drop off is measured in W/°C

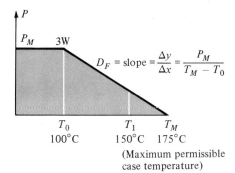

Figure 2.19 Maximum power rating vs. case temperature—the power derating curve.

and has the symbol D_F. In actuality, it is the inverse of the junction to case thermal resistance. That is, $D_F = 1/\Theta_{JC}$.

In the linear sloped region, the power rating at any temperature T_1 compared to the maximum power (P_M) at temperature T_o is determined by

$$\boxed{P_{T_1} = P_M - (T_1 - T_o)(D_F)} \quad \text{(watts)} \quad (2.7)$$

D_f can be determined by

$$\boxed{D_F = \frac{P_M}{T_M - T_o}} \quad (\text{W}/^\circ\text{C}) \quad (2.8)$$

although it is often provided simply as a change of so many watts per degree.

EXAMPLE 2.2

(a) Determine the maximum power rating for a device having the derating curve of Fig. 2.19 at a temperature of 150°C.
(b) Calculate Θ_{JC}.

Solution:

(a) From Eq. (2.8): $D_F = \dfrac{P_M}{T_M - T_o} = \dfrac{3}{175 - 100} = \dfrac{3}{75} = 0.04 \text{ W}/^\circ\text{C}$

From Eq. (2.7): $P_{T_1} = P_M - (T_1 - T_o)(D_F)$

$\qquad = 3 - (150 - 100)(0.04)$

$\qquad = 3 - (50)(0.04)$

$\qquad = 3 - 2 = 1 \text{ W}$ (a significant drop)

(b) $\Theta_{JC} = \dfrac{1}{D_F} = \dfrac{1}{0.04} = 25\,^\circ\text{C/W}$

2.7 DIODE SPECIFICATION SHEETS

Data on specific semiconductor devices is normally provided by the manufacturer in two forms. One is a very brief description of a device that will permit a quick review of all devices available within a few pages. The other is a thorough examination of a device, including graphs, applications, and so on, which usually appears as a separate entity. The latter is normally only provided when specifically requested.

There are certain pieces of data, however, that normally appear on either one. They are included below:

1. The maximum forward voltage $V_{F(\max)}$ (at a specified current and temperature).
2. The maximum forward current $I_{F(\max)}$ (at a specified temperature).
3. The maximum reverse current $I_{R(\max)}$ (at a specified temperature).
4. The reverse voltage rating (PIV) or PRV or V(BR), where BR comes from the term "breakdown" (at a specified temperature).
5. Maximum capacitance.
6. Maximum t_{rr}.
7. The maximum operating (or case) temperature.

Depending on the type of diode being considered, additional data may also be provided, such as frequency range, noise level, switching time, thermal resistance levels, and peak repetitive values. For the application in mind, the significance of the data will usually be self-apparent. If the maximum power or dissipation rating is also provided, it is understood to be equal to the following product:

$$\boxed{P_{D\max} = V_D I_D} \qquad (2.9)$$

where I_D and V_D are the diode current and voltage at a particular point of operation, each variable not to exceed its maximum value. The information in Table 2.1 was taken directly from a Texas Instruments, Inc., data book. Note that the forward voltage drop does not exceed 1 V, but the current has maximum values of 1 to 200 mA.

TABLE 2.1 General-Purpose Diodes

| Device Type | Forward Current | | V_{BR} (V) | Maximum I_R | | | |
| | | | | 25°C | | 150°C | |
	I_F (mA)	V_F (V)		V	μA	V	μA
1N463	1.0	1.0	200	175	0.5	175	30
1N462	5.0	1.0	70	60	0.5	60	30
1N459A	100.0	1.0	200	175	0.025	175	5
T151	200.0	1.0	20	10	1	—	—

For the 1N463, if we establish maximum forward voltage and current conditions:

$$P_D = V_D I_D = 1(1) = 1 \text{ mW} \quad \text{(a low-power device)}$$

Of course, a device may have a maximum dissipation less than that established by the maximum values. That is, if the voltage is a maximum, the current may have to be less than rated maximum value.

Note the increase in I_R for each device with temperature. For the 1N463, it is $30/0.5 = 60$ times larger.

An exact copy of the data provided by Fairchild Camera and Instrument Corporation for their BAY 73 and BA 129 high voltage/low leakage diodes appears in Figs. 2.20 and 2.21. This example would represent the expanded list of data and characteristics. Note that all but the average rectified current, peak repetitive

Figure 2.20 Electrical characteristics of the Bay73 · BA 129 high-voltage, low-leakage diodes. (Courtesy Fairchild Camera and Instrument Corporation.)

DIFFUSED SILICON PLANAR

- BV... 125 V (MIN) @ 100 µA (BAY73)
- BV... 200 V (MIN) @ 100 µA (BA129)

ABSOLUTE MAXIMUM RATINGS (Note 1)

Temperatures
- Storage Temperature Range — −65°C to +200°C
- Maximum Junction Operating Temperature — +175°C
- Lead Temperature — +260°C

Power Dissipation (Note 2)
- Maximum Total Power Dissipation at 25°C Ambient — 500 mW
- Linear Power Derating Factor (from 25°C) — 3.33 mW/°C

Maximum Voltage and Currents
- WIV Working Inverse Voltage BAY73 — 100 V
- BA129 — 180 V
- I_O Average Rectified Current — 200 mA
- I_F Continuous Forward Current — 500 mA
- if Peak Repetitive Forward Current — 600 mA
- if(surge) Peak Forward Surge Current
  - Pulse Width = 1 s — 1.0 A
  - Pulse Width = 1 µs — 4.0 A

DO-35 OUTLINE

NOTES:
Copper clad steel leads, tin plated
Gold plated leads available
Hermetically sealed glass package
Package weight is 0.14 gram

ELECTRICAL CHARACTERISTICS (25°C Ambient Temperature unless otherwise noted)

SYMBOL	CHARACTERISTIC	BAY73 MIN	BAY73 MAX	BA129 MIN	BA129 MAX	UNITS	TEST CONDITIONS
V_F	Forward Voltage	0.85	1.00			V	I_F = 200 mA
		0.81	0.94			V	I_F = 100 mA
		0.78	0.88	0.78	1.00	V	I_F = 50 mA
		0.69	0.80	0.69	0.83	V	I_F = 10 mA
		0.67	0.75			V	I_F = 5.0 mA
		0.60	0.68	0.60	0.71	V	I_F = 1.0 mA
				0.51	0.60	V	I_F = 0.1 mA
I_R	Reverse Current		500			nA	V_R = 20 V, T_A = 125°C
			5.0			nA	V_R = 100 V
			1.0			µA	V_R = 100 V, T_A = 125°C
					10	nA	V_R = 180 V
					5.0	µA	V_R = 180 V, T_A = 100°C
BV	Breakdown Voltage	125		200		V	I_R = 100 µA
C	Capacitance		8.0		6.0	pF	V_R = 0, f = 1.0 MHz
t_{rr}	Reverse Recovery Time		3.0			µs	I_f = 10 mA, V_r = 35 V, R_L = 1.0 to 100 KΩ, C_L = 10 pf, JAN 256

NOTES:
1. These ratings are limiting values above which the serviceability of the diode may be impaired.
2. These are steady state limits. The factory should be consulted on applications involving pulses or low duty-cycle operation.

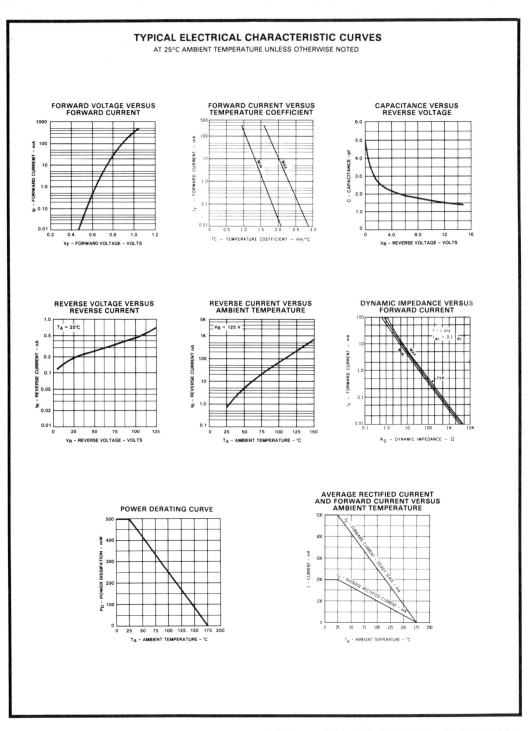

Figure 2.21 Terminal characteristics of the Fairchild Bay73 · BA 129 high-voltage diodes. (Courtesy Fairchild Camera and Instrument Corporation.)

forward current, and peak forward surge current have been defined in this chapter. The significance of these three quantities is as follows:

1. *Average Rectified Current:* The half-wave-rectified signal of Fig. 2.5 has an average value defined by $I_{av} = 0.318\, I_{peak}$. The average current rating is lower than the continuous forward current because a half-wave current waveform will have instantaneous values much higher than the average value.
2. *Peak Repetitive Forward Current:* This is the maximum instantaneous value of repetitive forward current. Note that since it is at this level for a brief period of time, its level can be higher than the continuous level.
3. *Peak Forward Surge Current:* On occasion during turn-on, malfunctions, and so on, there will be very high currents through the device for very brief intervals of time (that are not repetitive). This rating defines the maximum value and the time interval for such surges in current level.

Note the logarithmic scale appearing on some of the curves of Fig. 2.21. Each region is bisected such that the value of each horizontal line should be fairly obvious. For I_F versus V_F the horizontal lines between 1.0 and 10.0 mA are 2 mA, 4 mA, 6 mA, and 8 mA. Again, most of the axis variables on the graphs provided have been introduced, resulting in a set of curves that have some recognizable meaning. The temperature coefficient defines the change in voltage with temperature at different current levels. A range of values for the temperature coefficient is provided at each current level. The dynamic impedance (actually, simply the resistance of the device at that forward current) will be discussed in a later section. Note the effect of temperature in the power rating and current ratings of the device in the bottom right figure.

2.8 SEMICONDUCTOR DIODE NOTATION

The notation most frequently used for semiconductor diodes is provided in Fig. 2.22. For most diodes any marking such as a dot or band, as shown in Fig. 2.22, appears at the cathode end. The terminology anode and cathode is a carryover from vacuum-tube notation. The anode refers to the higher or positive potential and the cathode refers to the lower or negative terminal. This combination of bias levels will result in a forward-bias or "on" condition for the diode. In general, the maximum current-carrying capacity of the diodes of Fig. 2.22 increases from

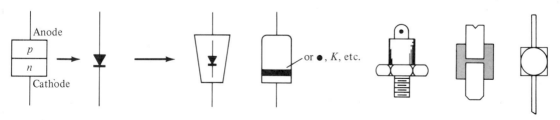

Figure 2.22 Semiconductor diode notation.

the left to the right. For each the size will increase with the current rating to ensure that it can handle the additional power dissipation. All but the stud type are limited to a few amperes.

2.9 DIODE OHMMETER CHECK

The condition of a semiconductor diode can be quickly determined by using an ohmmeter such as is found on the standard VOM. The internal battery (often 1.5 V) of the ohmmeter section will either forward- or reverse-bias the diode when applied. If the positive (normally the red) lead is connected to the anode and the negative (normally the black) lead to the cathode, the diode is forward-biased and the meter should indicate a low resistance. The R × 1000 or R × 10,000 setting should be suitable for this measurement. With the reverse polarity the internal battery will back bias the diode and the resistance should be very large. A small reverse-bias resistance reading indicates a "short" condition while a large forward-bias resistance indicates an "open" situation. The basic connections for the tests appear in Fig. 2.23.

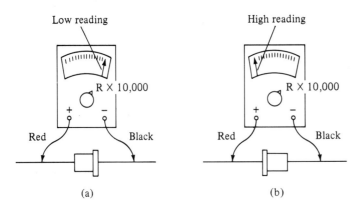

Figure 2.23 Ohmmeter testing of a semiconductor diode: (a) forward-bias; (b) reverse-bias.

2.10 SEMICONDUCTOR DIODE FABRICATION

Semiconductor diodes are normally one of the following types: grown junction, alloy, diffusion, epitaxial growth, or point contact. Each will be described in some detail in this section. A rereading of Section 1.6 is suggested before proceeding with the following description.

Grown Junction

Diodes of this type are formed during the Czochralski *crystal pulling* process. Impurities of *p*- and *n*-type can be alternately added to the molten semiconductor material in the crucible, resulting in a *p-n* junction, as indicated in Fig. 2.24, when the crystal is pulled. After slicing, the large-area device can then be cut into a large number (sometimes thousands) of smaller-area semiconductor diodes. The area of grown junction diodes is sufficiently large to handle high currents

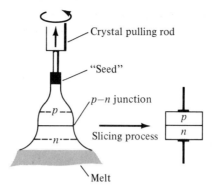

Figure 2.24 Grown junction diode.

(and therefore have high power ratings). The large area, however, will introduce undesired junction capacitive effects.

Alloy

The alloy process will result in a junction-type semiconductor diode that will also have a high current rating and large PIV rating. The junction capacitance is also large, however, due to the large junction area.

The *p-n* junction is formed by first placing a *p*-type impurity on an *n*-type substrate and heating the two until liquefaction occurs where the two materials meet (Fig. 2.25). An alloy will result that, when cooled, will produce a *p-n* junction at the boundary of the alloy and substrate. The roles played by the *n*- and *p*-type materials can be interchanged.

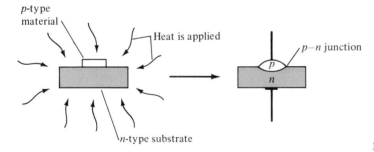

Figure 2.25 Alloy process diode.

Diffusion

The diffusion process of forming semiconductor junction diodes can employ either solid or gaseous diffusion. This process requires more time than the alloy process but it is relatively inexpensive and can be very accurately controlled. Diffusion is a process by which a heavy concentration of particles will "diffuse" into a surrounding region of lesser concentration. The primary difference between the diffusion and alloy process is the fact that liquefaction is not reached in the diffusion process. Heat is applied in the diffusion process only to increase the activity of the elements involved.

The process of solid diffusion commences with the "painting" of an acceptor impurity on an *n*-type substrate and heating the two until the impurity diffuses into the substrate to form the *p*-type layer (Fig. 2.26a).

In the process of gaseous diffusion, an *n*-type material is submerged in a gaseous atmosphere of acceptor impurities and then heated (Fig. 2.26b). The impurity diffuses into the substrate to form the *p*-type layer of the semiconductor diode. The roles of the *p*- and *n*-type materials can also be interchanged in each case. The diffusion process is the most frequently used today in the manufacture of semiconductor diodes.

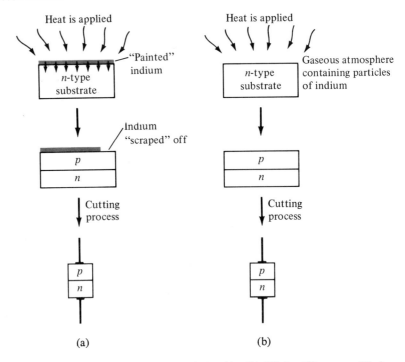

Figure 2.26 Diffusion process diodes: (a) solid diffusion; (b) gaseous diffusion.

Epitaxial Growth

The term "epitaxial" has its derivation from the Greek terms *epi* meaning "upon" and *taxis* meaning "arrangement." A base wafer of n^+ material is connected to a metallic conductor as shown in Fig. 2.27. The n^+ indicates a very high doping level for a reduced resistance characteristic. Its purpose is to act as a semiconductor extension of the conductor and not the *n*-type material of the *p-n* junction. The *n*-type layer is to be deposited on this layer as shown in Fig. 2.27 using a diffusion process. This technique of using an n^+ base gives the manufacturer definite design advantages. The *p*-type silicon is then applied by using a diffusion technique and the anode metallic connector added as indicated in Fig. 2.27.

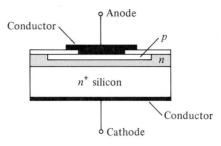

Figure 2.27 Epitaxial growth semiconductor diode.

Point Contact

The point-contact semiconductor diode is constructed by pressing a phosphor bronze spring (called a cat whisker) against an *n*-type substrate (Fig. 2.28). A high current is then passed through the whisker and substrate for a short period of time, resulting in a number of atoms passing from the wire into the *n*-type material to create a *p*-region in the wafer. The small area of the *p-n* junction results in a very small junction capacitance (typically 1 pF or less). For this reason, the point-contact diode is frequently used in applications where very high frequencies are encountered, such as in microwave mixers and detectors. The disadvantage of the small contact area is the resulting low current ratings and characteristics less ideal than those obtained from junction-type semiconductor diodes. The basic construction and photographs of point-contact diodes appear in Fig. 2.29. Various types of junction diodes appear in Fig. 2.30.

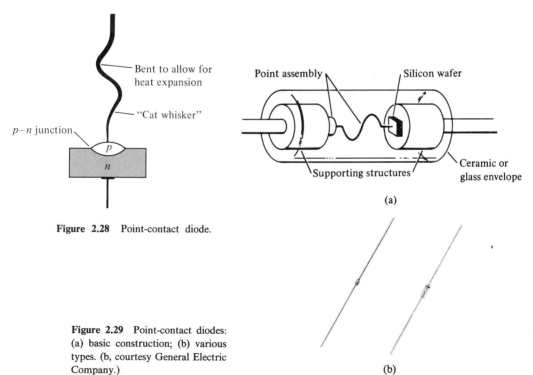

Figure 2.28 Point-contact diode.

Figure 2.29 Point-contact diodes: (a) basic construction; (b) various types. (b, courtesy General Electric Company.)

SEC. 2.10 SEMICONDUCTOR DIODE FABRICATION

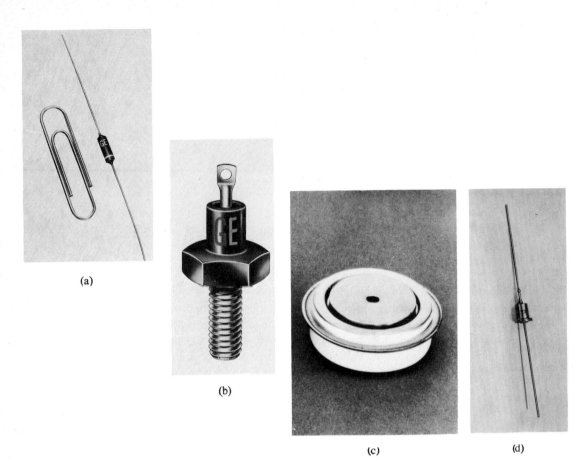

Figure 2.30 Various types of junction diodes. (Courtesy International Rectifier Corporation.)

2.11 DIODE ARRAYS—INTEGRATED CIRCUITS

The unique characteristics of integrated circuits will be introduced in Chapter 8. However, we have reached a plateau in our introduction to electronic circuits that permits at least a surface examination of diode arrays in the integrated circuit package. You will find that the integrated circuit is not a unique device with characteristics totally different from those we will examine in these introductory chapters. It is simply a packaging technique that permits a significant reduction in the size of electronic systems. In other words, internal to the integrated circuit are systems and discrete devices that were available long before the integrated circuit as we know it today became a reality.

One possible array appears in Fig. 2.31. Note that eight diodes are internal to the Fairchild FSA 1410M diode array. That is, in the container shown in Fig. 2.32 there are diodes set in a single silicon wafer that have all the anodes connected to pin 1 and the cathodes of each to pins 2 through 9. Note in the

CH. 2 DIODES

same figure that pin 1 can be determined as being to the left of the small projection in the case if we look from the bottom toward the case. The other numbers then follow in sequence. If only one diode is to be used, then only pins 1 and 2 (or any number from 3 to 9) would be used. The remaining diodes would be left hanging and not affect the network that pins 1 and 2 are connected to.

Another diode array appears in Fig. 2.33. In this case the package is different but the numbering sequence appears in the outline. Pin 1 is the pin directly above the small indentation as you look down on the device.

Figure 2.31 Monolithic diode array. (Courtesy Fairchild Camera and Instrument Corporation.)

FSA1410M
PLANAR AIR-ISOLATED MONOLITHIC DIODE ARRAY

- C...5.0 pF (MAX)
- ΔV_F...15 mV (MAX) @ 10 mA

CONNECTION DIAGRAM
FSA1410M

See Package Outline TO-96

ABSOLUTE MAXIMUM RATINGS (Note 1)

Temperatures
Storage Temperature Range	−55°C to +200°C
Maximum Junction Operating Temperature	+150°C
Lead Temperature	+260°C

Power Dissipation (Note 2)
Maximum Dissipation per Junction at 25°C Ambient	400 mW
per Package at 25°C Ambient	600 mW
Linear Derating Factor (from 25°C) Junction	3.2 mW/°C
Package	4.8 mW/°C

Maximum Voltage and Currents
WIV	Working Inverse Voltage	55 V
I_F	Continuous Forward Current	350 mA
$i_{f(surge)}$	Peak Forward Surge Current	
	Pulse Width = 1.0 s	1.0 A
	Pulse Width = 1.0 µs	2.0 A

ELECTRICAL CHARACTERISTICS (25°C Ambient Temperature unless otherwise noted)

SYMBOL	CHARACTERISTIC	MIN	MAX	UNITS	TEST CONDITIONS
B_V	Breakdown Voltage	60		V	I_R = 10 µA
V_F	Forward Voltage (Note 3)		1.5	V	I_F = 500 mA
			1.1	V	I_F = 200 mA
			1.0	V	I_F = 100 mA
I_R	Reverse Current		100	nA	V_R = 40 V
	Reverse Current (T_A = 150°C)		100	µA	V_R = 40 V
C	Capacitance		5.0	pF	V_R = 0, f = 1 MHz
V_{FM}	Peak Forward Voltage		4.0	V	I_f = 500 mA, t_r < 10 ns
t_{fr}	Forward Recovery Time		40	ns	I_f = 500 mA, t_r < 10 ns
t_{rr}	Reverse Recovery Time		10	ns	$I_f = I_r$ = 10−200 mA, R_L = 100 Ω, Rec. to 0.1 I_r
			50	ns	I_f = 500 mA, I_r = 50 mA, R_L = 100 Ω, Rec. to 5 mA
ΔV_F	Forward Voltage Match		15	mV	I_F = 10 mA

NOTES:
1. These ratings are limiting values above which life or satisfactory performance may be impaired.
2. These are steady state limits. The factory should be consulted on applications involving pulsed or low duty cycle operation.
3. V_F is measured using an 8 ms pulse.

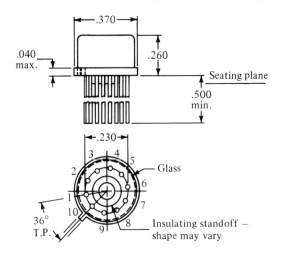

Notes:
Kovar leads, gold plated
Hermetically sealed package
Package weight is 1.32 grams

Figure 2.32 Package outline TO-96 for the FSA 1410M diode array. All dimensions are in inches.

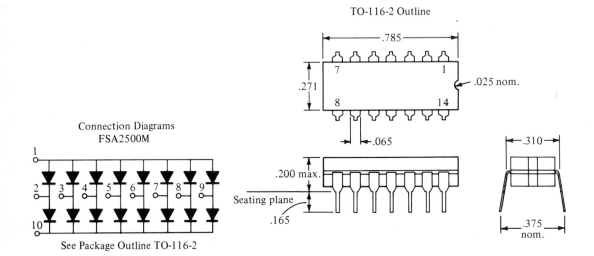

Notes:
Alloy 42 pins, tin plated
Gold plated pins available
Hermetically sealed ceramic package

Figure 2.33 Monolithic diode array. (Courtesy Fairchild Camera and Instrument Corporation.) All dimensions are in inches.

2.12 DC CONDITIONS

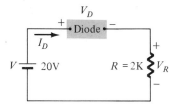

Figure 2.34 Fundamental diode circuit.

The analysis of earlier sections employed the ideal diode. We must now examine the effect of introducing a device that is less than perfect. Our first interest will be in the *dc* or *quiescent* conditions—the latter term meaning still, quiet, or inactive. The circuit of Fig. 2.34 is the simplest possible if the effects of the nonideal characteristics are to be examined. The block symbol was inserted to represent any diode that we may choose to use in this particular circuit.

Applying Kirchhoff's voltage law around the indicated loop will result in the following equation:

$$V = V_D + V_R \qquad (2.10)$$

Solving for V_D and substituting $V_R = I_D R$, we have

$$V_D = V - V_R$$

and

$$V_D = V - I_D R \qquad (2.11)$$

Equation (2.11) has two dependent variables (V_D and I_D) and two fixed values (V and R). Since a minimum of two equations is required to solve for two unknown dependent variables, Eq. (2.11) is not sufficient for a complete solution. The second equation necessary to determine the value of V_D and I_D determined by V and R is provided by the characteristics of the diode element in the enclosed container; that is, for the diode employed, we know that the current is a *function of* the voltage across the diode, or mathematically,

$$I_D = f(V_D) \qquad (2.12)$$

It is necessary, therefore, to find the common solution of the equation determined by the load circuit [Eq. (2.11)] and the characteristics of the diode. One method of finding this solution is the graphical method, which will now be outlined. It is extremely important that the procedure described in the next few paragraphs be fully understood since similar operations apply to other devices, such as the transistor, and FET.

Rewriting Eq. (2.11) in a slightly different form, we have

$$I_D = -\frac{1}{R} V_D + \frac{V}{R} \qquad (2.13)$$

$$y = \quad mx \quad + b \quad \text{(straight-line equation)}$$

Below this newly formed equation, the general equation for a straight line has been included. Note that the slope of the line is negative (I_D decreases in magnitude with increase in V_D) with a magnitude $1/R$, while the y-intercept is V/R and I_D and V_D are the y- and x-variables, respectively. The intercepts of this straight line with the axes of the graph of Fig. 2.35 can be found rather quickly by applying the following conditions. If we consider first that if $I_D = 0$ mA, we must be somewhere along the horizontal axis of Fig. 2.35 and if we apply this condition to Eq. (2.13), then

$$I_D = 0 = -\frac{V_D}{R} + \frac{V}{R}$$

and solving for V_D yields

$$\boxed{V_D = V\big|_{I_D=0}} \qquad (2.14)$$

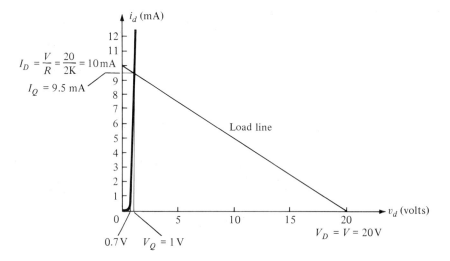

Figure 2.35 Sketching the load line and determining the Q point for the network of Fig. 2.34.

The intersection of the straight line with the horizontal axis is the applied voltage V. If we then consider that if $V_D = 0$, we must be somewhere along the vertical axis, and we must apply this condition to Eq. (2.13), then

$$I_D = -\frac{V_D}{R} + \frac{V}{R} = 0 + \frac{V}{R}$$

and

$$\boxed{I_D = \frac{V}{R}\bigg|_{V_D=0}} \qquad (2.15)$$

The intersection, therefore, of the straight line with the vertical axis is determined by the ratio of the applied voltage and load. Both intersections have been indicated in Fig. 2.35. All that remains is to connect these two points by a straight line to obtain a graphical representation of Eq. (2.11). This resulting line is called the *load line* since it represents the properties of the applied voltage and load and tells us nothing about the diode's characteristics.

In Fig. 2.35 the characteristics of a semiconductor diode have also been included. The intersection of the load line and the diode's characteristic curve will determine the point of operation for that diode. This point, due only to the dc input, is called the *quiescent* point. The voltage across and current through the diode can now be found by simply drawing a vertical and a horizontal line, respectively, to the voltage and current axis as indicated in Fig. 2.35. The subscript Q is used to denote quiescent values of current and voltage as shown in Fig. 2.35. The results are (to the accuracy possible)

$$V_Q = 1 \text{ V} \quad \text{and} \quad I_Q = 9.5 \text{ mA}$$

Substituting into Eq. (2.10), we get

$$V_R = V - V_D = V - V_Q = 20 - 1 = 19.0 \text{ V}$$

or

$$V_R = I_Q R = (9.5 \times 10^{-3})(2 \times 10^3) = 19.0 \text{ V}$$

The power delivered to the load is

$$P_L = I_Q^2 R = (9.5 \times 10^{-3})^2 (2 \times 10^3) = 180.5 \text{ mW}$$

or

$$P_L = P_S - P_D$$

where P_S is the power supplied by the source and P_D is the power dissipation of the diode, so that

$$P_L = VI_Q - V_Q I_Q = I_Q(V - V_Q)$$
$$= (9.5 \times 10^{-3})(20 - 1) = 180.5 \text{ mW}$$

Take special note of how closely the semiconductor diode approaches that of the ideal diode for the magnitudes of current and voltage indicated.

2.13 STATIC RESISTANCE

A second glance at Fig. 2.35 will reveal that the diode has a fixed voltage and current associated with the point of operation. Applying Ohm's law to this value will result in the *static* or *dc* resistance of the diode at the quiescent point.

$$\boxed{R_{dc} = \frac{V_D}{I_D}} \quad (2.16)$$

In this case,

$$R_{dc} = \frac{V_D}{I_D} = \frac{1}{9.5 \times 10^{-3}} = 105.2 \text{ }\Omega$$

For the reverse-bias region of a semiconductor diode with $V_D = -20$ V (for example) and $I_S = 1$ μA:

$$R_{dc} = \frac{V_D}{I_D} = \frac{20}{1 \text{ }\mu\text{A}} = 20 \text{ M}\Omega \gg 105.2 \text{ }\Omega \quad \text{(obtained above)}$$

Once the dc resistance has been determined, the diode can be replaced by a resistor of this value. Any change in the applied voltage or load resistor, however, will result in a different Q-point and therefore different dc resistance.

2.14 DYNAMIC RESISTANCE

It is obvious from Fig. 2.35 that the dc resistance of a diode is independent of the shape of the characteristic in the region surrounding the point of interest. If a sinusoidal rather than dc input is applied to the circuit of Fig. 2.35, the situation will change completely. Consider the circuit of Fig. 2.36a, which has as its input a sinusoidal signal on a dc level. Since the magnitude of the dc level is much greater than that of the sinusoidal signal at any instant of time, the diode will always be forward-biased and current will exist continuously in the circuit in the direction shown.

The dc load line resulting from the dc input of 20 V is shown in Fig. 2.36b. The effect of the ac signal is also demonstrated pictorially in the same figure. Note that two additional load lines have been drawn at the positive and negative peaks of the input signal. At the instant the sinusoidal signal is at its positive peak value the input could be replaced by a dc battery with a magnitude of

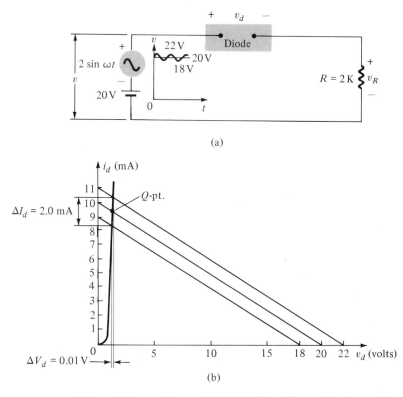

Figure 2.36 ac resistance: (a) circuit; (b) resulting region of operation.

22 V and the resultant load line drawn as shown. For the negative peak, $V_{dc} = 18$ V. A moment of thought, however, should reveal the relative simplicity of superimposing the sinusoidal signal on the dc load line and drawing the load lines coinciding with the positive and negative peaks of the sinusoidal signal.

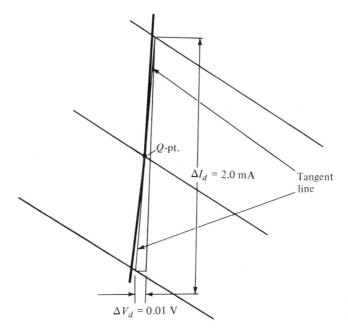

Figure 2.37 Semiconductor diode ac resistance.

Note that we are now interested in a region of the diode characteristics as determined by the sinusoidal signal rather than a single point, as was the case for purely dc inputs. Since the resistance will vary from point to point along this region of interest, which value should we use to represent this portion of the characteristic curve? The value chosen is determined by drawing a straight line tangent to the curve at the quiescent point as shown in Fig. 2.37 for the vacuum-tube diode. The tangent line should "best fit" the characteristics in the region of interest as shown. The resultant resistance, called the *dynamic* or *ac* resistance, is then calculated on an approximate basis, in the following manner:

$$r_d = \frac{\Delta V_d}{\Delta I_d}\bigg|_{\text{tangent line}} \quad (2.17)$$

and

$$r_d = \frac{\Delta V_d}{\Delta I_d} \cong \frac{0.01}{2 \times 10^{-3}} = 5 \, \Omega$$

There is a basic definition in differential calculus that states that *the derivative of a function at a point is equal to the slope of a tangent line drawn at that point.* Equation (2.17), as defined by Fig. 2.37 is, therefore, essentially finding the derivative of the function at the Q-point of operation. If we find the derivative of the general equation [Eq. (2.1)] for the semiconductor diode with respect to the applied forward bias and then invert the result, we will have an equation for the dynamic or ac resistance in that region. That is, taking the derivative of Eq. (2.1) with respect to the applied bias will result in

$$\frac{dI}{dV} = \frac{k}{T_K}(I + I_S)$$

For values of $I \gg I_s$, $I + I_s \cong I$ and, as indicated earlier, $\eta = 1$ for Ge and Si in the vertical rise section of the characteristics. Therefore,

$$k = \frac{11{,}600}{\eta} = \frac{11{,}600}{1} = 11{,}600$$

with (at room temperature)

$$T_K = T_C + 273° = 25° + 273° = 298°$$

and

$$\frac{dI}{dV} = \frac{11{,}600}{298} I \cong 38.93 I$$

or

$$\frac{dV}{dI} = \frac{1}{38.93 I} \cong \frac{0.026}{I}$$

and

$$r_d' = \frac{dV}{dI} = \frac{0.026\ \text{V}}{I_D} = \frac{26\ \text{mV}}{I_D} \bigg|_{\text{Ge, Si}} \qquad (2.18)$$

The significance of Eq. (2.18) must be understood. It implies that the dynamic resistance can be found by simply substituting the quiescent value of the diode current into the equation. There is no need to have the characteristics available or to worry about sketching tangent lines as defined by Eq. (2.17). Its use will be demonstrated below.

We already realize from Eq. (2.17) that the shape of the curve will have an effect on the dynamic resistance. The fact that the silicon and germanium curves in Fig. 2.11 are almost identical after they begin their vertical rise would suggest that the equation for the dynamic resistance of each might be the same as indicated by Eq. (2.18).

It was already noted on Fig. 2.10 that the characteristics of the commercial unit are slightly different from those determined by Eq. (2.1) because of the bulk and contact resistance of the semiconductor device. This additional resistance level must be included in Eq. (2.18) by adding a factor denoted r_B as appearing in Eq. (2.19).

$$r_d = \frac{26\ \text{mV}}{I_D\ (\text{mA})} + r_B \qquad (\text{ohms}) \qquad (2.19)$$

The factor r_B (measured in ohms) can range from typically 0.1 for high-power devices to 2 for some low-power, general-purpose diodes. As construction techniques improve, this additional factor will continue to decrease in importance until it can be dropped and Eq. (2.18) applied. For values of I_D in mA, the units of the first term are like those of r_B: ohms. For low levels of current, the first factor of Eq. (2.19) will certainly predominate. Consider

$$I_D = 1 \text{ mA}$$

with
$$r_B = 2 \text{ }\Omega$$

Then
$$r_d = \frac{26}{1} + 2 = \mathbf{28 \text{ }\Omega}$$

At higher levels of current the second factor may predominate. Consider

$$I_D = 52 \text{ mA}$$

with
$$r_B = 2 \text{ }\Omega$$

Then
$$r_d = \frac{26}{52} + 2 = 0.5 + 2 = \mathbf{2.5 \text{ }\Omega}$$

For the example provided earlier where r_d was graphically determined to be 5 Ω, if we choose $r_B = 2$ Ω, then

$$r_d = \frac{26}{9.8} + 2 = 2.65 + 2 = 4.65 \text{ }\Omega$$

which is very close to the graphically determined value.

The question of how one is to determine which value to choose for r_B will probably arise. For some devices 2 Ω will be an excellent choice, while for others the approximate average of 1 Ω will perhaps be more appropriate. Certainly, 2 Ω could always be used as a worst-case design approach. However, it would appear that technology is reaching the point where an average value of 1 Ω would, in general, be more appropriate. Of course, the problem of choosing a correct value only arises in the intermediate range of current levels. At low levels of current either choice of r_B would be an insignificant factor. At higher levels the resistance level is so low in comparison to the other series elements that it can probably be ignored. For the purposes of this text the value of r_B chosen for an example will be directly related to the current level; it will extend from a minimum value at high currents of 0.1 Ω to a maximum value of 2 Ω at low levels. Experience will develop a sense for what value to choose, and, indeed, whether it is a factor of significance at all.

In summary, keep in mind that the static or dc resistance of a diode is determined solely by the point of operation, while the dynamic resistance is determined by the shape of the curve in the region of interest.

2.15 AVERAGE AC RESISTANCE

If the input signal is sufficiently large to produce the type of swing indicated in Fig. 2.38, the resistance associated with the device for this region is called the *average ac resistance*. The average ac resistance is, by definition, the resistance

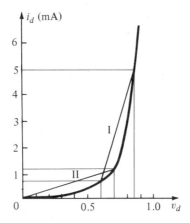

Figure 2.38 Average ac resistance.

determined by a straight line drawn between the two intersections determined by the maximum and minimum values of input voltage. In equation form (note Fig. 2.38)

$$r_{av} = \frac{\Delta V_d}{\Delta I_d}\bigg|_{\text{pt. to pt.}} \quad (2.20)$$

For the situation indicated by Fig. 2.38 in region I,

$$r_{av} = \frac{\Delta V_d}{\Delta I_d}\bigg|_{\text{pt. to pt.}} = \frac{0.85 - 0.6}{(5 - 0.75) \times 10^{-3}} = \frac{0.25}{4.25 \times 10^{-3}} = 58.8 \, \Omega$$

For region II:

$$r_{av} = \frac{\Delta V_d}{\Delta I_d}\bigg|_{\text{pt. to pt.}} = \frac{0.7 - 0}{(1.2 - 0) \times 10^{-3}} = 583 \, \Omega$$

Note the significant increase in resistance as you progress down the curve. For curves in which the current is the vertical axis and the voltage the horizontal, it is useful to remember that the more horizontal the region, the higher the resistance. Or, if preferred, the more vertical the region, the less the resistance.

It is important to note in this discussion of average ac resistance that the resistance to be associated with the element is determined *only* by the region of interest, *not* by the entire characteristic.

2.16 EQUIVALENT CIRCUITS

An equivalent circuit is a combination of elements properly chosen to best represent the actual terminal characteristics of a device, system, and so on. That is, once the equivalent circuit is determined, the device symbol can be removed from a schematic and the equivalent circuit inserted in its place without severely affecting the behavior of the overall system.

One technique for obtaining an equivalent circuit for a diode is to approximate the characteristics of the device by straight-line segments, as shown in Fig. 2.39.

This type of equivalent circuit is called a *piecewise-linear equivalent circuit*. It should be obvious from each curve that the straight-line segments do not result in an exact equivalence between the characteristics and the equivalent circuit. It will, however, at least provide a *first approximation* to its terminal behavior. In each case the resistance chosen is the average ac resistance as defined by Eq. (2.20). The equivalent circuit appears below the curve in Fig. 2.39. The ideal diode was included to indicate that there is only one direction of conduction through the device and that the reverse-bias state is an open-circuit state.

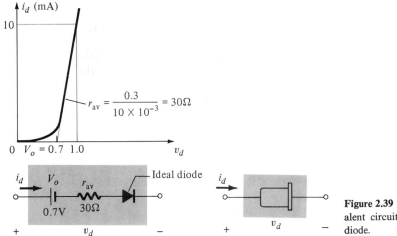

Figure 2.39 Piecewise linear equivalent circuits for a semiconductor diode.

Since a silicon semiconductor diode does not reach the conduction state until approximately 0.7 V, an opposing battery V_o of this value must appear in the equivalent circuit. This indicates that the total forward voltage V_D across the diode must be greater than V_o before the ideal diode in the equivalent circuit will be forward-biased.

Keep in mind, however, that V_o is not an independent source of energy in a system. You will not measure a voltage $V_o = 0.7$ V across an isolated silicon diode using simply a voltmeter. It is simply a useful way of representing the horizontal offset of the semiconductor diode.

The value of r_{av} can usually be determined purely from a few numerical values given on a specification sheet. The complete characteristics, therefore, are usually unnecessary for this calculation. For instance, for a semiconductor diode, if $I_F = 10$ mA, at 1 V, we know that for silicon a shift of 0.7 V is required before the characteristics rise and

$$r_{av} = \frac{1 - 0.7}{10 \text{ mA}} = \frac{0.3}{10 \text{ mA}} = 30 \text{ }\Omega$$

For a germanium diode it would be

$$\frac{1 - 0.3}{10 \text{ mA}} = \frac{0.7}{10 \text{ mA}} = 70 \text{ }\Omega$$

The use of the derived equivalent circuit can best be demonstrated by a few examples.

EXAMPLE 2.3 For the network of Fig. 2.34, determine the voltage across R, the total diode drop V_D, and the equivalent dc resistance of the diode.

Solution: The complete equivalent is substituted in Fig. 2.40.

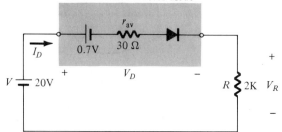

Figure 2.40

Since the applied *emf* of 20 V is much greater than 0.7 V, the ideal diode is forward-biased and the short-circuit equivalent can be substituted. Then, using the voltage divider rule, we get

$$V_R = \frac{(2\,\text{K})(20 - 0.7)}{2\,\text{K} + 30} = \frac{(2\,\text{K})(19.3)}{2030} = 19.0\,\text{V}$$

This compares exactly with the value obtained in Section 2.12.

The dc current through the circuit is

$$I_D = \frac{20}{2030} = 9.85\,\text{mA}$$

as compared to 9.5 mA and

$$V_D = 0.7 + I_D(r_{av}) = 0.7 + (9.85 \times 10^{-3})(30) = 0.7 + 0.296 \cong 1\,\text{V}$$

as determined earlier.

Finally,

$$R_{dc} = \frac{V_D}{I_D} = \frac{1}{9.85\,\text{mA}} = 101.5\,\Omega \text{ versus } 102\,\Omega$$

The results in this case were excellent. It must be realized, however, that when such an equivalent is used, this type of accuracy cannot always be expected, although a good first approximation is usually provided.

Examining the results above, we see that it should be obvious that the 30-Ω forward resistance of the diode is swamped by the 2-K resistor and could be effectively eliminated from the equivalent circuit and still obtain a good first approximate solution to the circuit. That is,

$$V_R = E - V_0 = 20 - 0.7 = 19.3\,\text{V}$$

$$I_D = \frac{20}{2\,\text{K}} = 10\,\text{mA versus } 9.85\,\text{mA} \quad \text{and} \quad V_D = 0.7\,\text{V}$$

This removal of r_{av} from the equivalent circuit is the same as implying that the characteristics of the diode appear as shown in Fig. 2.41. Indeed, this approxi-

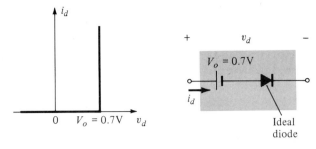

Figure 2.41 Approximate equivalent circuit for the silicon semiconductor diode.

mation is frequently employed in semiconductor circuit analysis. The reduced equivalent circuit appears in the same figure. It states that a forward-biased silicon diode in an electronic system under dc conditions has a drop of 0.7 V across it in the conduction state no matter what the diode current (within rated values, of course).

In fact, we can now go a step further and say that the 0.7 V in comparison to the applied 20 V can be ignored leaving only the ideal diode as an equivalent for the semiconductor device. It is for this very reason that a great deal of the applications to follow in later sections use ideal diodes rather than the complete equivalent. Except for small applied voltages or series resistances, it is never too far from the actual response and it does not cloud the application with a great deal of mathematical exercises.

EXAMPLE 2.4 For the input shown in Fig. 2.42, determine the output voltage by using the semiconductor diode of Fig. 2.39. Use the complete equivalent circuit.

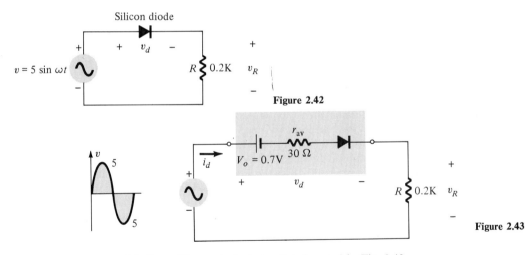

Figure 2.42

Figure 2.43

Solution: The equivalent circuit is inserted in Fig. 2.43.

The diode will not enter its conduction state (i clockwise through the circuit corresponding with the arrow in the diode symbol) until the applied voltage is greater than 0.7 V. This is shown in the output solution of Fig. 2.44. With the input at its maximum value of 5 V, the output is determined by the voltage divider rule.

$$V_o = \frac{(200)(5 - 0.7)}{200 + 30} = \frac{200}{230}(4.3) = 3.74 \text{ V}$$

SEC. 2.16 EQUIVALENT CIRCUITS

For an intermediate value such as 3 V:

$$V_o = \frac{(200)(3-0.7)}{230} = \frac{200}{230}(2.3) = 2 \text{ V}$$

For v less than 0 V, the ideal diode is certainly reverse-biased, and $v_o = 0$ V, as shown in Fig. 2.44. In this type of application in which the input swing extends throughout a wide range of the characteristics, the piecewise equivalent circuit that employs r_{av} will give excellent results as a first approximation. However, let us now consider the small-signal situation in which the region of operation is very limited.

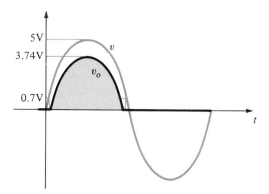

Figure 2.44

At high currents, such as 10 mA, if we use Eq. (2.19), we find the resistance to be

$$r_d = \frac{26}{I_D} + 2 = \frac{26}{10} + 2 = 4.65 \text{ }\Omega$$

while at 0.5 mA it is

$$r_d = \frac{26}{I_D} + 2 = \frac{26}{0.5} + 2 = 52 + 2 = 54 \text{ }\Omega$$

Depending on the region of operation, therefore, the average value of 30 Ω may be far from accurate. In this case, since the dc diode current will probably be discernible by first using the approximation of Fig. 2.41, it would be best to substitute this value into Eq. (2.19) and use this value of resistance. For small-signal applications in which the signal rides on a dc level such as in Fig. 2.36, the diode will always be forward-biased and the ideal diode can be replaced by its short-circuit equivalent. The dc level of V_o can be removed from the equivalent for the ac response since it will only determine the "riding" dc level and not affect the peak-to-peak (p–p) ac response. We will find in the analysis of electronic systems that the dc and ac response can normally be determined separately. That is, the theorem of superposition can usually be applied. This will be demonstrated in Example 2.5.

EXAMPLE 2.5 Determine the voltage v_d across the diode for the input shown in Fig. 2.45.

Solution: The 2-V dc level will ensure that the diode is always forward-biased. The "dc" equivalent appears in Fig. 2.46 using the approximation of Fig. 2.41.

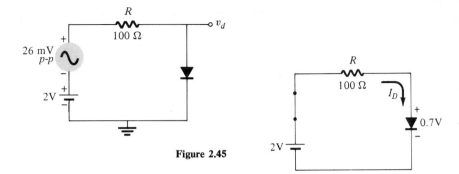

Figure 2.45

Figure 2.46

The dc diode current is found to be

$$I_D \cong \frac{2-0.7}{100} = \frac{1.3}{100} = 13 \text{ mA}$$

Substituting into Eq. (2.19), we get

$$r_d = \frac{26}{13} + 2 = 2 + 2 = 4 \text{ }\Omega$$

The "ac" equivalent is then drawn in Fig. 2.47 and the ac voltage across the diode is determined.

$$v_{d_{ac}} = \frac{4(26 \text{ mV}_{p-p})}{4+100} = 1 \text{ mV}_{(p-p)}$$

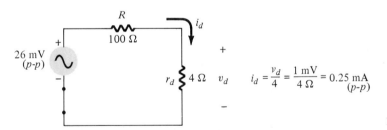

Figure 2.47

The complete solution (applying the superposition theorem) appears in Fig. 2.48. The technique demonstrated above will be used frequently in the analysis in later chapters.

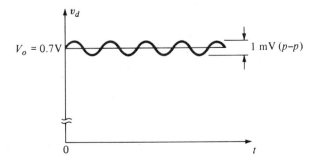

Figure 2.48

SEC. 2.16 EQUIVALENT CIRCUITS

2.17 CLIPPERS AND CLAMPERS

Clippers and clampers are diode waveshaping circuits. Each performs the function indicated by its name. The output of clipping circuits appears as if a portion of the input signal were clipped off. Clamping circuits simply clamp the waveform to a different dc level.

Clippers

A clipping circuit requires at least two fundamental components, a diode and a resistor. A dc battery, however, is also frequently used. The output waveform can be clipped at different levels simply by interchanging the position of the various elements and changing the magnitude of the dc battery. Only ideal diodes appear in the examples to follow. As indicated in the preceding section, however, the response would not be severely altered if semiconductor devices were used.

For networks of this type it is often helpful to consider particular instants of the time-varying input signal to determine the state of the diode. Keep in mind that *at any instant of time* a varying signal can simply be replaced by a dc source of the same value. This is clearly shown in Example 2.6.

EXAMPLE 2.6

Clipper: Find the output voltage waveshape (V_o) for the inputs shown in Fig. 2.49.

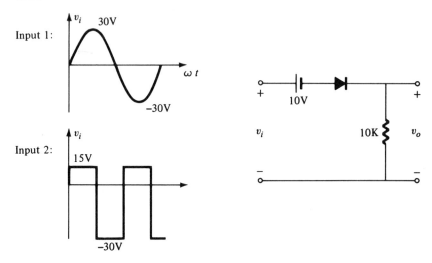

Figure 2.49 Clipping circuit and inputs for Example 2.6.

Solution:

Input 1: For any value of $v_i > 10$ V the ideal diode is forward-biased and $v_o = v_i - 10$. For example, at $v_i = 15$ V (Fig. 2.50), the result is $v_o = 15 - 10 = 5$ V.

For any value of $v_i < 10$ V the ideal diode is reverse-biased and $v_o = 0$ since the current in the circuit is zero. For example, note v_o with $v_i = 5$ V (Fig. 2.51).

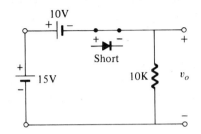

Figure 2.50 Clipping circuit of Fig. 2.49 at instant $v_i = 15$ V of input 1.

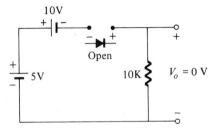

Figure 2.51 Clipping circuit of Fig. 2.49 at instant $v_i = 5$ V of input 1.

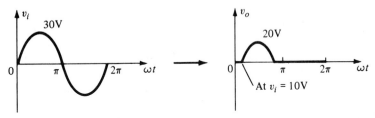

Figure 2.52 Output wave form (v_o) for input 1 to the clipping circuit of Fig. 2.49.

The output waveform v_o appears as if the entire input were clipped off except the positive peak (Fig. 2.52).

Input 2: The diode will change state at the same levels indicated for the first input. The output waveform appears as shown in Fig. 2.53.

Another technique, other than treating instantaneous values of the input as dc levels, is to redraw the applied voltage as shown in Fig. 2.54. Note that the 10-V dc level has only shifted the sinusoidal input down 10 V (Fig. 2.54).

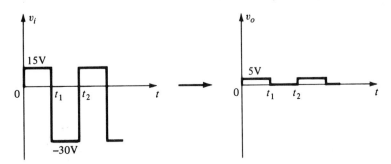

Figure 2.53 Output wave form (v_o) for input 2 to the clipping circuit of Fig. 2.49.

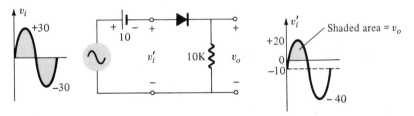

Figure 2.54

SEC. 2.17 CLIPPERS AND CLAMPERS

In order for the diode to be forward-biased, the input v_i' must be positive. This region as clearly shown in the same figure represents the only region that will pass through to the load. A number of clipping circuits and their effect on the applied signal appear in Fig. 2.55.

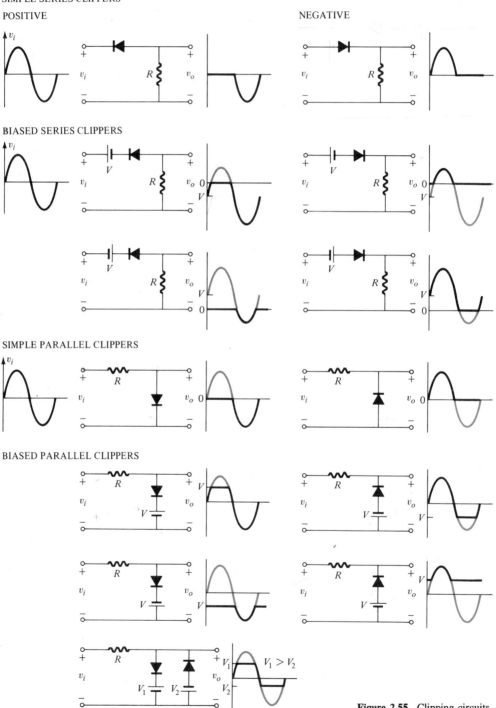

Figure 2.55 Clipping circuits.

Clampers

The clamping circuit has a minimum requirement of three elements: a diode, a capacitor, and a resistor. The clamping circuit may also be augmented by a dc battery. The magnitudes of R and C must be chosen such that the time constant $\tau = RC$ is large enough to ensure that the voltage across the capacitor does not change significantly during the interval of time, determined by the input, that both R and C affect the output waveform. The need for this condition will be demonstrated in Example 2.7. Throughout the discussion, we shall assume that for all practical purposes a capacitor will charge to its final value in five time constants.

It is usually advantageous when examining clamping circuits to first consider the conditions that exist when the input is such that the diode is forward-biased.

EXAMPLE 2.7
Clamper: Draw the output voltage waveform (v_o) for the input shown (Fig. 2.56).

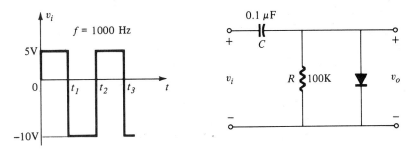

Figure 2.56 Clamping circuit and input for Example 2.7.

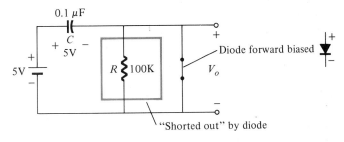

Figure 2.57 Clamping circuit of Fig. 2.56, when $v_i = 5$ V ($0 \rightarrow t_1$).

Solution: At the instant the input switches to the +5-V state the circuit will appear as shown in Fig. 2.57. The input will remain the +5-V state for an interval of time equal to one-half the period of the waveform since the time interval $0 \rightarrow t_1$ is equal to the interval $t_1 \rightarrow t_2$.

The period of v_i is $T = 1/f = 1/1000 = 1$ ms and the time interval of the +5-V state is $T/2 = 0.5$ ms.

Since the output is taken from directly across the diode it is 0 V for this interval of time. The capacitor, however, will rapidly charge to 5 V, since the time constant of the network is now $\tau = RC \cong 0C = 0$.

When the input switches to -10 V, the circuit of Fig. 2.58 will result.

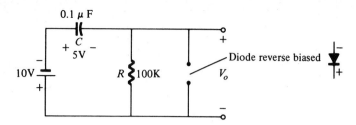

Figure 2.58 Clamping circuit of Fig. 2.56, when $v_i = -10$ V ($t_1 \rightarrow t_2$).

The time constant for the circuit of Fig. 2.58 is

$$\tau = RC = 100 \times 10^3 \times 0.1 \times 10^{-6} = 10 \text{ ms}$$

Since it takes approximately five time constants or 50 ms for a capacitor to discharge, and the input is only in this state for 0.5 ms, to assume the voltage across the capacitor does not change appreciably during this interval of time is certainly a reasonable approximation. The output is therefore

$$V_o = \underset{\text{supply}}{-10} - \underset{\text{capacitor}}{5} = -15 \text{ V}$$

The resulting output waveform (v_o) is provided in Fig. 2.59. As indicated, the output is clamped to the negative region and will repeat itself at the same frequency as the input signal. Note that the swing of the input and output voltages is the same: 15 V. *For all clamping circuits the voltage swing of the input and output waveforms will be the same.* This is certainly not the case for clipping circuits.

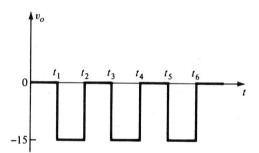

Figure 2.59 Output (v_o) for clamping circuit and input of Fig. 2.56.

If, for discussion sake, the 100-K resistor were replaced by a 1-K resistor, the time constant

$$\tau = RC = (10^3)(0.1 \times 10^{-6}) = 0.1 \text{ ms}$$

and $\quad\quad 5\tau = 0.5$ ms (approximate total discharge time)

The capacitor, therefore, would discharge during the interval in which the voltage is 10 V since the time intervals match. The output waveform would then appear as shown in Fig. 2.60. Clamping networks must have time constants determined by the product RC, which will result in $5RC$ being significantly greater than the time interval in which the diode is reverse-biased or the waveform will be severely distorted.

A number of clamping circuits and their effects on the input signal appear in Fig. 2.61.

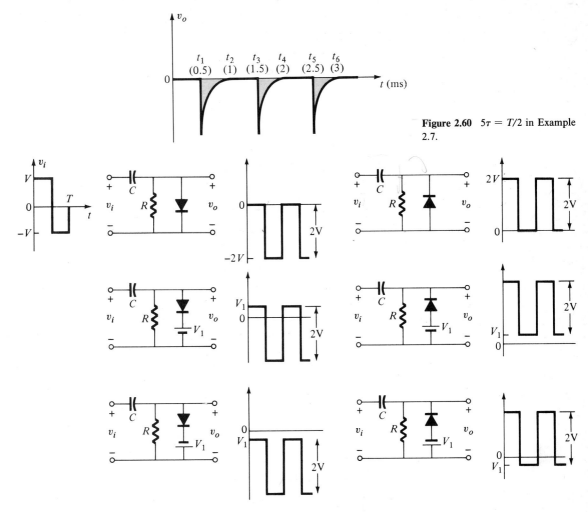

Figure 2.60 $5\tau = T/2$ in Example 2.7.

Figure 2.61 Clamping circuits ($5\tau = 5RC \gg T/2$).

PROBLEMS

§ 2.2

1. Describe, in your own words, the characteristics of the *ideal* diode and how they determine the on and off state of the device. That is, describe why the short-circuit and open-circuit equivalents are appropriate.

2. (a) For the network of Fig. 2.4, sketch the waveform across the resistor R if its value is 6 K and the input signal has a peak value of 120 V.
   (b) Repeat part (a) for the voltage across the diode.

(c) Sketch the waveform for the current in the network for one full cycle of the input voltage.

3. Repeat Problem 2 if the diode is reversed.

4. If a second resistor of 2 K is added in series with the 6-K resistor of Problem 2(a), sketch the waveform of the voltage across the 2-K resistor for the same 120-V input.

5. For the network of Fig. 2.62, determine the voltage across the resistor R_L for the input signals shown in the same figure. Assume an ideal diode.

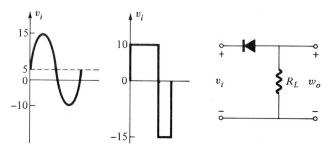

Figure 2.62

6. Repeat Problem 5 for the voltages across the diode.

7. Determine the voltage across the diode in Fig. 2.63 for the inputs appearing in Fig. 2.62. Assume an ideal diode.

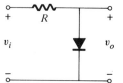

Figure 2.63

§ 2.3

8. Describe in your own words the conditions established by a forward- and reverse-bias condition on a p-n junction diode and how it effects the resulting current flow.

9. Describe how you will remember the forward- and reverse-bias states of the p-n junction diode. That is, how will you remember which potential (positive or negative) is applied to which terminal?

10. Referring to Fig. 2.10, determine the average difference in voltage between the typical commercially available GE unit and the characteristic determined by Eq. (2.1) for the range $i_d = 20$ mA to 100 mA.

11. Using Eq. (2.1), determine the diode current at 20°C for a silicon diode with $I_s = 5$ μA and an applied forward bias of 0.4 V.

12. Repeat Problem 2.11 for $T = 100°C$ (boiling point of water).

13. In the reverse-bias region the saturation current of a silicon diode is about 3 μA ($T = 20°C$). Determine its approximate value if the temperature is increased 40°C.

14. Compare the characteristics of a silicon and germanium diode and determine which you would prefer to use for most practical applications. Give some detail. Refer to a manufacturer's listing and compare the characteristics of a germanium and silicon diode of similar maximum ratings.

§ 2.4

15. (a) Referring to Fig. 2.12, determine the transition capacitance at reverse-bias potentials of -20 V and -5 V. What is the ratio of the change in capacitance to the change in voltage?
    (b) Repeat part (a) for reverse-bias potentials of 10 V and 1 V. Determine the ratio of the change in capacitance to the change in voltage.
    (c) How do the ratios determined in parts (a) and (b) compare? What does it tell you about which range may have more areas of practical application?

16. (a) Referring to Fig. 2.12, determine the diffusion capacitance at 0 V and 0.3 V.
    (b) What is the ratio of the change in level of capacitance to the change in voltage level?

17. Describe in your own words how diffusion and transition capacitances differ.

18. Determine the reactance offered by the diode of Fig. 2.12 at a forward potential of 0.2 V and a reverse potential of -20 V if the applied frequency is 6 MHz.

§ 2.5

19. Sketch the waveform for i of the network of Fig. 2.64 if $t_t = 2t_s$ and the total reverse recovery time is 9 ns.

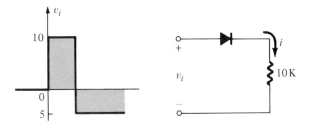

Figure 2.64

§ 2.6

20. Determine the forward voltage drop across the diode whose characteristics appear in Fig. 2.15 at temperatures of $-75°C$, $25°C$, $100°C$, and $200°C$ and a current of 10 mA. For each temperature, determine the level of saturation current. Compare the extremes of each and comment on the ratio of the two.

21. Determine the maximum power dissipation for a semiconductor diode under the following operating conditions: $T_J = 125°C$, $T_A = 25°C$, $\Theta_{JC} = 2.0°C/W$, $\Theta_{CS} = 0.5°C/W$, and $\Theta_{SA} = 2.0°C/W$.

22. Repeat Problem 21 if the heat sink was not employed and $\Theta_{JA} = 20°C/W$.

23. (a) If the maximum permissible power dissipation of a diode with a heat sink is 15 W and $T_A = 25°C$, $T_J = 150°C$, and $\Theta_{CA} = 3.0°C/W$, determine Θ_{JC}.
    (b) What is Θ_{JA}?

24. Determine the heat-sink requirement Θ_{SA} for the device of Problem 21 if the power demand had to be increased by 25%.

25. (a) If P_M were increased to 5 W in Fig. 2.19 at $T_o = 100°C$ and T_M remained the same, determine the maximum power rating of the device at 150°C. How does it compare to the results of Example 2.2?
    (b) What is the new value of D_F and Θ_{JC}?
    (c) At what temperature T_1 will the power rating drop to 4 W?
    (d) If $\Theta_{SA} = 4\Theta_{CS}$ and $\Theta_{JA} = 25°C/W$, determine Θ_{CS}, Θ_{SA}, and Θ_{CA}. Use the level of Θ_{JC} obtained in part (b).

§ 2.7

26. Determine the maximum power dissipation for the T151 diode. What is the maximum reverse-bias dissipation at $V = -10$ V ($T = 25°C$)?

27. Repeat Problem 26 for the 1N459A diode.

28. Using the data of Fig. 2.20, sketch the power derating curve for the BAY73 diode and find its rated power level at 50°C.

29. Plot a curve of V_F(max) versus I_F for the BAY73 diode from the data of Fig. 2.20 and make any noteworthy comments.

30. Repeat Problem 29 for the reverse current.

31. What is the capacitive reactance of the BA129 diode at a frequency of 1 MHz?

32. If $t_t = t_s$, sketch a curve of the reverse recovery period for the BAY73 diode.

33. Determine the peak current associated with a rated average rectified level of $I_0 = 200$ mA (half-wave rectified signal).

34. How does the curve of V_F versus I_F in Fig. 2.21 compare with the results of Problem 29?

35. (a) Referring to Fig. 2.21, determine the temperature coefficient at a forward current of 1.0 mA under maximum conditions.
    (b) Using the results of part (a), determine the change in forward voltage if the temperature should increase 20°C.

36. Compare the power derating curve of Fig. 2.21 with the results of Problem 28.

37. What is the change in dynamic impedance if the current should be reduced from 10 mA to 0.1 mA under maximum conditions? Refer to Fig. 2.21

§ 2.10

38. List the types of diode fabrication introduced in Section 2.10 and discuss the disadvantages and advantages when applied to various practical applications.

§ 2.11

39. What is the maximum power dissipation at 75°C of each diode in the FSA 1410M array? Sketch the power derating curve.

40. Referring to Fig. 2.31, list the detrimental effects of increasing the forward current beyond 100 mA and the temperature above room temperature (25°C).

41. If each diode of the FSA 1410M has a current of 40 mA, what is the current through terminal 1? What is the forward voltage drop from pin 1 to 9 if diodes 1, 5, and 9 are active as forward currents (I_F) of 100 mA?

42. If each diode in Fig. 2.33 has a V_F of 0.7 V and an I_F equal to 30 mA, determine the current through terminals 1 and 10 and the voltage from across pins 1 and 10.

§ 2.12

43. Sketch the load line on the characteristics of Fig. 2.35 if the supply voltage is 18 V and $R = 3$ K. Determine the Q-point for the diode and the corresponding levels of current and voltage. What is the dc dissipation of the diode at this quiescent point of operation?

44. Repeat Problem 43 by cutting the voltage down to 9 V and the resistor down to 1.5 K. Are the results one-half of those obtained in Problem 43? If not, why?

45. How would you expect the results obtained in Problem 43 to change if a germanium diode were inserted in place of the silicon device?

§ 2.13

46. Determine the static or dc resistance of the diode of Fig. 2.35 at a forward current of 5 mA.

47. Repeat Problem 46 at a forward current of 10 mA. Has the dc resistance doubled? If not, why?

48. Determine the dc resistance of the diode of Problem 43 at the Q-point.

§ 2.14

49. Repeat Problem 46 for the dynamic resistance. How do the static and dynamic resistances compare? (Make the necessary approximations, due to the size of the characteristic, for the semiconductor diode.)

50. Determine the dynamic resistance for the conditions of Problem 49 using Eq. (2.19) if r_B is 2 Ω.

51. Using Eq. (2.19), determine the dynamic resistance at 10 mA for the diode of Fig. 2.35 and compare with the results you obtain using Eq. (2.17). Use $r_B = 2$ Ω, and assume the curve changes at a rate of 0.05 V/mA in this region.

§ 2.15

52. Determine the average ac resistance for the diode of Fig. 2.35 for the region greater than 0.7 V. How does this value compare with the dc and dynamic values at a forward current of 10 mA (as required in Problems 47 and 51)?

§ 2.16

53. Find the piecewise-linear equivalent circuit for the diode of Fig. 2.35. Assume that the straight-line segment for the semiconductor diode intersects the horizontal axis at 0.7 V (Si).

54. Determine V_R, I_D, V_D, and R_{dc} for the network of Fig. 2.40 using the equivalent circuit obtained in Problem 53.

55. If V in Problem 54 was increased to 50 V, would it be a reasonable approximation to assume that we are working with an ideal diode as in Fig. 2.1? Why?

56. Using the approximate characteristics of Fig. 2.41 for a silicon semiconductor diode, determine the level and appearance of the output voltage for the network of Fig. 2.42 and compare with the results of Example 2.4. Sketch the waveform of v_d using the approximate and piecewise-linear equivalent models for the diode.

57. If the supply voltage and resistance R of Fig. 2.43 were increased to $40 \sin \omega t$ and 2 K, respectively, sketch the appearance of v_R and v_d using the approximate and piecewise-linear equivalent models for the diode. Is it reasonable to assume that we are simply working with an ideal diode as in Fig. 2.1? Why?

58. Determine the peak-to-peak value of v_d in Fig. 2.45 if the dc level is increased to 4 V. Sketch the waveform of the ac sinusoidal voltage across the 100-Ω resistor.

59. Repeat Problem 58 for a dc level of 4 V and the 100-Ω resistor replaced by one of 1 K.

§ 2.17

60. Assuming an ideal diode in the circuit of Fig. 2.65, determine the output waveform for each of the input signals of Fig. 2.66.

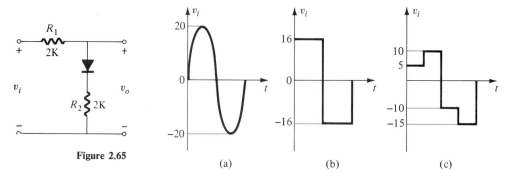

Figure 2.65

Figure 2.66

61. Repeat Problem 60 with the diode reversed.
62. Repeat Problem 60 with the diode and resistor R_1 interchanged.
63. Draw the output waveform for the circuit of Fig. 2.67a for each of the input signals of Fig. 2.66.

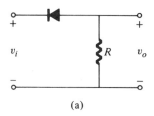
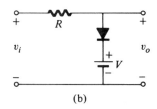

Figure 2.67

64. Draw the output waveform for the circuit of Fig. 2.67b for each of the input signals of Fig. 2.66. Use $V = 5$ V.

65. Repeat Problem 64 for $V = -10$ V.

66. Sketch the output waveform for the network of Fig. 2.68a for the input of Fig. 2.66b.

67. Sketch the output waveform for the network of Fig. 2.68b for the input of Fig. 2.66b. Use $V = 5$ V.

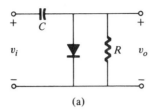 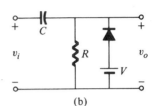

(a)　　　　　　　　　　(b)　　　　　　Figure 2.68

68. Repeat Problem 67 for $V = -10$ V.

69. Design a network that will only permit the +5- to +10-V swing of the input of Fig. 2.66c to pass through (riding on a +5-V level).

70. Design a network that will shift the input of Fig. 2.66b to a −10- to −42-V swing.

GLOSSARY

Ideal Diode A two-terminal device having a short-circuit and an open-circuit state.

Depletion Region A region of uncovered positive and negative ions between the p and n materials of a semiconductor diode.

Reverse Bias The condition established by an external bias that limits the flow of charge to a very small minority-carrier level.

Forward Bias The condition established by biasing conditions that results in a heavy flow of majority carriers through a semiconductor diode.

Saturation Current (I_s) The minority-carrier current flow through a semiconductor diode under reverse-bias conditions.

Zener Region The region in which the reverse-bias current of a semiconductor diode increases very sharply from the saturation level.

Peak Inverse Voltage (PIV) The maximum reverse-bias potential that can be applied across the diode before entering the Zener region.

Transition Capacitance (C_T) The capacitance associated with the depletion region of a semiconductor diode in the reverse-bias state.

Depletion Region Capacitance (C_T) A term often applied to transition capacitance.

Diffusion Capacitance (C_D) The capacitance associated with a forward-biased semiconductor diode that is heavily dependent on the rate at which charge is injected into the regions just outside the depletion region.

Storage Time (t_s) The time required for the minority carriers on one side of a junction to return to their majority-carrier state in the opposite material when a semiconductor diode switches from a forward-bias to a reverse-bias condition.

Transition Interval (t_t) The period of time associated with the current dropping from the storage level to the accepted nonconducting level during an on-to-off state transition.

Reverse Recovery Time (t_{rr}) The sum of the storage and transition intervals described above.

Heat Sink A metallic structure designed to remove the heat from the device in an effort to increase its power-handling capabilities.

Thermal Conductivity The ease with which heat can be transferred from one element to another.

Thermal Resistance A measure of the opposition of a medium to the flow of heat.

Power Derating Curve A plot often provided on specification sheets to demonstrate the effect of temperature on the power-handling capabilities of the device.

Average Rectified Current The level of current associated with the average value of the rectified signal.

Peak Repetitive Forward Current The maximum instantaneous value of repetitive forward current.

Peak Forward Surge Current The maximum permissible level of current flow for a short interval of time.

Grown Junction Diode A diode formed during the Czochralski crystal pulling process.

Alloy Process Diode A diode formed by placing a *p*-type impurity on an *n*-type substrate and heating the two until liquefaction occurs.

Diffusion Diode A diode manufactured through the diffusion of a heavy concentration of particles into a surrounding region of lesser concentration.

Epitaxial Growth The introduction of a n^+ base in the design process to achieve improved diode characteristics.

Point-Contact Diode A diode constructed by pressing a phosphorus bronze spring against an *n*-type substrate.

Diode Array The pattern of available diodes in an integrated circuit.

Quiescent Conditions (**Q**-*point*) Those fixed levels of voltage and current determined by the applied dc bias and resistor values.

Load Line A straight line superimposed on the characteristics of a device that defines the possible range of *Q*-points as determined by the applied bias and load.

Static Resistance The resistance determined by a dc voltage and current at a particular point of operation.

Dynamic Resistance The resistance determined by the ratio of a small change in voltage to the corresponding change in current along a line tangent to the point of operation.

Average ac Resistance The resistance determined by the ratio of a change in voltage to the corresponding change in current between two points on the characteristic curve.

Piecewise-Linear Equivalent Circuit A circuit derived from the straight-line approximation of the characteristics of a device to permit an examination of the response of a device to a first approximation.

Clipper A combination of elements, including at least one diode and resistor, that will clip off a portion of the applied signal.

Clamper A combination of elements having at least one diode, capacitor, and resistor that will clamp an input signal to a different level but not change its peak-to-peak swing.

CHAPTER 3

Zener and Other Devices

3.1 INTRODUCTION

There are a number of two-terminal devices having a single *p-n* junction like the semiconductor diode but with different modes of operation, terminal characteristics, and areas of application. A number, including Zener, Schottky, tunnel, and varicap diodes, photodiodes, LEDs, and solar cells will be introduced in this chapter. In addition, two-terminal devices of a different construction, such as the photoconductive cell, LCD (liquid crystal display), and thermistor will be examined.

3.2 ZENER DIODES

The Zener and avalance region of the semiconductor diode were discussed in detail in Section 2.3. It occurs at a reverse-bias potential of V_Z for the diode of Fig. 3.1a. The Zener diode is a device that is designed to make full use of this Zener region. If we present the characteristics as shown in Fig. 3.1b (the mirror image of Fig. 3.1a) to emphasize the region of interest by placing it in the first quadrant, a similarity appears between the characteristics and those of the ideal diode introduced in Section 2.2. The vertical rise approaches the ideal, although it is offset by a voltage V_Z. Any voltage from 0 to V_Z will result in an open-circuit equivalent as occurred below zero for the ideal diode. In the third quadrant the ideal diode remains an open-circuit while the Zener diode approaches the short-circuit state once again. In total, the almost vertical rise at V_Z has found numerous areas of application that make it a valuable device in electronic design. The first quadrant

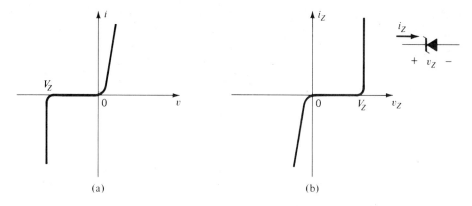

Figure 3.1 Zener diodes: (a) Zener potential; (b) characteristics and notation.

of Fig. 3.1b is defined by the polarities and current direction appearing next to the Zener diode symbol in the same figure.

The location of the Zener region can be controlled by varying the doping levels. An increase in doping, producing an increase in the number of added impurities, will decrease the Zener potential. Zener diodes are available having Zener potentials of 2.4 to 200 V with power ratings from $\frac{1}{4}$ to 50 W. Because of its higher temperature and current capability, silicon is usually preferred in the manufacture of Zener diodes.

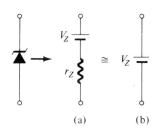

Figure 3.2 Zener equivalent circuit: (a) complete; (b) approximate.

The complete equivalent circuit of the Zener diode in the Zener region includes a small dynamic resistance and dc battery equal to the Zener potential, as shown in Fig. 3.2. For all applications to follow, however, we shall assume as a first approximation that the external resistors are much larger in magnitude than the Zener-equivalent resistor and that the equivalent circuit is simply that indicated in Fig. 3.2b.

A larger drawing of the Zener region is provided in Fig. 3.3 to permit a description of the Zener nameplate data appearing in Table 3.1 for a 1N961, Fairchild, 500 mW, 20% diode.

TABLE 3.1 Electrical Characteristics (25°C Ambient Temperature Unless Otherwise Noted)

Jedec Type	Zener Voltage Nominal, V_Z (V)	Test Current, I_{ZT} (mA)	Max Dynamic Impedance, $Z_{ZT}@I_{ZT}$ (Ω)	Maximum Knee Impedance, $Z_{ZK}@I_{ZK}$ (Ω) (mA)		Maximum Reverse Current, $I_R@V_R$ (μA)	Test Voltage, V_R (V)	Maximum Regulator Current, I_{ZM} (mA)	Typical Temperature Coefficient (%/°C)
1N961	10	12.5	8.5	700	0.25	10	7.2	32	+0.072

SEC. 3.2 ZENER DIODES

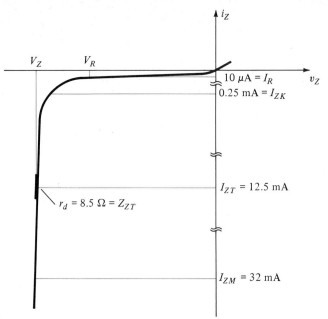

Figure 3.3 Zener test characteristics (Fairchild 1N961).

The term "nominal" associated with V_Z indicates that it is a typical average value. Since this is a 20% diode, the Zener potential can be expected to vary as 10 V ± 20% or from 8 to 12 V in its range of application. Also available are 10% and 5% diodes with the same specifications. The test current I_{ZT} is a typical operating level and Z_{ZT} is the dynamic impedance at this current level. The maximum knee impedance occurs at the knee current of I_{ZK}. The reverse saturation current is provided at a particular potential level and I_{ZM} is the maximum current for the 20% unit.

The temperature coefficient reflects the percent change in V_Z with temperature. It is defined by the equation

$$T_C = \frac{\Delta V_Z}{V_Z(T_1 - T_0)} \times 100\% \qquad (\%/°C) \qquad (3.1)$$

where ΔV_Z is the resulting change in Zener potential due to the temperature variation. Note in Fig. 3.4a that the temperature coefficient can be positive, negative, or even zero for different Zener levels. A positive value would reflect an increase in V_Z with an increase in temperature, while a negative value would result in a decrease in value with increase in temperature. The 24-V, 6.8-V, and 3.6-V levels refer to three Zener diodes having these nominal values within the same family of Zeners as the IN961. The curve for the 10-V IN961 Zener would naturally lie between the curves of the 6.8-V and 24-V devices. Note that it has a positive temperature coefficient for the entire region. Returning to Eq. (3.1), T_0 is the temperature at which V_Z is provided (normally room temperature — 25°C) and T_1 is the new level. Example 3.1 will demonstrate the use of Eq. (3.1).

EXAMPLE 3.1 Determine the nominal voltage for a IN961 Fairchild Zener diode at a temperature of 100°C.

Solution:
From Eq. (3.1):

$$\Delta V_Z = \frac{T_C V_Z}{100}(T_1 - T_0)$$

Substituting yields

$$\Delta V_Z = \frac{(0.072)(10)}{100}(100 - 25)$$

$$= (0.0072)(75)$$

$$= 0.54 \text{ V}$$

and because of the positive temperature coefficient, the new Zener potential, defined by V'_Z, is

$$V'_Z = V_Z + 0.54$$

$$= \mathbf{10.54 \text{ V}}$$

Figure 3.4 Electrical characteristics for a 500 mW Fairchild Zener diode. (Courtesy Fairchild Camera and Instrument Corporation.)

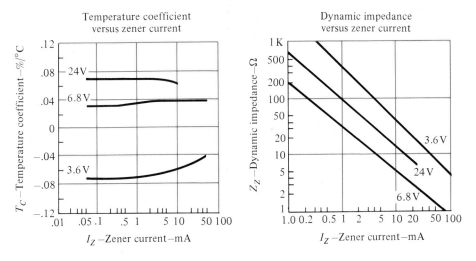

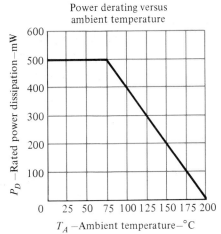

The variation in dynamic impedance (fundamentally, its series resistance) with current appears in Fig. 3.4b. Again, the 10-V Zener appears between the 6.8-V and 24-V Zeners. Note that the heavier the current (or the farther up the vertical rise you are in Fig. 3.1b), the less the resistance value. And also note that as you approach the knee of the curve and beyond, the resistance increases to significant levels.

The power derating curve of Fig. 3.4c is very similar to that described for the semiconductor diode. The linear power derating factor for Zener diodes in this family is 3.33 W/°C. The power rating at the 100°C employed in Example 3.1 can be determined from the derating factor in the following manner:

$$P_{100°C} = P_{75°C} - (D_F)(\Delta T)$$
$$= 500 \times 10^{-3} - (3.33 \times 10^{-3})(25)$$
$$= 500 \times 10^{-3} - 83.25 \times 10^{-3}$$
$$= \mathbf{416.75\ mW}$$

while the graph would provide a value of approximately 400 mW.

The terminal identification and the casing for a variety of Zener diodes appear in Fig. 3.5.

Figure 3.6 is an actual photograph of a variety of Zener devices. Note that their appearance is very similar to the semiconductor diode. A few areas of application for the Zener diode will not be examined.

It is often necessary to have a fixed *reference* voltage in a network for biasing and comparison purposes. This can be accomplished using a Zener diode as shown in Fig. 3.7. The variation in dc supply voltage due to any number of reasons has been included as a small sinusoidal signal.

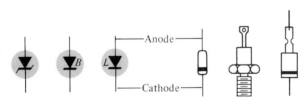

Figure 3.5 Zener terminal identification and symbols.

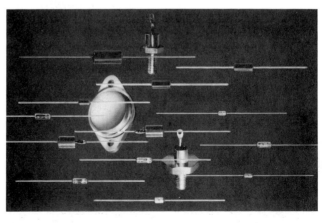

Figure 3.6 Zener diodes. (Courtesy Siemens Corporation.)

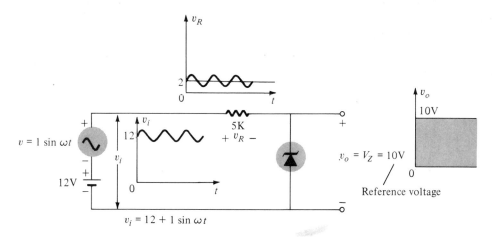

Figure 3.7 Reference voltage.

Since v_i is always greater than 10 V, the Zener diode will always be in the "on" state, a condition defined by the vertical-rise region of Fig. 3.1b. Our analysis here will employ only the reduced equivalent circuit of Fig. 3.2. The output voltage V_o, therefore, will remain fixed at the Zener potential of 10 V, our *reference* potential.

The voltage appearing across the 5-K resistor is then the difference between the two as defined by Kirchhoff's voltage law:

$$v_R = v_i - v_o$$
$$= (12 + 1 \sin \omega t) - 10$$
$$= 2 + 1 \sin \omega t$$

as shown in Fig. 3.2.

Two reference levels can be established by the network of Fig. 3.8. Two back-to-back Zeners can be used as an ac regulator (Fig. 3.9). For the sinusoidal signal v_i the circuit will appear as shown in Fig. 3.9b at the instant $v_i = 10$ V. The region of operation for each diode is indicated in the adjoining figure. Note that the impedance associated with Z_1 is very small, or essentially a short, since it is in series with 5 K, while the impedance of Z_2 is very large corresponding to the open-circuit representation. Since Z_2 is an open circuit, $v_o = v_i = 10$ V. This will continue to be the case until v_i is slightly greater than 20 V. Then Z_2 will enter the low-resistance region (Zener region) and Z_1 will for all practical purposes be a short circuit and Z_2 will be replaced by $V_Z = 20$ V. The resultant output waveform is indicated in the same figure. Note that the waveform is not purely sinusoidal, but its rms (effective) value is closer to the desired 20-V peak sinusoidal waveform than the sinusoidal input having a peak value of 22 V.

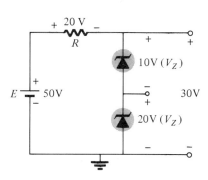

Figure 3.8 Two reference voltages.

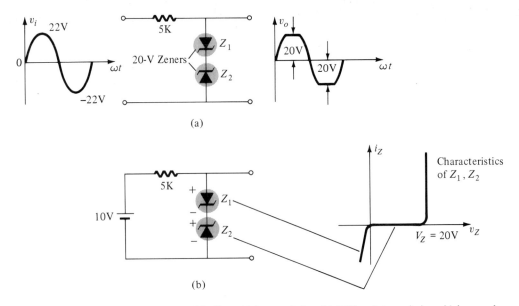

Figure 3.9 Sinusoidal ac regulation: (a) 40-V peak-to-peak sinusoidal ac regulator; (b) circuit operation at $V_i = 10$ V.

The circuit of Fig. 3.9a can be extended to that of a simple square-wave generator (due to its clipping action) if the signal v_i is increased to perhaps a 40-V peak with 10-V Zeners. The resultant waveform is indicated in Fig. 3.10.

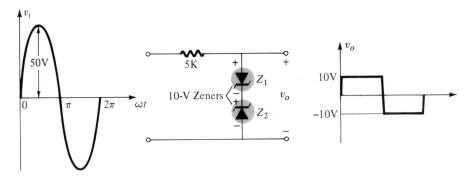

Figure 3.10 Simple square-wave generator.

Finally, let us use the Zener diode to maintain a fixed voltage across a load for a variation in the load (a voltage regulator). The basic Zener regulator appears in Fig. 3.11. In order to maintain V_Z at 10 V (the Zener potential), the diode must first be in the "on" state. Certainly, for values of v_i less than 10 V the Zener device has not reached its Zener potential. Will it "fire" at 12 V? To determine its state, first remove the Zener diode from the network and find the voltage across its open-circuit terminals using the voltage divider rule (Fig. 3.12).

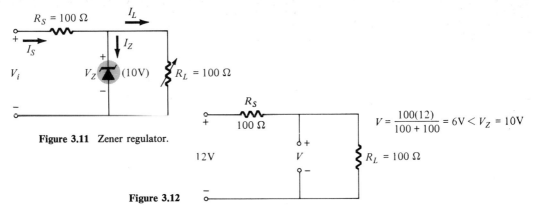

Figure 3.11 Zener regulator.

$$V = \frac{100(12)}{100 + 100} = 6\text{V} < V_Z = 10\text{V}$$

Figure 3.12

The resulting voltage is still less than the needed 10 V and the diode remains open. By setting V_Z to 10 V, we can calculate the required applied voltage in the following manner:

$$V_Z = 10 = \frac{100(V_i)}{100 + 100} \quad \text{and} \quad V_i = 20 \text{ V}$$

Let us now examine the behavior of the network with $V_i = 100$ V (Fig. 3.13).

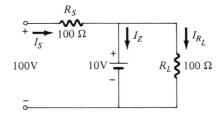

Figure 3.13

The voltage

$$V_{R_S} = 100 - 10 = 90 \text{ V}$$

$$I_S = \frac{90}{100} = 900 \text{ mA}$$

with

$$I_{R_L} = \frac{10}{100} = 100 \text{ mA}$$

and

$$I_Z = I_S - I_L = 900 - 100 = 800 \text{ mA}$$

EXAMPLE 3.2 For the network of Fig. 3.14, determine the range of I_L that will result in V_{R_L} being maintained at 10 V.

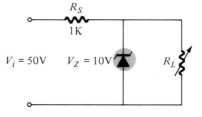

Figure 3.14 Network for Example 3.2.

SEC. 3.2 ZENER DIODES

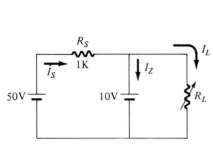

Figure 3.15

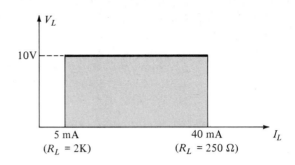

Figure 3.16 V_{R_L} vs. I_L for the network of Fig. 3.14.

Solution: The minimum value of R_L that will ensure that the diode is in the on state can be determined through Fig. 3.15.

$$V_{R_L} = V_Z = 10 = \frac{R_{L(\min)}(50\text{ V})}{R_{L(\min)} + 1000}$$

and
$$10{,}000 + 10R_{L(\min)} = 50R_{L(\min)}$$

or
$$40R_{L(\min)} = 10{,}000$$

and
$$R_{L(\min)} = 250\ \Omega$$

The minimum load resistance corresponds with the maximum I_L and

$$I_{L(\max)} = \frac{V_Z}{R_{L(\min)}} = \frac{10}{250} = 40\text{ mA}$$

$I_{L(\min)}$ is determined by examining the network with maximum I_Z (I_{ZM}). When the diode is in the "on" state, $V_S = 50 - 10 = 40$ V and I_S is maintained at $I_S = 40/1000 = 40$ mA. Since $I_L = I_S - I_Z$, I_L will be a minimum when I_Z is a maximum, or I_{ZM}. Therefore,

$$I_{L(\min)} = I_S - I_{ZM} = 40 - 35 = 5\text{ mA}$$

and incidentally

$$R_{L(\max)} = \frac{V_Z}{I_{L(\min)}} = \frac{10}{5\text{ mA}} = 2\text{ K}$$

A plot of V_L versus I_L appears in Fig. 3.16.

3.3 SCHOTTKY BARRIER (HOT-CARRIER) DIODE

In recent years there has been increasing interest in a two-terminal device referred to as a *Schottky-barrier, surface-barrier,* or *hot-carrier* diode. Its areas of application were first limited to the very high frequency range as a competitor for the point-contact diode. It succeeded in this test because it was significantly more rugged and had a quicker response time (especially important at high frequencies) and a lower noise figure (a quantity of real importance in high-frequency applications). In recent years, however, it is appearing more and more in low voltage/high current power supplies and ac-to-dc convertors. Other areas of application of the device include radar systems, Schottky TTL logic for computers, mixers and detectors in communication equipment, instrumentation, and analog-to-digital convertors.

Its construction is very different from the conventional *p-n* junction in that a metal-semiconductor junction is created such as shown in Fig. 3.17. The semiconductor is normally *n*-type silicon (although *p*-type silicon is sometimes used), while a host of different metals, such as molybdenum, platinum, chrome, or tungsten, are used. Different construction techniques will result in a different set of characteristics for the device, such as increased frequency range, lower forward bias, and so on. Priorities do not permit an examination of each technique here, but information will usually be provided by the manufacturer. In general, however, Schottky diode construction results in a more uniform junction region and increased ruggedness compared to the point-contact diode.

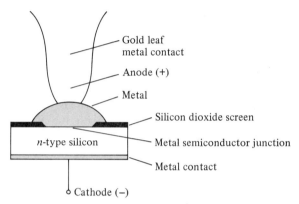

Figure 3.17 Passivated hot-carrier diode.

In both materials, the electron is the majority carrier. In the metal, the level of minority carriers (holes) is insignificant. When the materials are joined the electrons in the *n*-type silicon semiconductor material immediately flow into the adjoining metal, establishing a heavy flow of majority carriers. Since the injected carriers have a very high kinetic energy level compared to the electrons of the metal, they are commonly called "hot carriers." In the conventional *p-n* junction there was the injection of minority carriers into the adjoining region. Here the electrons are injected into a region of the same electron plurality. Schottky diodes are therefore unique, in that conduction is entirely by majority carriers. The heavy flow of electrons into the metal creates a region near the junction surface depleted of carriers in the silicon material—much like the depletion region in the *p-n* junction diode. The additional carriers in the metal establish a negative wall in the metal at the boundary between the two materials. The net result is a "surface barrier" between the two materials, preventing any further current flow. That is, any electrons (negatively charged) in the silicon material face a carrier-free region and a negative wall at the surface of the metal.

The application of a forward bias as shown in Fig. 3.17 will reduce the strength of the negative barrier through the attraction of the applied positive potential for electrons from this region. The result is a return to the heavy flow of electrons across the boundary, the magnitude of which is controlled by the level of the applied bias potential. The barrier at the junction for a Schottky diode is less than that of the *p-n* junction device in both the forward- and reverse-bias regions. The result is therefore a higher current at the same applied bias in the forward- and reverse-bias regions. This is a desirable effect in the forward-bias region but highly undesirable in the reverse-bias region.

The exponential rise in current with forward bias is described by Eq. (2.1) but with η dependent on the construction technique (1.05 for the metal whisker type of construction, which is somewhat similar to the point-contact diode). In the reverse-bias region the current I_s is due primarily to those electrons in the metal passing into the semiconductor material. One of the areas of continuing research on the Schottky diode centers on reducing the high leakage currents that result with temperatures over 100°C. Through design, improvement units are now becoming available that have a temperature range from $-65°C$ to $+150°C$. At room temperature, I_s is typically in the microampere range for low-power units and milliampere range for high-power devices, although it is typically larger than that encountered using conventional *p-n* junction devices with the same current limits. In addition, even though Schottky diodes exhibit better characteristics than the point-contact diode in the reverse-bias region as shown in Fig. 3.18, the PIV of these diodes is usually significantly less than that of a comparable *p-n* junction unit. Typically, for a 50-A unit, the PIV of the Schottky diode would be about 50 V as compared to 150 V for the *p-n* junction variety. Recent advances, however, have resulted in Schottky diodes with PIVs greater than 100 V at this current level. It is obvious from the characteristics of Fig. 3.18 that the Schottky diode is closer to the ideal set of characteristics than the point contact and has levels of V_o less than the typical silicon semiconductor *p-n* junction. The level of V_o for the "hot-carrier" diode is controlled to a large measure by the metal employed. There exists a required trade-off between temperature range and level of V_o.

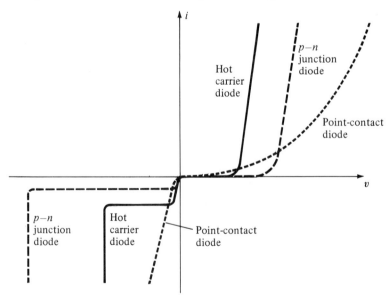

Figure 3.18 Comparison of characteristics of hot-carrier, point-contact, and *p-n* junction diodes.

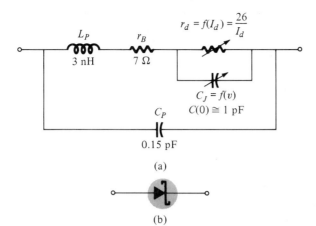

Figure 3.19 Schottky (hot-carrier) diode: (a) equivalent circuit; (b) symbol.

An increase in one appears to correspond to a resulting increase in the other. In addition, the lower the range of allowable current levels, the lower the value of V_o. For some low-level units, the value of V_o can be assumed to be essentially zero on an approximate basis. For the middle and high range, however, a value of 0.2 V would appear to be a good representative value.

The maximum current rating of the device is presently limited to about 75 A, although 100-A units appear to be on the horizon. One of the primary areas of application of this diode is in *switching power supplies* that operate in the frequency range of 20 kHz or more. A typical unit at 25°C may be rated at 50 A at a forward voltage of 0.6 V with a recovery time of 10 ns for use in one of these supplies. A *p-n* junction device with the same current limit of 50 A may have a forward voltage drop of 1.1 V and a recovery time of 30 to 50 ns. The difference in forward voltage may not appear significant, but consider the power dissipation difference: $P_{\text{hot carrier}} = (0.6)(50) = 30$ W compared to $P_{p\text{-}n} = (1.1)(50) = 55$ W, which is a measurable difference when efficiency criteria must be met. There will, of course, be a higher dissipation in the reverse-bias region for the Schottky diode due to the higher leakage current, but the total power loss in the forward- and reverse-bias regions is still significantly improved as compared to the *p-n* junction device.

Recall from our discussion of reverse recovery time for the semiconductor diode that the injected minority carriers accounted for the high level of t_{rr} (the reverse recovery time). The absence of minority carriers at any appreciable level in the Schottky diode results in a reverse recovery time of significantly lower levels, as indicated above. This is the primary reason Schottky diodes are so effective at frequencies approaching 20 GHz, where the device must switch states at a very high rate. For higher frequencies the point-contact diode, with its very small junction area, is still employed.

The equivalent circuit for the device (with typical values) and a commonly used symbol appear in Fig. 3.19. A number of manufacturers prefer to use the standard diode symbol for the device, since its function is essentially the same. The inductance L_P and capacitance C_P are package values. R_s is the series resistance, which includes the contact and bulk resistance. The resistance r_d and capacitance C_J are values defined by equations introduced in earlier sections. For many applica-

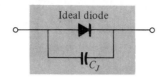

Figure 3.20 Approximate equivalent circuit for the Schottky diode.

tions, an excellent approximate equivalent circuit simply includes an ideal diode in parallel with the junction capacitance as shown in Fig. 3.20.

A number of hot-carrier rectifiers manufactured by Motorola Semiconductor Products, Inc., appear in Fig. 3.21 with their specifications and terminal identification. Note that the maximum forward, voltage drop V_F does not exceed 0.65 V for any of the devices, while this was essentially V_o for a silicon diode.

Three sets of curves for the Hewlett-Packard 5082-2300 series of general-purpose Schottky barrier diodes are provided in Fig. 3.22. Note at $T = 100°C$ in Fig. 3.22a that V_F is only 0.1 V at a current of 0.01 mA. Note also that the reverse current has been limited to nanoamperes in Fig. 3.22b and the capacitance to 1 pF in Fig. 3.22c to ensure a high switching rate.

3.4 VARICAP (VARACTOR) DIODES

Varicap [also called varactor, VVC (voltage-variable-capacitance), or tuning] diodes are semiconductor, voltage-dependent, variable capacitors. Their mode of operation depends on the capacitance that exists at the *p-n* junction when the element is reverse-biased. Under reverse-bias conditions, it was established in Section 2.3 that there is a region of uncovered charge on either side of the junction that together make up the depletion region and define the depletion width W_d. The transition capacitance (C_T) established by the isolated uncovered charges is determined by

$$C_T = \frac{\epsilon A}{W_d} \tag{3.2}$$

where ϵ is the permittivity of the semiconductor materials, A is the *p-n* junction area, and W_d is the depletion width.

As the reverse-bias potential increases, the width of the depletion region increases, which in turn reduces the transition capacitance. The characteristics of a typical commercially available varicap diode appear in Fig. 3.23. Note the initial sharp decline in C_T with increase in reverse bias. The normal range of V_r for VVC diodes is limited to about 20 V. In terms of the applied reverse bias, the transition capacitance is given by

$$C_T = \frac{K}{(V_o + V_r)^n} \tag{3.3}$$

where $K =$ constant determined by the semiconductor material and construction technique
$V_o =$ knee potential as defined in Section 2.3
$V_r =$ applied reverse-bias potential
$n = \frac{1}{2}$ for alloy junctions and $\frac{1}{3}$ for diffused junctions

I_O, Average rectified forward current (amperes)														
	0.5	1.0		3.0		3.0	5.0	15		25		40		
Case	51-02 (DO-7) Glass	59-04 Plastic		267 Plastic		60 Metal		245 (DO-4) Metal				257 (DO-5) Metal		430-2 (DO-21) Metal
Anode: / Cathode: 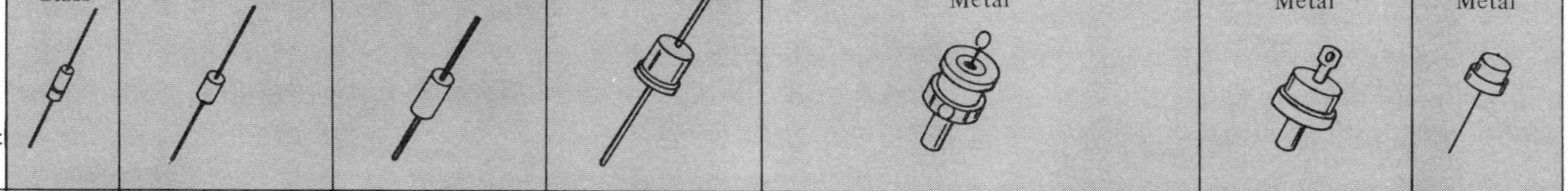														
V_{RRM} (Volts)														
20	MBR020	1N5817	MBR120P	1N5820	MBR320P	MBR320M	1N5823	1N5826	MBR1520	1N5829	MBR2520	1N5832	MBR4020	MBR4020PF
30	MBR030	1N5818	MBR130P	1N5821	MBR330P	MBR330M	1N5824	1N5827	MBR1530	1N5830	MBR2530	1N5833	MBR4030	MBR4030PF
35			MBR135P		MBR335P	MBR335M			MBR1535		MBR2535		MBR4035	MBR4035PF
40		1N5819	MBR140P	1N5822	MBR340P	MBR340M	1N5825	1N5828	MBR1540	1N5831	MBR2540	1N5834	MBR4040	
I_{FSM} (Amps)	5.0	100	50	250	200	500	500	500	500	800	800	800	800	800
T_C @ Rated I_O (°C)								85	80	85	80	75	70	50
T_J Max (°C)	125	125	125	125	125	125	125	125	125	125	125	125	125	125
Max V_F @ $I_{FM} = I_O$	0.50	*0.60	0.65	*0.525	0.60	0.45 @ 5 A	*0.38	*0.50	0.55	*0.48	0.55	*0.59	0.63	0.63

. . . Schottky barrier devices, ideal for use in low-voltage, high-frequency power supplies and as free-wheeling diodes. These units feature very low forward voltages and switching times estimated at less than 10 ns. They are offered in current ranges of 0.5 to 5.0 amperes and in voltages to 40.

V_{RRM}—respective peak reverse voltage
I_{FSM}—forward current, surge peak
I_{FM}—forward current, maximum

Figure 3.21 Motorola Schottky barrier devices. (Courtesy Motorola Semiconductor Products, Incorporated.)

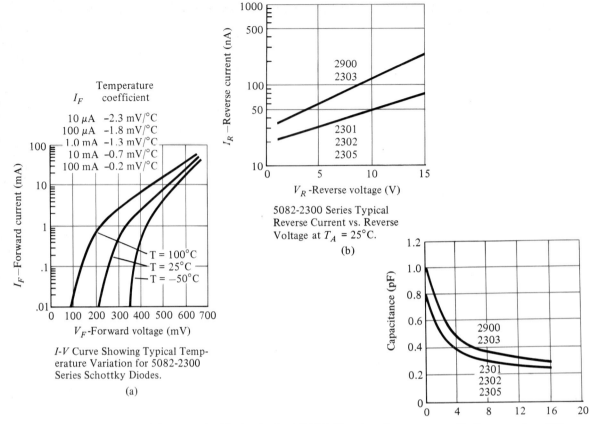

Figure 3.22 Characteristic curves for Hewlett-Packard 5082-2300 series of general-purpose Schottky barrier diodes. (Courtesy Hewlett-Packard Corporation.)

(a) I-V Curve Showing Typical Temperature Variation for 5082-2300 Series Schottky Diodes.

(b) 5082-2300 Series Typical Reverse Current vs. Reverse Voltage at $T_A = 25°C$.

(c) 5082-2300 Series Typical Capacitance vs. Reverse Voltage at $T_A = 25°C$.

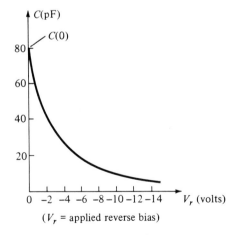

(V_r = applied reverse bias)

Figure 3.23 Varicap characteristics: C(pF) vs. V_r.

In terms of the capacitance at the zero-bias condition $C(0)$, the capacitance as a function of V_r is given by

$$C_T(V_r) = \frac{C(0)}{\left(1 + \dfrac{V_r}{V_o}\right)^n} \qquad (3.4)$$

The symbols most commonly used for the varicap diode and a first approximation for its equivalent circuit in the reverse-bias region are shown in Fig. 3.24. Since we are in the reverse-bias region, the resistor in the equivalent circuit is very large in magnitude—typically 1 M or larger—while R_S, the geometric resistance of the diode, is, as indicated in Fig. 3.24, very small. The magnitude of C

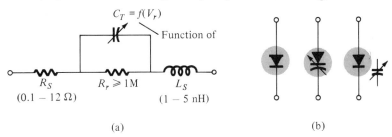

Figure 3.24 Varicap diode: (a) equivalent circuit in the reverse-bias region; (b) symbols.

will vary from about 2 to 100 pF depending on the varicap considered. To ensure that R_r is as large (for minimum leakage current) as possible, silicon is normally used in varicap diodes. The fact that the device will be employed at very high frequencies requires that we include the inductance L_s even though it is measured in nanohenries. Recall that $X_L = 2\pi fL$ and a frequency of 1 GHz with $L_s = 1$ nH will result in an $X_{L_s} = 2\pi fL = (6.28)(10^{12})(10^{-9}) = 6.28$ K—a large value. There is obviously, therefore, a frequency limit associated with the use of each varicap diode.

Assuming the proper frequency range, a low value of R_s, and X_{L_s} compared to the other series elements, then the equivalent circuit for the varicap of Fig. 3.24a can be replaced by the variable capacitor alone. The complete data sheet and its characteristic curves appear in Figs. 3.25 and 3.26, respectively. The C_3/C_{25} ratio in Fig. 3.25 is the ratio of capacitance levels at reverse-bias potentials of 3 and 25 V. It provides a quick estimate of how much the capacitance will change with reverse-bias potential. The figure of merit is a quantity of consideration in the application of the device and is a measure of the ratio of energy stored by the capacitive device per cycle to the energy dissipated (or lost) per cycle. Since energy loss is seldom considered a positive attribute, the higher its relative value the better. The resonant frequency of the device is determined by $f_0 = 1/2\pi\sqrt{LC}$ and affects the range of application of the device.

In Fig. 3.26, most quantities are self-explanatory. However, the *capacitance temperature coefficient* is defined by

$$TC_C = \frac{\Delta C}{C_0(T_1 - T_0)} \times 100 \qquad (\%/°C) \qquad (3.5)$$

where ΔC is the change in capacitance due to the temperature change $T_1 - T_0$ and C_0 is the capacitance at T_0 for a particular reverse-bias potential. For example, Fig. 3.25 indicates that $C_0 = 29$ pF with $V_R = 3$ V and $T_0 = 25°C$. A change

Figure 3.25 Electrical characteristics for a VHF/FM Fairchild Varactor Diode. (Courtesy Fairchild Camera and Instrument Corporation.)

BB139
VHF/FM VARACTOR DIODE
DIFFUSED SILICON PLANAR

- C_3/C_{25} ... 5.0–6.5
- **MATCHED SETS** (Note 2)

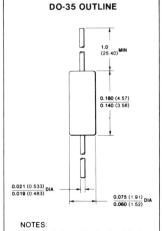

ABSOLUTE MAXIMUM RATINGS (Note 1)

Temperatures
Storage Temperature Range   −55°C to +150°C
Maximum Junction Operating Temperature   +150°C
Lead Temperature   +260°C

Maximum Voltage
WIV   Working Inverse Voltage   30 V

NOTES:
Copper clad steel leads, tin plated
Gold plated leads available
Hermetically sealed glass package
Package weight is 0.14 gram

ELECTRICAL CHARACTERISTICS (25°C Ambient Temperature unless otherwise noted)

SYMBOL	CHARACTERISTIC	MIN	TYP	MAX	UNITS	TEST CONDITIONS
BV	Breakdown Voltage	30			V	$I_R = 100$ μA
I_R	Reverse Current		10	50	nA	$V_R = 28$ V
			0.1	0.5	μA	$V_R = 28$ V, $T_A = 60°C$
C	Capacitance		29		pF	$V_R = 3.0$ V, $f = 1$ MHz
		4.3	5.1	6.0	pF	$V_R = 25$ V, $f = 1$ MHz
C_3/C_{25}	Capacitance Ratio	5.0	5.7	6.5		$V_R = 3$ V/25 V, $f = 1$ MHz
Q	Figure of Merit		150			$V_R = 3.0$ V, $f = 100$ MHz
R_S	Series Resistance		0.35		Ω	$C = 10$ pF, $f = 600$ MHz
L_S	Series Inductance		2.5		nH	1.5 mm from case
f_0	Series Resonant Frequency		1.4		GHz	$V_R = 25$ V

NOTES:
1. These ratings are limiting values above which the serviceability of the diode may be impaired.
2. The capacitance difference between any two diodes in one set is less than 3% over the reverse voltage range of 0.5 V to 28 V.

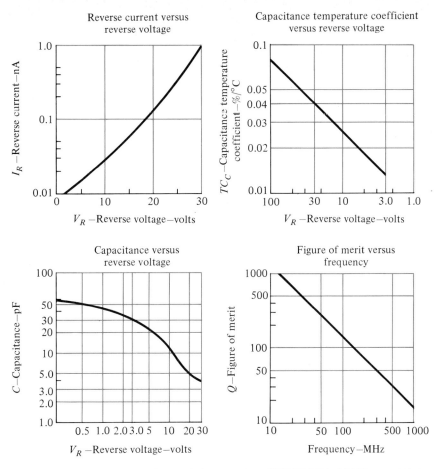

Figure 3.26 Characteristic curves for a VHF/FM Fairchild Varactor Diode. (Courtesy Fairchild Camera and Instrument Corporation.)

in capacitance ΔC could then be determined using Eq. (3.5) by simply substituting the new temperature T_1 and the TC_C as determined from the graph (= 0.013). At a new V_R the value of TC_C would change accordingly. Returning to Fig. 3.25, note that the maximum frequency appearing is 600 MHz. At this frequency,

$$X_L = 2\pi f L = (6.28)(600 \times 10^6)(2.5 \times 10^{-9}) = 9.42 \ \Omega$$

normally a quantity of sufficiently small magnitude to be ignored.

Some of the high-frequency (as defined by the small capacitance levels) areas of application include FM modulators, automatic-frequency-control devices, adjustable band-pass filters, and parametric amplifiers.

3.5 POWER DIODES

There are a number of diodes designed specifically to handle the high-power and high-temperature demands of some applications. The most frequent use of power diodes occurs in the rectification process, in which ac signals (having zero average value) are converted to one having an average or dc level. When used in this capacity, diodes are normally referred to as *rectifiers*.

The majority of the power diodes are constructed using silicon because of its higher current, temperature, and PIV ratings. The higher current demands require that the junction area be larger, to ensure that there is a low forward diode resistance. If the forward resistance were too large, the I^2R losses would be excessive. The current capability of power diodes can be increased by placing two or more in parallel and the PIV rating can be increased by stacking the diodes in series.

Various types of power diodes and their current rating have been provided in Fig. 3.27a. The high temperatures resulting from the heavy current flow require, in many cases, that heat sinks be used to draw the heat away from the element. A few of the various types of heat sinks available are shown in Fig. 3.27b. If heat sinks are not employed, stud diodes are designed to be attached directly to the chassis, which in turn will act as the heat sink. The analysis associated with the temperature limitations is exactly the same as presented for semiconductor diodes in Section 2.6.

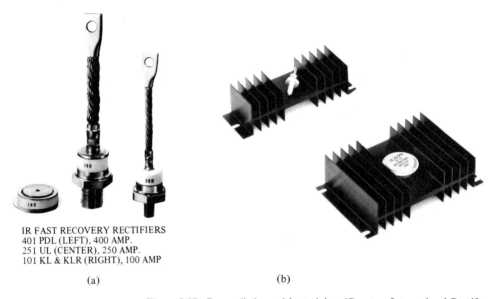

IR FAST RECOVERY RECTIFIERS
401 PDL (LEFT), 400 AMP.
251 UL (CENTER), 250 AMP.
101 KL & KLR (RIGHT), 100 AMP

(a)

(b)

Figure 3.27 Power diodes and heat sinks. (Courtesy International Rectifier Corporation.)

3.6 TUNNEL DIODES

The tunnel diode was first introduced by Leo Esaki in 1958. Its characteristics, shown in Fig. 3.28, are different from any diode discussed thus far in that it has a negative resistance region. In this region, an increase in terminal voltage results in a reduction in diode current.

The tunnel diode is fabricated by doping the semiconductor materials that will form the *p-n* junction at a level one hundred to several thousand times that

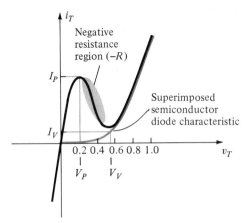

Figure 3.28 Tunnel diode characteristic.

of a typical semiconductor diode. This will result in a greatly reduced depletion region, of the order of magnitude of 10^{-6} cm, or typically about $\frac{1}{100}$ the width of this region for a typical semiconductor diode. It is this thin depletion region that many carriers can "tunnel" through, rather than attempt to surmount, at low forward-bias potentials that accounts for the peak in the curve of Fig. 3.28. For comparison purposes, a typical semiconductor diode characteristic has been superimposed on the tunnel diode characteristic of Fig. 3.28.

This reduced depletion region results in carriers "punching through" at velocities that far exceed that available with conventional diodes. The tunnel diode can therefore be used in high-speed applications such as in computers, where switching times in the order of nanoseconds or picoseconds are desirable.

You will recall from a previous section that an increase in the doping level will drop the Zener potential. Note the effect of a very high doping level on this region in Fig. 3.28. The semiconductor materials most frequently used in the manufacture of tunnel diodes are germanium and gallium arsenide. The ratio I_p/I_v is very important for computer applications. For germanium it is typically 10:1, while for gallium arsenide it is closer to 20:1.

The peak current, I_p, of a tunnel diode can vary from a few microamperes to several hundred amperes. The peak voltage, however, is limited to about 600 mV. For this reason, a simple VOM with an internal dc battery potential of 1.5 V can severely damage a tunnel diode if used improperly.

The tunnel diode equivalent circuit in the negative-resistance region is provided in Fig. 3.29, with the symbols most frequently employed for tunnel diodes.

Figure 3.29 Tunnel diode: (a) equivalent circuit; (b) symbols.

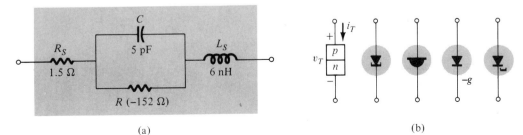

(a)   (b)

The values for each parameter are for the 1N2939 GE tunnel diode whose specifications appear in Table 3.2. The inductor L_S is due mainly to the terminal leads. The resistor R_S is due to the leads, ohmic contact at the lead-semiconductor junction, and the semiconductor materials themselves. The capacitance C is the junction diffusion capacitance and the R is the negative resistance of the region. The negative resistance finds application in oscillators to be described later.

TABLE 3.2 Specifications: Ge 1N2939

	Min.	Typ.	Max.	
Absolute maximum ratings (25°C)				
Forward current (−55 to + 100°C)		5 mA		
Reverse current (−55 to + 100°C)		10 mA		
Electrical characteristics (25°C)($\frac{1}{8}$ in. leads)				
I_P	0.9	1.0	1.1	mA
I_V		0.1	0.14	mA
V_P	50	60	65	mV
V_V		350		mV
Reverse voltage ($I_R = 1.0$ mA)			30	mV
Forward peak point current voltage, V_{fp}	450	500	600	mV
I_p/I_v		10		
$-R$		−152		Ω
C		5	15	pF
L_s		6		nH
R_S		1.5	4.0	Ω

Note the lead length of $\frac{1}{8}$ in. included in the specifications. An increase in this length will cause L_s to increase. In fact, it was given for this device that L_s will vary 1 to 12 nH, depending on lead length. At high frequencies ($X_{L_s} = 2\pi f L_s$) this factor can take its toll.

The fact that $V_{fp} = 500$ mV (typ.) and $I_{forward}$ (max.) $= 5$ mA indicates that tunnel diodes are low-power devices [$P_D = (0.5)(5 \times 10^{-3}) = 2.5$ mW], which is also excellent for computer applications. A rendering of the device appears in Fig. 3.30.

Figure 3.30 A GE IN2939 tunnel diode. (Rendered from specifications courtesy General Electric Corporation.)

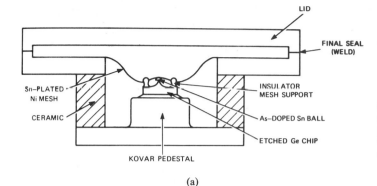

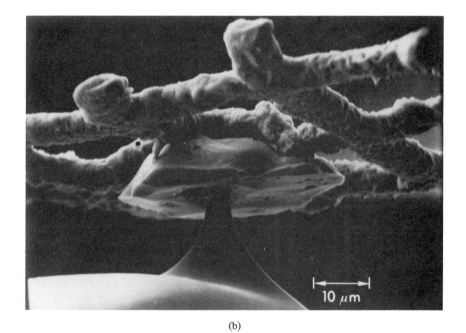

Figure 3.31 Tunnel diode: (a) construction; (b) photograph. (Courtesy *COMSAT Technical Review*, Vol. 3, No. 1, Spring 1973, P. F. Varadi and T. D. Kirkendall.)

Although the use of tunnel diodes in present-day high-frequency systems has been dramatically stalled by manufacturing techniques that suggest alternatives to the tunnel diode, its simplicity, linearity, low power drain, and reliability ensure its continued life and application. The basic construction of an advance design tunnel diode appears in Fig. 3.31 with a photograph of the actual junction.

3.7 PHOTODIODE

The interest in light-sensitive devices has been increasing at an almost exponential rate in recent years. The resulting field of *optoelectronics* will be receiving a great deal of research interest as efforts are made to improve efficiency levels.

Through the advertising media, the layperson has become quite aware that light sources offer a unique source of energy. This energy, transmitted as discrete packages called *photons,* has a level directly related to the frequency of the traveling light wave as determined by the following equation:

$$W = hf \quad \text{(joules)} \tag{3.6}$$

where h is called Planck's constant and is equal to 6.624×10^{-34} joule-seconds. It clearly states that since h is a constant, the energy associated with incident light waves is directly related to the frequency of the traveling wave.

The frequency is, in turn, directly related to the wavelength (distance between successive peaks) of the traveling wave by the following equation:

$$\lambda = \frac{v}{f} \tag{3.7}$$

where
 λ = wavelength, meters
 v = velocity of light, 3×10^8 m/s
 f = frequency, hertz, of the traveling wave

The wavelength is usually measured in angstrom units (Å) or microns (μ), where

$$1 \text{ Å} = 10^{-10} \text{ m} \quad \text{and} \quad 1 \mu = 10^{-6} \text{ m}$$

The wavelength is important because it will determine the material to be used in the optoelectronic device. The relative spectral response for Ge, Si, and selenium is provided in Fig. 3.32. The visible light spectrum has also been indicated with an indication of the wavelength associated with the various colors.

The number of free electrons generated in each material is proportional to the *intensity* of the incident light. Light intensity is a measure of the amount of *luminous flux* falling in a particular surface area. Luminous flux is normally measured in *lumens* (lm) or watts. The two units are related by

$$1 \text{ lm} = 1.496 \times 10^{-3} \text{ W}$$

The light intensity is normally measured in lm/ft², foot-candles (fc), or W/m², where

$$1 \text{ lm/ft}^2 = 1 \text{ fc} = 1.609 \times 10^{-12} \text{ W/m}^2$$

The photodiode is a semiconductor *p-n* junction device whose region of operation is limited to the reverse-bias region. The basic biasing arrangement, construction, and symbol for the device appear in Fig. 3.33.

Recall from Chapter 2 that a reverse saturation current will flow under reverse-bias conditions whose magnitude is normally limited to a few microamperes. It is due solely to the thermally generated minority carriers in the *n*- and *p*-type materials. The application of light to the junction will result in a transfer of energy from the incident traveling light waves (in the form of photons) to the atomic structure, resulting in an increased number of minority carriers and an increased level of reverse current. This is clearly shown in Fig. 3.34 for different

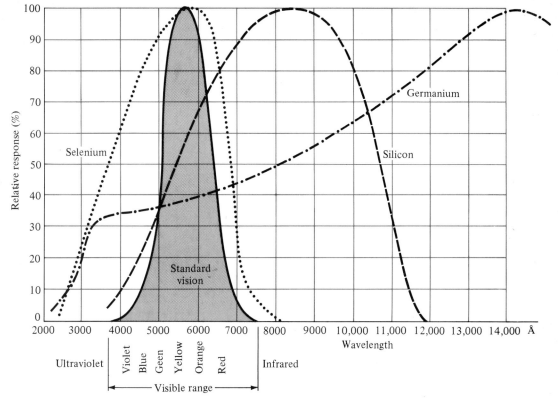

Figure 3.32 Relative spectral response for Si, Ge, and selenium as compared to the human eye.

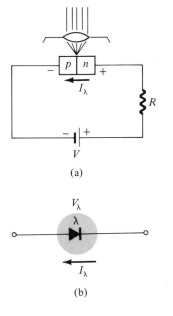

Figure 3.33 Photodiode: (a) basic biasing arrangement and construction; (b) symbol.

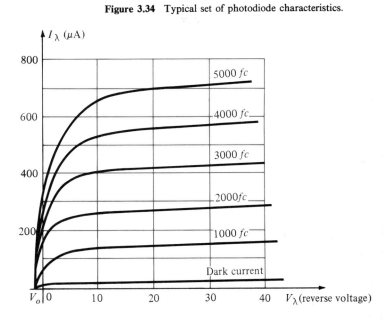

Figure 3.34 Typical set of photodiode characteristics.

intensity levels. The *dark* current is that current that will flow with no applied illumination. Note that the current will only return to zero with a positive applied bias equal to V_o. In addition, Fig. 3.33 demonstrates the use of a lens to concentrate the light on the junction region. An actual device showing the lens in the cap appears in Fig. 3.35.

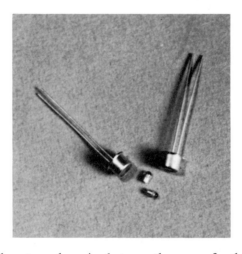

Figure 3.35 Photograph of Hewlett-Packard 5082-4200 S pin photodiode. (Courtesy Hewlett-Packard Corporation.)

The almost equal spacing between the curves for the same increment in luminous flux reveals that the reverse current and luminous flux are almost linearly related. In other words, an increase in light intensity will result in a similar increase in reverse current. A plot of the two to show this linear relationship appears in Fig. 3.36 for a fixed voltage V_λ of 20 V. On a relative basis we can assume that the reverse current is essentially zero in the absence of incident light. Since the rise and fall times (change-of-state parameters) are very small for this device (in the nanosecond range), the device can be used for high-speed counting or switching applications. Returning to Fig. 3.32, we note that Ge encompasses a wider spectrum

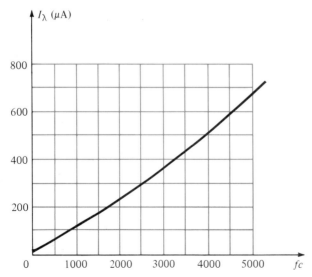

Figure 3.36 I_λ vs. fc (at $V_A = 20$ V) for the photodiode of Fig. 3.34.

CH. 3 ZENER AND OTHER DEVICES

of wavelengths than Si. This would make it suitable for incident light in the infrared region as provided by lasers and IR (infrared) light sources to be described shortly. Of course, Ge has a higher dark current than silicon, but it also has a higher level of reverse current. The level of current generated by the incident light on a photodiode is not such that it could be used as a direct control, but it can be amplified for this purpose.

3.8 PHOTOCONDUCTIVE CELL

The photoconductive cell is a two-terminal semiconductor device whose terminal resistance will vary (linearly) with the intensity of the incident light. For obvious reasons, it is frequently called a photoresistive device. A typical photoconductive cell and the most widely used graphic symbol for the device appear in Fig. 3.37.

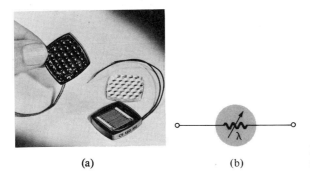

Figure 3.37 Photoconductive cell: (a) appearance; (b) symbol.

The photoconductive materials most frequently used include cadmium sulfide (Cds) and cadmium selenide (CdSe). The peak spectral response of CdS occurs at approximately 5100 Å and for CdSe at 6150 Å, as shown in Fig. 3.32. The response time of CdS units is about 100 ms and 10 ms for CdSe cells.

The photoconductive cell does not have a junction like the photodiode. A thin layer of the material connected between terminals is simply exposed to the incident light energy.

As the illumination on the device increases in intensity, the energy state of a larger number of electrons in the structure will also increase, because of the increased availability of the photon packages of energy. The result is an increasing number of relatively "free" electrons in the structure and a decrease in the terminal resistance. The sensitivity curve for a typical photoconductive device appears in Fig. 3.38. Note the linearity (when plotted using a log-log scale) of the resulting curve and the large change in resistance (100 K → 100 Ω) for the indicated change in illumination.

One rather simple, but interesting, application of the device appears in Fig. 3.39. The purpose of the system is to maintain V_o at a fixed level even through V_i may fluctuate from its rated value. As indicated in the figure, the photoconductive cell, bulb, and resistor, all form part of this voltage regulator system. If V_i should drop in magnitude for any number of reasons, the brightness of the bulb would also decrease. The decrease in illumination would result in an increase in

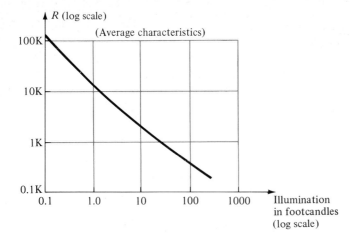

Figure 3.38 Photoconductive cell —terminal characteristics (GE type B425).

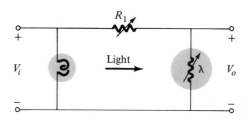

Figure 3.39 Voltage regulator employing a photoconductive cell.

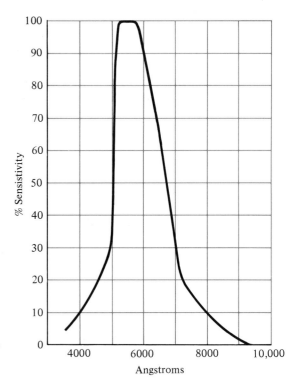

Figure 3.40 Characteristics of a Clairex CdS photoconductive cell. (Courtesy Clairex Electronics.)

**Variation of Conductance
With Temperature and Light**

Foot Candles	.01	0.1	1.0	10	100
Temperature			% Conductance		
−25°C	103	104	104	102	106
0	98	102	102	100	103
25°C	100	100	100	100	100
50°C	98	102	103	104	99
75°C	90	106	108	109	104

Response Time Versus Light

Foot Candles	.01	0.1	1.0	10	100
Rise (seconds)*	0.5	.095	.022	.005	.002
Decay (seconds)**	.125	.021	.005	.002	.001

* Time to $(1 - 1/e)$ of final reading after 5 seconds Dark adaptation.
** Time to $1/e$ of initial reading.

Figure 3.40 (continued)

the resistance (R_λ) of the photoconductive cell to maintain V_o at its rated level as determined by the voltage divider rule; that is,

$$V_o = \frac{R_\lambda V_i}{R_\lambda + R_i} \qquad (3.8)$$

In an effort to demonstrate the wealth of material available on each device from manufacturers, consider the CdS (cadmium sulfide) photoconductive cell described in Fig. 3.40. Note again the concern with temperature and response time.

3.9 IR EMITTERS

Infrared-emitting diodes are solid-state gallium arsenide devices that emit a beam of radiant flux when forward-biased. The basic construction of the device is shown in Fig. 3.41. When the junction is forward-biased, electrons from the n-region will recombine with excess holes of the p-material in a specially designed recombina-

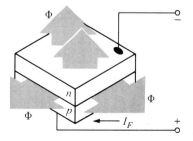

Figure 3.41 General structure of a semiconductor IR-emitting diode. (Courtesy RCA Solid State Division.)

tion region sandwiched between the *p*- and *n*-type materials. During this recombination process, energy is radiated away from the device in the form of photons. The generated photons will either be reabsorbed in the structure or leave the surface of the device as radiant energy, as shown in Fig. 3.41.

The radiant flux in mW versus the dc forward current for a typical device appears in Fig. 3.42. Note the almost linear relationship between the two. An interesting pattern for such devices is provided in Fig. 3.43. Note the very narrow

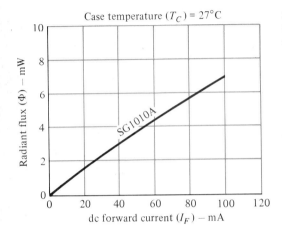

Figure 3.42 Typical radiant flux vs. dc forward current for an IR-emitting diode. (Courtesy RCA Solid State Division.)

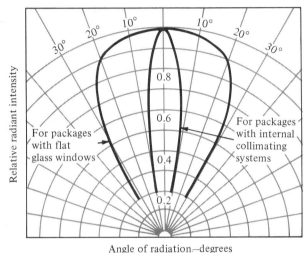

Figure 3.43 Typical radiant intensity patterns of RCA IR-emitting diodes. (Courtesy RCA Solid State Division.)

Figure 3.44 RCA IR-emitting diode: (a) construction; (b) photo; (c) symbol. (*b*, courtesy RCA Solid State Division.)

104

pattern for devices with an internal collinating system. One such device appears in Fig. 3.44, with its internal construction and graphic symbol. A few areas of application for such devices include card and paper-tape readers, shaft encoders, data transmission systems, and intrusion alarms.

3.10 LIGHT-EMITTING DIODES

The light-emitting diode (LED) is, as the name implies, a diode that will give off visible light when it is energized. In any forward-biased *p-n* junction there is, within the structure and primarily close to the junction, a recombination of holes and electrons. This recombination requires that the energy possessed by the unbound free electron be transferred to another state. In all semiconductor *p-n* junctions some of this energy will be given off as heat and some in the form of photons. In silicon and germanium the greater percentage is given up in the form of heat and the emitted light is insignificant. In other materials, such as gallium arsenide phosphide (GaAsP) or gallium phosphide (GaP), the number of photons of light energy emitted is sufficient to create a very visible light source. The process of giving off light by applying an electrical source of energy is called *electroluminescence*. As shown in Fig. 3.45, the conducting surface connected to the *p*-material is much smaller, to permit the emergence of the maximum number of photons of light energy. Note in the figure that the recombination of the injected carriers due to the forward-biased junction is resulting in emitted light at the site of recombination. There may, of course, be some absorption of the packages of photon energy in the structure itself, but a very large percentage are able to leave, as shown in the figure.

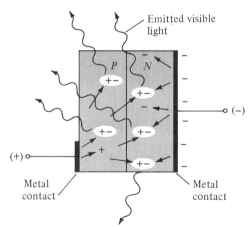

Figure 3.45 The process of electroluminescence in the LED.

The appearance and characteristics of a subminiature high-efficiency solid-state lamp manufactured by Hewlett-Packard appears in Fig. 3.46. Two quantities yet undefined appear under the heading electrical/optical characteristics at $T_A = 25°C$. They are the *axial luminous intensity* (I_V) and the *luminous efficacy* (η_v). Light intensity is measured in *candella*. One candella emits a light flux of 4π lumens and establishes an illumination of 1 foot-candle on a 1-ft² area 1 ft from the

(a)

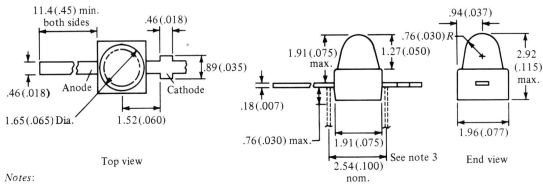

(b)

Absolute Maximum Ratings at $T_A = 25°C$					
Parameter	Red 4100/4101	High Eff. Red 4160	Yellow 4150	Green 4190	Units
Power dissipation	100	120	120	120	mW
Average forward current	50[1]	20[1]	20[1]	30[2]	mA
Peak forward current	1000	60	60	60	mA
Operating and storage temperature range	−55°C to 100°C				
Lead soldering temperature [1.6mm (0.063 in.) from body]	230°C for 3 seconds				

[1]. Derate from 50°C at 0.2mA/°C
[2]. Derate from 50°C at 0.4mA/°C

(c)

Figure 3.46 Hewlett-Packard subminiature high-efficiency red solid-state lamp: (a) appearance; (b) package dimensions; and (c) absolute maximum ratings. (Continued on pp. 107 and 108.) (Courtesy Hewlett-Packard Corporation.)

Electrical/Optical Characteristics at $T_A = 25°C$

Symbol	Description	5082–4100/4101 Min.	Typ.	Max.	5082–4160 Min.	Typ.	Max.	5082–4150 Min.	Typ.	Max.	5082–4190 Min.	Typ.	Max.	Units	Test Conditions
I_V	Axial luminous intensity	–/0.5	.7/1.0		1.0	3.0		1.0	2.0		0.8	1.5		mcd	$I_F = 10$mA,
$2\Theta_{1/2}$	Included angle between half luminous intensity points		45			80			90		At $I_F = 20$mA 70			deg.	Note 1
λ_{peak}	Peak wavelength		655			635			583			565		nm	Measurement at peak
λ_d	Dominant wavelength		640			628			585			572		nm	Note 2
τ_S	Speed of response		15			90			90			200		ns	
C	Capacitance		100			11			15			13		pF	$V_F = 0; f = 1$ MHz
Θ_{JC}	Thermal resistance		125			120			100			100		°C/W	Junction to cathode lead at 0.79mm (.031 in) from body
V_F	Forward voltage		1.6	2.0		2.2	3.0		2.2	3.0		2.4	3.0	V	$I_F = 10$mA,
BV_R	Reverse breakdown voltage	3.0	10		5.0			5.0			At $I_F = 20$mA 5.0			V	$I_F = 100\mu$A
η_v	Luminous efficacy		55			147			570			665		lm/W	Note 3

NOTES:
1. $\Theta_{1/2}$ is the off-axis angle at which the luminous intensity is half the axial luminous intensity.
2. The dominant wavelength, λ_d, is derived from the CIE chromaticity diagram and represents the single wavelength which defines the color of the device.
3. Radiant intensity, I_e, in watts/steradian, may be found from the equation $I_e = I_v/\eta_v$, where I_v is the luminous intensity in candellas and η_v is the luminous efficacy in lumens/watt.

(d)

Figure 3.46 (continued) (d) Electrical/optical characteristics.

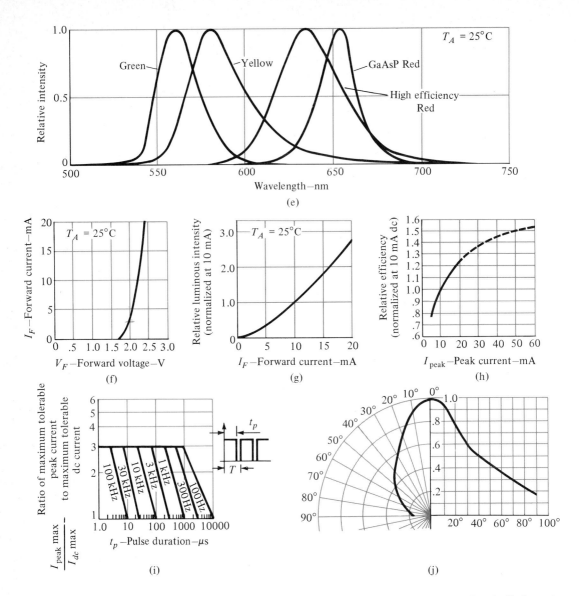

Figure 3.46 (continued) (e) Relative intensity vs. wavelength; (f) forward current vs. forward voltage; (g) relative luminous intensity vs. forward current; (h) relative efficiency vs. peak current; (i) maximum peak current vs. pulse duration; and (j) relative luminous intensity vs. angular displacement.

light source. Even though this description may not provide a clear understanding of the candela as a unit of measurement, its level can certainly be compared between similar devices. The term "efficacy" is, by definition, a measure of the ability of a device to produce a desired effect. For the LED this is the ratio of the number of lumens generated per applied watt of electrical energy. Note the high efficacy for the high-efficiency red LED. The relative efficiency is defined by the luminous intensity per unit current, as shown in Fig. 3.46h. Note also how close the peak wavelength of the light waves produced by each LED (red, yellow, green) approaches the defined wavelength (λ_d) for each color. The relative intensity of each color versus wavelength appears in Fig. 3.46e.

Since the LED is a *p-n*-junction device, it will have a forward-biased characteristic (Fig. 3.46f) similar to the diode response curves introduced in Chapter 2. Note, however, that a change from Ge to a Si base has resulted in a much larger V_o. Note the almost linear increase in relative luminous intensity with forward current (Fig. 3.46g). Figure 3.46i reveals that the longer the pulse duration at a particular frequency, the lower the permitted peak current (after you pass the break value of t_p). Figure 3.46j simply reveals that the intensity is greater at 0° (or head on) and the least at 90° (when you view the device from the side).

LED displays are available today in many different sizes and shapes. The light-emitting region is available in lengths from 0.1 to 1 in. Numbers can be created by segments such as shown in Fig. 3.47. By applying a forward bias to the proper *p*-type material segment, any number from 0 to 9 can be displayed.

The display of Fig. 3.48 is used in calculators and will provide eight digits. There are also two-lead LED lamps (Fig. 3.49) that contain two LEDs, so that

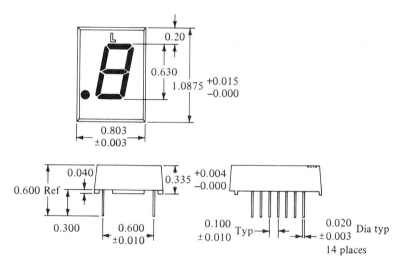

Figure 3.47 Litronix segment display.

Figure 3.48 Eight-digit and a sign calculator display. (Courtesy Hewlett-Packard Corporation.)

a reversal in biasing will change the color from green to red, or vice versa. LEDs are presently available in red, green, yellow, orange, and white. It would appear that the introduction of the color blue is a possibility in the very near future. In general, LEDs operate at voltage levels from 1.7 to 3.3 V, which makes them completely compatible with solid-state circuits. They have a fast response time (nanoseconds) and offer good contrast ratios for visibility. The power requirement is typically from 10 to 150 mW with a lifetime of 100,000+ hours. Their semiconductor construction adds a significant ruggedness factor.

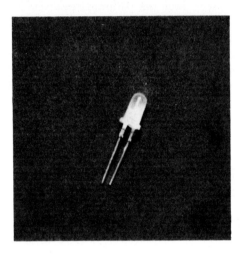

Figure 3.49 RED/GREEN LED diode. (Courtesy General Instrument Corporation.)

3.11 LIQUID CRYSTAL DISPLAYS

The liquid crystal display (LCD) has the distinct advantage of having a lower power requirement than the LED. It is typically in the order of microwatts for the display, as compared to the same order of milliwatts for LEDs. It does, however, require an external or internal light source, is limited to a temperature range of about 0° to 60°C, and lifetime is an area of concern because LCDs can chemically degrade. The types receiving the major interest today are the field-effect and dynamic scattering units. Each will be covered in some detail in this section.

A liquid crystal is a material (normally organic for LCDs) that will flow like a liquid but whose molecular structure has some properties normally associated with solids. For the light-scattering units, the greatest interest is in the *nematic liquid crystal,* having the crystal structure shown in Fig. 3.50. The individual molecules have a rodlike appearance as shown in the figure. The indium oxide conducting surface is transparent and, under the condition shown in the figure, the incident light will simply pass through and the liquid crystal structure will appear clear. If a voltage (for commercial units the threshold level is usually between 6 and 20 V) is applied across the conducting surfaces, as shown in Fig. 3.51, the molecular arrangement is disturbed, with the result that regions will be established with different indices of refraction. The incident light is, therefore, reflected in different directions at the interface between regions of different indices of refraction (referred to as *dynamic scattering*—first studied by RCA in 1968)

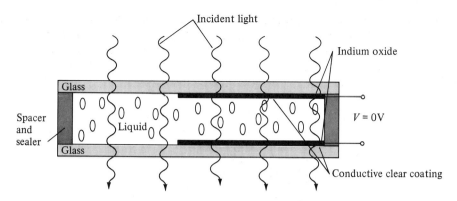

Figure 3.50 Nematic liquid crystal with no applied bias.

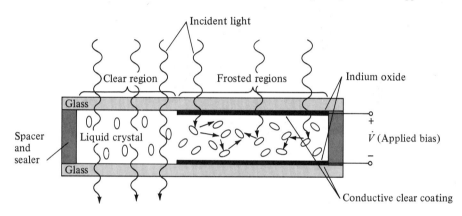

Figure 3.51 Nematic liquid crystal with applied bias.

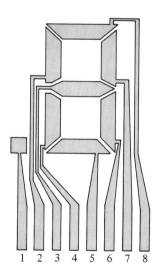

Figure 3.52 LCD 8-segment digit display.

with the result that the scattered light has a frosted glass appearance. Note in Fig. 3.51, however, that the frosted look occurs only where the conducting surfaces are opposite each other and that the remaining areas remain translucent.

A digit on an LCD display may have the segment appearance shown in Fig. 3.52. The black area is actually a clear conducting surface connected to the terminals below for external control. Two similar masks are placed on opposite sides of a sealed thin layer of liquid crystal material. If the number 2 were required, the terminals 8, 7, 3, 4, and 5 would be energized and only those regions would be frosted while the other areas would remain clear.

As indicated earlier, the LCD does not generate its own light but depends on an external or internal source. Under dark conditions it would be necessary for the unit to have its own internal light source either behind or to the side of the LCD. During the day, or in lighted areas, a reflector can be put behind the LCD to reflect the light back through the display for maximum intensity. For optimum operation, current watch manufacturers are using a combination of the transmissive (own light source) and reflective modes called *transflective*.

The *field-effect* or *twisted nematic* LCD has the same segment appearance and thin layer of encapsulated liquid crystal, but its mode of operation is very different. Similar to the dynamic scattering LCD, the field effect can be operated in the reflective mode or transmissive mode with an internal source. The transmissive display appears in Fig. 3.53. The internal light source is on the right and the viewer is on the left. This figure is most noticeably different from Fig. 3.50 in that there is an addition of a *light polarizer*. Only the vertical component of the entering light on the right can pass through the vertical light polarizer on the right. In the field-effect LCD, either the clear conducting surface to the right is

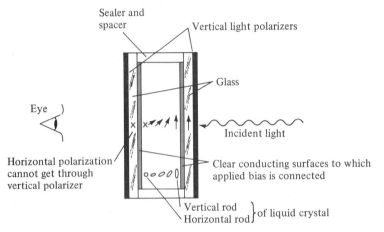

Figure 3.53 Transmissive field-effect LCD with no applied bias.

chemically etched or an organic film is applied to orient the molecules in the liquid crystal in the vertical plane, parallel to the cell wall. Note the rods to the far right in the liquid crystal. The opposite conducting surface is also treated to ensure that the molecules are 90° out of phase in the direction shown (horizontal) but still parallel to the cell wall. In between the two walls of the liquid crystal there is a general drift from one polarization to the other, as shown in the figure. The left-hand light polarizer is also such that it permits the passage of only the vertically polarized incident light. If there is no applied *emf* to the conducting surfaces, the vertically polarized light enters the liquid crystal region and follows the 90° bending of the molecular structure. Its horizontal polarization at the left-hand vertical light polarizer does not allow it to pass through, and the viewer sees a uniformly dark pattern across the entire display. When a threshold voltage is applied (for commercial units from 2 to 8 V), the rodlike molecules align themselves with the field (perpendicular to the wall) and the light passes directly through without the 90° shift. The vertically incident light can then pass directly through the second vertically polarized screen and a light area is seen by the viewer. Through proper excitation of the segments of each digit the pattern will appear as shown in Fig. 3.54. The reflective type field effect is shown in Fig. 3.55.

Figure 3.54 Reflective-type LCD. (Courtesy RCA Solid State Division.)

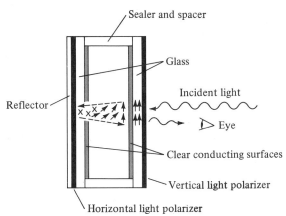

Figure 3.55 Reflective field-effect LCD with no applied bias.

Figure 3.56 Transmissive-type LCD. (Courtesy RCA Solid State Division.)

In this case the horizontally polarized light at the far left encounters a horizontally polarized filter and passes through to the reflector, where it is reflected back into the liquid crystal, bent back to the other vertical polarization, and returned to the observer. If there is no *emf,* there is a uniformly lit display. The application of a voltage results in a vertically incident light encountering a horizontally polarized filter at the left which will not be able to pass through and be reflected. A dark area results on the crystal, and the pattern as shown in Fig. 3.56 appears.

Field-effect LCDs are normally used when a source of energy is a prime factor (e.g., in watches, portable instrumentation, etc.) since they absorb considerably less power than the light-scattering types—the microwatt range compared to the low-milliwatt range. The cost is typically higher for field-effect units, and their height is limited to about 2 in., while light-scattering units are available up to 8 in. in height.

A further consideration in displays is turn-on and turn-off time. LCDs are characteristically much slower than LEDs. LCDs typically have response times in the range 100 to 300 ms, while LEDs are available with response times below 100 ns. However, there are numerous applications, such as in a watch, where the difference between 100 ns and 100 ms ($\frac{1}{10}$ of a second) is of little consequence. For such applications the lower power demand of LCDs is a very attractive characteristic. The lifetime of LCD units is steadily increasing beyond the 10,000+ hours limit. Since the color generated by LCD units is dependent on the source of illumination, there is a greater range of color choice.

SEC. 3.11 LIQUID CRYSTAL DISPLAYS

3.12 SOLAR CELLS

In recent years there has been increasing interest in the solar cell as an alternative source of energy. When we consider that the power density received from the sun at sea level is about 100 mW/cm² (SI units: 1 kW/m²), it is certainly an energy source that requires further research and development to maximize the conversion efficiency from solar to electrical energy.

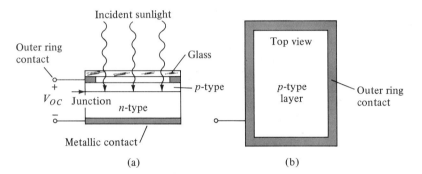

Figure 3.57 Solar cell: (a) cross section; (b) top view.

The basic construction of a silicon *p-n* junction solar cell appears in Fig. 3.57. As shown in the top view, every effort is made to ensure that the surface area perpendicular to the sun is a maximum. Also, note that the metallic conductor connected to the *p*-type material and the thickness of the *p*-type material are such that they ensure that a maximum number of photons of light energy will reach the junction. A photon of light energy in this region may collide with a valence electron and impart to it sufficient energy to leave the parent atom. The result is a generation of free electrons and holes. This phenomenon will occur on each side of the junction. In the *p*-type material the newly generated electrons are minority carriers and will move rather freely across the junction as explained for the basic *p-n* junction with no applied bias. A similar discussion is true for the holes generated in the *n*-type material. The result is an increase in the minority carrier flow which is opposite in direction to the conventional forward current of a *p-n* junction. This increase in reverse current is shown in Fig. 3.58. Since $V=0$ anywhere on the vertical axis and represents a short-circuit condition, the current at this intersection is called the *short-circuit current* and is represented by the notation I_{SC}. Under open-circuit conditions ($i_d = 0$) the *photovoltaic* voltage V_{oc} will result. This is a logarithmic function of the illumination, as shown in Fig. 3.59. V_{oc} is the terminal voltage of a battery under no-load (open-circuit) conditions. Note, however, in the same figure that the short-circuit current is a linear function of the illumination. That is, it will double for the same increase in illumination (f_{c_1} and $2f_{c_1}$ in Fig. 3.59) while the change in V_{oc} is less for this region. The major increase in V_{oc} occurs for lower-level increases in illumination. Eventually, a further increase in illumination will have very little effect on V_{oc}, although I_{sc} will increase, causing the power capabilities to increase.

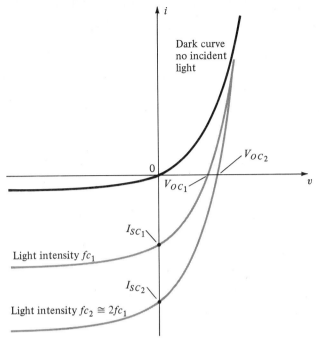

Figure 3.58 Short-circuit current and open-circuit voltage vs. light intensity for a solar cell.

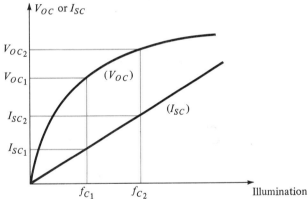

Figure 3.59 V_{oc} and I_{SC} vs. illumination for a solar cell.

Selenium and silicon are the most widely used materials for solar cells, although gallium arsenide, indium arsenide, and cadmium sulfide, among others, are also used. The wavelength of the incident light will affect the response of the *p-n* junction to the incident photons. Note in Fig. 3.60 how closely the selenium cell response curve matches that of the eye. This fact has widespread application in photographic equipment such as exposure meters and automatic exposure diaphragms. Silicon also overlaps the visible spectrum but has its peak at the 0.8 μ (8000 Å) wavelength, which is in the infrared region. In general, silicon has a higher conversion efficiency and greater stability and is less subject to fatigue. Both materials have excellent temperature characteristics. That is, they can withstand extreme high or low temperatures without a significant drop-off in efficiency. A typical solar cell, with its electrical characteristics, appears in Fig. 3.61.

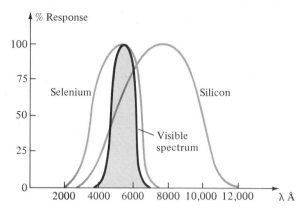

Figure 3.60 Spectral response of Se, Si, and the naked eye.

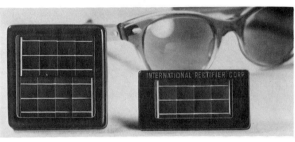

Electrical Characteristics*

IR Number	Load Voltage (volts) (min.)	Load Current (milliamps) (min.)	Power (milliwatts) (min.)
SP2A40B	1.6	36	58
SP2B48B	1.6	40	64
SP4C40B	3.2	36	115
SP2C80B	1.6	72	115
SP4D48B	3.2	40	128
SP2D96B	1.6	80	129
S2900E5M	.4	60	24
S2900E7M	.4	90	36
S2900E9.5M	.4	120	48

* Current Voltage characteristics are based on an illumination level of 100 mW/cm² (bright average sunlight).

Figure 3.61 Typical solar cell and its electrical characteristics. (Courtesy International Rectifier Corporation.)

A very recent innovation in the use of solar cells appears in Fig. 3.62. The series arrangement of solar cells permits a voltage beyond that of a single element. The performance of a typical four-cell array appears in the same figure. At a current of approximately 2.6 mA the output voltage is about 1.6 V, resulting in

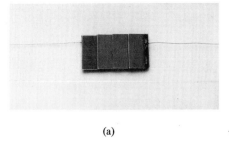

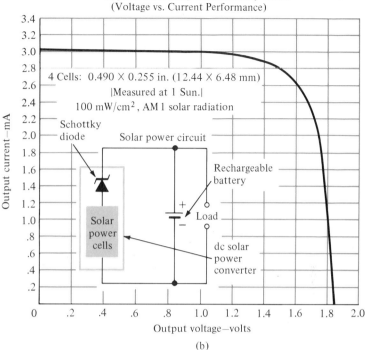

Figure 3.62 International Rectifier 4-cell array: (a) appearance; (b) characteristics. (Courtesy International Rectifier Corporation.)

an output power of 4.16 mW. The Schottky barrier diode is included to prevent battery current drain through the power converter. That is, the resistance of the Schottky diode is so high to current flowing down through (+ to −) the power converter that it will appear as an open circuit to the rechargeable battery and not draw current from it.

It might be of interest to note that the Lockheed Missiles and Space Company has been awarded a grant from the National Aeronautics and Space Administration to develop a massive solar-array wing for the space shuttle. The wing will measure 13.5 ft by 105 ft when extended and will contain 41 panels, each carrying 3060 silicon solar cells. The wing can generate a total of 12.5 kW of electrical energy.

The efficiency of operation of a solar cell is determined by the electrical power output divided by the power provided by the light source. That is,

$$\eta\% = \frac{P_{o(\text{electrical})}}{P_{i(\text{light energy})}} \times 100\% = \frac{P_{\max(\text{device})}}{(\text{area in cm}^2)(100 \text{ mW/cm}^2)} \times 100\% \quad (3.9)$$

Typical levels of efficiency range from 10 to 14%—a level that should improve measurably if the present interest continues. A typical set of output characteristics for silicon solar cells of 10% efficiency with an active area of 1 cm² appears in Fig. 3.63. Note the optimum power locus and the almost linear increase in output current with luminous flux for a fixed voltage.

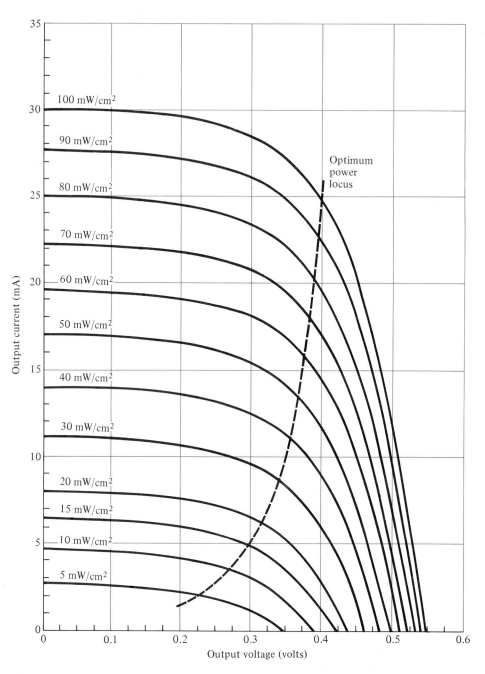

Figure 3.63 Typical output characteristics for silicon solar cells of 10% efficiency having an active area of 1 cm². Cell temperature is 30°C.

3.13 THERMISTORS

The thermistor is, as the name implies, a temperature-sensitive resistor; that is, its terminal resistance is related to its body temperature. It has a negative temperature coefficient, indicating that its resistance will decrease with an increase in its body temperature. It is not a junction device and is constructed of Ge, Si, or a mixture of oxides of cobalt, nickel, strontium, or manganese.

The characteristics of a representative thermistor are provided in Fig. 3.64, with the commonly used symbol for the device. Note, in particular, that at room temperature (20°C) the resistance of the thermistor is approximately 5000 Ω, while at 100°C (212°F) the resistance has decreased to 100 Ω. A temperature span of 80°C has therefore resulted in a 50:1 change in resistance. It is typically 3% to 5% per degree change in temperature. There are, fundamentally, two ways to change the temperature of the device: internally and externally. A simple change in current through the device will result in an internal change in temperature.

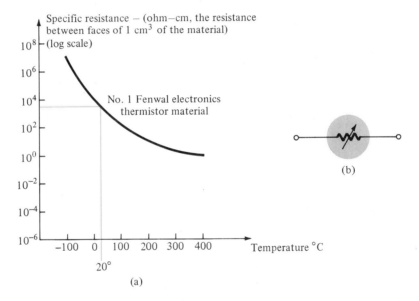

Figure 3.64 Thermistor: (a) typical set of characteristics; (b) symbol.

A small applied voltage will result in a current too small to raise the body temperature above that of the surroundings. In this region, as shown in Fig. 3.65, the thermistor will act like a resistor and have a positive temperature coefficient. However, as the current increases, the temperature will rise to the point where the negative temperature coefficient will appear as shown in Fig. 3.65. The fact that the rate of internal flow can have such an effect on the resistance of the device introduces a wide vista of applications in control, measuring techniques, and so on. An external change would require changing the temperature of the surrounding medium or immersing the device in a hot or cold solution.

A photograph of a number of commercially available thermistors is provided in Fig. 3.66.

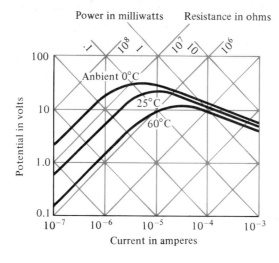

Figure 3.65 Steady-state voltage-current characteristics of Fenwal Electronics BK65V1 Thermistor. (Courtesy Fenwal Electronics, Incorporated.)

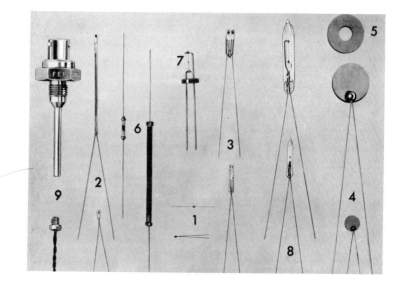

Figure 3.66 Various types of thermistors: (1) beads; (2) glass probes; (3) iso-curve interchangeable probes and beads; (4) discs; (5) washers; (6) rods; (7) specially mounted beads; (8) vacuum and gas-filled probes; (9) special probe assemblies. (Courtesy Fenwal Electronics, Incorporated.)

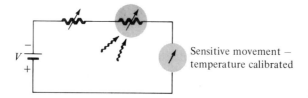

Figure 3.67 Temperature-indicating circuit.

A simple temperature-indicating circuit appears in Fig. 3.67. Any increase in the temperature of the surrounding medium will result in a decrease in the resistance of the thermistor and an increase in the current I_T. An increase in I_T will produce an increased movement deflection, which when properly calibrated will accurately indicate the higher temperature. The variable resistance was added for calibration purposes.

PROBLEMS

§ 3.2

1. The following characteristics are specified for a particular Zener diode: $V = 20$ V, $I_{ZT} = 10$ mA, $I_R = 20$ μA, and $I_{ZM} = 40$ mA. Sketch the characteristic curve in the manner displayed in Fig. 3.3.

2. At what temperature will the IN961 10-V Fairchild Zener diode have a nominal voltage of 10.75 V? (Hint: Note data in Table 3.1.)

3. Determine the temperature coefficient of a 5-V Zener diode (rated 25°C value) if the nominal voltage drops to 4.8 V at a temperature of 100°C.

4. Using the curves of Fig. 3.4a, what level of temperature coefficient would you expect for a 20-V diode? Repeat for a 5-V diode. Assume a linear scale between nominal voltage levels.

5. Determine the dynamic impedance for the 24-V diode at $I_Z = 10$ mA from Fig. 3.4b. Note that it is a log scale.

6. Compare the levels of dynamic impedance for the 24-V diode of Fig. 3.4b at current levels of 0.2 mA, 1 mA, and 10 mA. How do the results relate to the shape of the characteristics in this region?

7. Using the curve of Fig. 3.4c, determine the temperature at which the rated power drops to 100 mW. How does this value compare with the results obtained using $D_F = 3.33$ W/°C?

8. Sketch the output (v_o) of Fig. 3.7 if the 12-V battery is reduced to 10 V. Sketch v_R also.

9. Sketch the output of Fig. 3.9a if v_i is reduced to peak values of 10 V. Assume straight-line segments for the diode with zero dynamic impedance in the forward-bias region of Fig. 3.1. Repeat for peak values of 20 V.

10. For the Zener diode network of Fig. 3.11, if $V_i = 40$ V, $R_S = 200$ Ω, and $V_Z = 22$ V:
    (a) Find the minimum value of R_L to ensure that the Zener diode has fired.
    (b) From part (a) determine the maximum load current under Zener regulation.
    (c) Find the minimum I_L if the maximum $R_L = 5$ K.
    (d) Sketch the terminal voltage V_L versus I_L for the range of I_L.
    (e) At $R_L = 1$ K determine I_Z and the current through the 40-V source.
    (f) From the results of part (e), determine the power dissipated by the Zener diode.

11. Using a 10-V Zener, R_L with a range from 50 Ω to 2 K, $V_i = 30$ V, find a suitable value of R_S to ensure that the Zener diode maintains an "on" condition (use the circuit configuration in Fig. 3.11).

§ 3.3

12. (a) Describe in your own words how the construction of the hot-carrier diode is significantly different from the conventional semiconductor diode.
    (b) In addition, describe its mode of operation.

13. (a) Consult Fig. 3.18. How would you compare the dynamic resistances of the diodes in the forward-bias regions?
    (b) How do they compare at any level of reverse current more negative than I_S?

14. Referring to Fig. 3.21, how does the maximum surge current I_{FSM} relate to the average rectified forward current? Is it typically greater than 20:1? Why is it possible to have such high levels of current? What noticeable difference is there in construction as the current rating increases?

15. Referring to Fig. 3.22a, at what temperature is the forward voltage drop 300 mV at a current of 1 mA? Which current levels have the highest levels of temperature coefficients? Assume a linear progression between temperature levels.

16. For the curve of Fig. 3.22b denoted 2900/2303, determine the percent change in I_R for a change in reverse voltage from 5 to 10 V. At what reverse voltage would you expect to reach a reverse current of 1 μA? Note the log scale for I_R.

17. Determine the percent change in capacitance between 0 and 2 V for the 2900/2303 curve of Fig. 3.22c. How does this compare to the change between 8 and 16 V?

§ 3.4

18. (a) Determine the transition capacitance of a diffused junction varicap diode at a reverse potential of 4.2 V if $C(0) = 80$ pF and $V_o = 0.7$ V.
    (b) From the information of part (a), determine the constant K in Eq. (3.3).

19. (a) For a varicap diode having the characteristics of Fig. 3.23, determine the difference in capacitance between reverse-bias potentials of -3 and -12 V.
    (b) Determine the incremental rate of change ($\Delta C/\Delta V_r$) at $V = -8$ V. How does this value compare with the incremental change determined at -2 V?

20. (a) The resonant frequency of a series RLC network is determined by $f_o = 1/2\pi\sqrt{LC}$. Using the value of f_o and L_s provided in Fig. 3.25, determine the value of C.
    (b) How does the value calculated in part (a) compare with that determined by the curve in Fig. 3.26 at $V_R = 25$ V?

21. Referring to Fig. 3.26, determine the ratio of capacitance at $V_R = 3$ V to $V_R = 25$ V and compare to the value of C_3/C_{25} given in Fig. 3.25 (maximum = 6.5).

22. Determine T_1 for a varactor diode if $C_0 = 22$ pF, $TC_C = 0.02$, and due to an increase in temperature above $T_0 = 25°C$, $\Delta C = 11$ pF.

23. What region of V_R would appear to have the greatest change in capacitance per change in reverse voltage? Be aware that the scales are nonlinear.

24. If $Q = X_L/R = 2\pi fL/R$ and R_S is 0.35 Ω, determine the value of L_S at $V_R = 3.0$ V, $f = 100$ MHz. How does it compare to the value provided in the table of Fig. 3.25? If different, give some reasons why.

§ 3.5

25. Consult a manufacturer's data book and compare the general characteristics of a high-power device (>10 A) to a low-power unit (<100 mA). Is there a significant change in the data and characteristics provided? Why?

§ 3.6

26. What are the essential differences between a semiconductor junction diode and a tunnel diode?

27. Note in the equivalent circuit of Fig. 3.29 that the capacitor appears in parallel with the negative resistance. Determine the reactance of the capacitor at 1 MHz and 100 MHz if $C = 5$ pF, and determine which frequency would most affect the net impedance of the parallel combination if $R = -152$ Ω. Is the magnitude of the inductive reactance anything to be overly concerned about at either of these frequencies if $L_S = 6$ nH?

28. Why do you believe the maximum reverse current rating for the tunnel diode can be greater than the forward current rating. (*Hint:* Note the characteristics and consider the power rating.)

29. Determine the energy associated with the photons of green light if the wavelength is 5000 Å. Give your answer in joules and electron volts.

30. (a) Referring to Fig. 3.32, what would appear to be the frequencies associated with the upper and lower limits of the visible spectrum?
    (b) What is the wavelength in microns associated with the peak relative response of silicon?
    (c) If we define the bandwidth of the spectral response of each material to occur at 70% of its peak level, which material has the widest bandwidth?

31. Referring to Fig. 3.34, determine I_λ if a reverse voltage of 30 V is applied and the light intensity is 4×10^{-9} W/m².

32. (a) Which material of Fig. 3.32 would appear to provide the best response to red, green, and infrared (smaller wavelengths) light sources?
    (b) At a frequency of 0.5×10^{15} Hz, which color has the maximum spectral response?

33. Determine the voltage drop across the resistor of Fig. 3.33 if the incident flux is 3000 fc, $V_\lambda = 25$ V, and $R = 10$ K. Use the characteristics of Fig. 3.34.

§ 3.8

34. What is the approximate rate of change of resistance with illumination for a photoconductive cell with the characteristics of Fig. 3.38 for the ranges (a) 0.1 → 1 K; (b) 1 → 10 K; (c) 10 → 100 K. (Note that this is a log scale.) Which region has the greatest rate of change in resistance with illumination?

35. What is the "dark current" of a photodiode?

36. If the illumination on the photoconductive diode in Fig. 3.39 is 10 fc, determine the magnitude of V_i to establish 6 V across the diode if R_1 is equal to 5 K. Use the characteristics of Fig. 3.38.

37. Using the data provided in Fig. 3.40, sketch a curve of percent conductance versus temperature for 0.01, 1.0, and 100 fc. Are there any noticeable effects?

38. (a) Sketch a curve of rise time versus illumination using the data from Fig. 3.40.
    (b) Repeat part (a) for the decay time.
    (c) Discuss any noticeable effects of illumination in parts (a) and (b).

39. Which colors is the CdS unit of Fig. 3.40 most sensitive to?

§ 3.9

40. (a) Determine the radiant flux at a dc forward current of 70 mA for the device of Fig. 3.42.
    (b) Determine the radiant flux in lumens at a dc forward current of 45 mA.

41. (a) Through the use of Fig. 3.43, determine the relative radiant intensity at an angle of 25° for a package with a flat glass window.
    (b) Plot a curve of relative radiant intensity versus degrees for the flat package.

42. If 60 mA of dc forward current is applied to an SG1010A IR emitter, what will be the incident radiant flux in lumens 5° off the center if the package has an internal collimating system? Refer to Figs. 3.42 and 3.43.

§ 3.10

43. (a) Convert the horizontal scale of Fig. 3.46e to angstrom units.
    (b) How does the peak value of the relative intensity fall into the color bands provided in Fig. 3.32?

44. Referring to Fig. 3.46f, what would appear to be an appropriate value of V_o for this device? How does it compare to the value of V_o for silicon and germanium?

45. Using the information provided in Fig. 3.46, determine the forward voltage across the diode if the relative luminous intensity is 1.5.

46. (a) What is the percent increase in relative efficiency of the device of Fig. 3.46 if the peak current is increased from 5 to 10 mA?
    (b) Repeat part (a) for 30 to 35 mA (the same increase in current).
    (c) Compare the percent increase from parts (a) and (b). At what point on the curve would you say there is little gained by further increasing the peak current?

47. (a) Referring to Fig. 3.46i, determine the maximum tolerable peak current if the period of the pulse duration is 1 ms, the frequency is 300 Hz, and the maximum tolerable dc current is 20 mA.
    (b) Repeat part (a) for a frequency of 100 Hz.

48. (a) If the luminous intensity at 0° angular displacement is 3.0 mcd for the device of Fig. 3.46, at what angle will it be 0.75 mcd?
    (b) At what angle does the loss of luminous intensity drop below the 50% level?

49. Sketch the current derating curve for the average forward current as determined by temperature. (Note the absolute maximum ratings.)

§ 3.11

50. Referring to Fig. 3.52, which terminals must be energized to display number 7?

51. In your own words, describe the basic operation of an LCD.

52. Discuss the relative differences in mode of operation between an LED and an LCD display.

53. What are the relative advantages and disadvantages of an LCD display as compared to an LED display?

§ 3.12

54. A 1 cm by 2 cm solar cell has a conversion efficiency of 9%. Determine the maximum power rating of the device.

55. If the power rating of a solar cell is determined on a very rough scale by the product $V_{OC} \cdot I_{SC}$, is the greatest rate of increase obtained at lower or higher levels of illumination? Explain your reasoning.

56. (a) Referring to Fig. 3.63, what power density is required to establish a current of 24 mA at an output voltage of 0.25 V?
    (b) Why is 100 mW/cm^2 the maximum power density in Fig. 3.63?
    (c) Determine the output current if the power is 40 mW/cm^2 and the output voltage is 0.3 V.

57. (a) Sketch a curve of output current versus power density at an output voltage of 0.15 V.
    (b) Sketch a curve of output voltage versus power density at a current of 19 mA.
    (c) Is either of the curves from (a) and (b) linear within the limits of the maximum power limitation?

§ 3.13

58. For the thermistor of Fig. 3.64, determine the dynamic rate of change in specific resistance with temperature at $T = 20°C$. How does this compare to the value determined at $T = 300°C$? From the results, determine whether the greatest change in resistance per unit change in temperature occurs at lower or higher levels of temperature. Note the vertical log scale.

59. Using the information provided in Fig. 3.64, determine the total resistance of a 2-cm length of the material having a perpendicular surface area of 1 cm^2 at a temperature of 0°C. Note the vertical log scale.

60. (a) Referring to Fig. 3.65, determine the current at which a 25°C sample of the material changes from a positive to negative temperature coefficient. (Figure 3.65 is a log scale.)
    (b) Determine the power and resistance levels of the device (Fig. 3.65) at the peak of the 0°C curve.
    (c) At a temperature of 25°C, determine the power rating if the resistance level is 1 M.

61. In Fig. 3.67, $V = 0.2$ V, and $R_{\text{variable}} = 10\ \Omega$. If the current through the sensitive movement is 2 mA and the voltage drop across the movement is 0 V, what is the resistance of the thermistor?

GLOSSARY

Zener Diode   A device designed around the avalanche breakdown region of a semiconductor diode.

Zener Temperature Coefficient   Reflects the percent change in Zener potential per degree Celsius.

Schottky-Barrier (Hot-Carrier) Diode   A two-terminal device having a metal semiconductor junction and improved characteristics in the forward-bias region as compared to a typical semiconductor diode.

Varicap (Varactor) Diode   A semiconductor diode with a terminal capacitance heavily dependent upon the reverse-bias potential.

Capacitance Temperature Coefficient   Reflects the percent change in capacitance per degree Celsius.

Power Diode   A semiconductor diode designed to have higher current, temperature, and PIV ratings.

Tunnel Diodes   A heavily doped semiconductor diode having a negative resistance region for use in oscillators and a fast-switching capability for computer applications.

Photodiode   A semiconductor diode whose state is determined by the level of incident light on its junction.

Photon   A discrete package of light energy.

Wavelength   The distance between successive peaks of a traveling light wave.

Angstrom (Å)   A unit of measure equal to the 10^{-10} m that is used in wavelength measurements.

Lumen   A measure of luminous flux equal to 1.496×10^{-3} W.

Photoconductive Cell   A two-terminal semiconductor device whose resistance will vary (linearly) with the intensity of the incident light.

Infrared-Emitting Diodes   A solid-state gallium arsenide device that emits a beam of radiant infrared light flux when forward-biased.

Light-Emitting Diode (LED)   A diode that will give off visible yellow, green, or red light when forward-biased.

Axial Luminous Intensity   Measured in candellas; 1 candella emits a light flux of 4π lumens and establishes an illumination of 1 foot-candle (fc) on a 1-ft² area 1 foot from the light source.

Luminous Efficacy (η_v)   A measure of the ability of a device to produce a desired effect. For the LED it is the ratio of the number of lumens generated per applied watt of electrical energy.

Nematic Liquid Crystal Display (LCD)   A display whose performance is dependent upon the effect an applied bias will have on an organic liquid crystal located between two conducting surfaces.

Field-Effect LCD A display whose performance is dependent upon the effect an applied bias will have on an organic light crystal located between two light polarizers.

Solar Cell A semiconductor *p-n*-junction device capable of converting solar energy to electrical energy.

Thermistor A two-terminal device whose resistance is sensitive to its body temperature and that of the surrounding medium.

CHAPTER 4
Bipolar Transistors

4.1 INTRODUCTION

During the period 1904–1947 the vacuum tube was undoubtedly the electronic device of interest and development. In 1904 the vacuum-tube diode was introduced by J. A. Fleming. Shortly thereafter, in 1906, Lee De Forest added a third element, called the *control grid,* to the vacuum diode, resulting in the first amplifier, the *triode.* In the following years, radio and television provided great stimulation to the tube industry. Production rose from about 1 million tubes in 1922 to about 100 million in 1937. In the early 1930s the four-element tetrode and five-element pentode gained prominence in the electron-tube industry. In the years to follow, the industry became one of primary importance and rapid advances were made in design, manufacturing techniques, high-power and high-frequency applications, and miniaturization.

On December 23, 1947, however, the electronic industry was to experience the advent of a completely new direction of interest and development. It was on the afternoon of this day that Walter H. Brattain and John Bardeen demonstrated the amplifying action of the first transistor at the Bell Telephone Laboratories. The original transistor (a point-contact transistor) is shown in Fig. 4.1. The advantages of this three-terminal solid-state device over the tube were immediately obvious: it was smaller and lightweight; had no heater requirement or heater loss; had rugged construction; and was more efficient since less power was absorbed by the device itself; it was instantly available for use, requiring no warm-up period; and lower operating voltages were possible. Note in the discussion above that

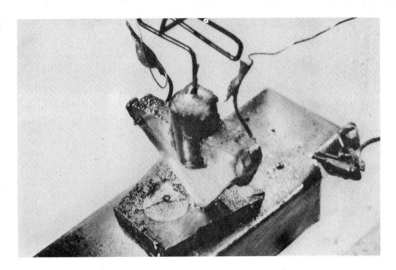

Figure 4.1 The first transistor. (Courtesy Bell Telephone Laboratories.)

this chapter is our first discussion of devices with three or more terminals. You will find that an amplifier (any device that increases the voltage, current, or power level) must have at least three terminals—one for the control of the flow between the other two.

4.2 TRANSISTOR CONSTRUCTION

The transistor is a three-layer semiconductor device consisting of either two *n*- and one *p*-type layers of material or two *p*- and one *n*-type layers of material. The former is called an *npn transistor,* while the latter is called a *pnp transistor.* Both are shown in Fig. 4.2 with the proper dc biasing. We will find in Chapter 5 that the dc biasing is necessary to establish the proper region of operation for ac amplification. The outer layers of the transistor are heavily doped semiconductor materials having widths much greater than that of the sandwiched *p*- or *n*-type material. For the transistors shown in Fig. 4.2 the ratio of the total width to

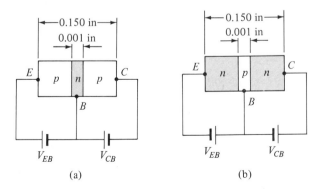

Figure 4.2 Types of transistors: (a) *pnp;* (b) *npn.*

that of the center layer is 0.150/0.001 = 150:1. The doping of the sandwiched layer is also considerably less than that of the outer layers (typically 10:1 or less). This lower doping level decreases the conductivity (increases the resistance) of this material by limiting the number of "free" carriers.

For the biasing shown in Fig. 4.2 the terminals have been indicated by the capital letters *E* for *emitter, C* for *collector,* and *B* for *base*. An appreciation for this choice of notation will develop when we discuss the basic operation of the transistor.

The abbreviation BJT, from *bipolar junction transistor,* is often applied to this three-terminal device. The term "bipolar" reflects the fact that holes *and* electrons participate in the injection process into the oppositely polarized material. If only one carrier is employed (electron or hole), it is considered a *unipolar* device. Recall that the Schottky diode was such a device.

4.3 TRANSISTOR OPERATION

The basic operation of the transistor will now be described using the *pnp* transistor of Fig. 4.2a. The operation of the *npn* transistor is exactly the same if the roles played by the electron and hole are interchanged.

In Fig. 4.3 the *pnp* transistor has been redrawn without the base-to-collector bias. Note the similarities between this situation and that of the *forward-biased* diode in Chapter 2. The depletion region has been reduced in width due to the applied bias, resulting in a heavy flow of majority carriers from the *p-* to the *n*-type material.

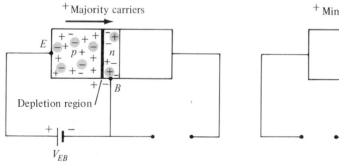
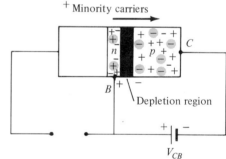

Figure 4.3 Forward-biased junction of a *pnp* transistor.

Figure 4.4 Reverse-biased junction of a *pnp* transistor.

Let us now remove the base-to-emitter bias of the *pnp* transistor of Fig. 4.2a as shown in Fig. 4.4. Consider the similarities between this situation and that of the *reverse-biased* diode of Section 2.3. Recall that the flow of majority carriers is zero, resulting in only a minority-carrier flow, as indicated in Fig. 4.4. *In summary, therefore, one* p-n *junction of a transistor is reverse-biased, while the other is forward-biased.*

In Fig. 4.5 both biasing potentials have been applied to a *pnp* transistor, with the resulting majority- and minority-carrier flow indicated. Note in Fig. 4.5 the

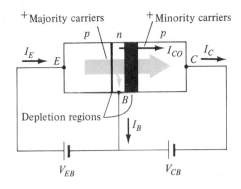

Figure 4.5 Majority and minority carrier flow of a *pnp* transistor.

widths of the depletion regions, indicating clearly which junction is forward-biased and which is reverse-biased. As indicated in Fig. 4.5, a large number of majority carriers will diffuse across the forward-biased *p-n* junction into the *n*-type material. The question then is whether these carriers will contribute directly to the base current I_B or pass directly into the *p*-type material. Since the sandwiched *n*-type material is very thin and has a low conductivity, a very small number of these carriers will take this path of high resistance to the base terminal. The magnitude of the base current is typically on the order of microamperes as compared to milliamperes for the emitter and collector currents. The larger number of these majority carriers will diffuse across the reverse-biased junction into the *p*-type material connected to the collector terminal as indicated in Fig. 4.5. The reason for the relative ease with which the majority carriers can cross the reverse-biased junction is easily understood if we consider that for the reverse-biased diode the injected majority carriers will appear as minority carriers in the *n*-type material. In other words, there has been an *injection* of minority carriers into the *n*-type base region material. Combining this with the fact that all the minority carriers in the depletion region will cross the reverse-biased junction of a diode accounts for the flow indicated in Fig. 4.5.

Applying Kirchhoff's current law to the transistor of Fig. 4.5 as if it were a single node, we obtain

$$\boxed{I_E = I_C + I_B} \quad (4.1)$$

and find that the emitter current is the sum of the collector and base currents. The collector current, however, is comprised of two components—the majority and minority carriers as indicated in Fig. 4.5. The minority current component is called the *leakage current* and is given the symbol I_{CO} (I_C current with emitter terminal *O*pen). The collector current, therefore, is determined in total by Eq. (4.2).

$$\boxed{I_C = I_{C_{\text{majority}}} + I_{CO_{\text{minority}}}} \quad (4.2)$$

For general-purpose transistors, I_C is measured in milliamperes, while I_{CO} is measured in microamperes or nanoamperes. I_{CO}, like I_S for a reverse-biased diode, is temperature-sensitive and must be examined carefully when applications of wide temperature ranges are considered. It can severely affect the stability of a system at high temperatures if not considered properly.

Improvements in construction techniques have resulted in significantly lower levels of I_{CO}, to the point where its effect can often be ignored. However, higher-power devices still typically have values of I_{CO} in the microampere range.

The configuration shown in Fig. 4.2 for the *pnp* and *npn* transistors is called the *common-base* configuration since the base is common to both the emitter and collector terminals. For fixed values of V_{CB} in the common-base configuration the ratio of a small change in I_C to a small change in I_E is commonly called the *common-base, short-circuit amplification factor* and is given the symbol α (alpha).

In equation form, the magnitude of α is given by

$$\alpha = \left.\frac{\Delta I_C}{\Delta I_E}\right|_{V_{CB}=\text{constant}} \quad (4.3)$$

The term "short circuit" indicates that the load is short-circuited when α is determined. More will be said about the necessity for shorting the load and the operations involved with using equations of the type indicated by Eq. (4.3) when we consider equivalent circuits in Chapter 5. Typical values of α vary from 0.90 to 0.998. For most practical applications, a first approximation for the magnitude of α, usually correct to within a few percent, can be obtained using the following equation:

$$\alpha \cong \frac{I_C}{I_E} \quad (4.4)$$

where I_C and I_E are the magnitude of the collector and emitter currents, respectively, at a particular point on the transistor characteristics.

Equations (4.3) and (4.4) are employed to determine α from the device characteristics or network conditions. However, in the strictest sense, α is only a measure of the percentage of holes (majority carriers) originating in the emitter *p*-material of Fig. 4.5 that reach the collector terminal. As defined by Eq. (4.2), therefore,

$$I_C = \alpha I_{E_{\text{majority}}} + I_{CO_{\text{minority}}} \quad (4.5)$$

4.4 TRANSISTOR AMPLIFYING ACTION

The basic voltage-amplifying action of the common-base configuration can now be described using the circuit of Fig. 4.6. For the common-base configuration, the input resistance between the emitter and the base of a transistor will typically

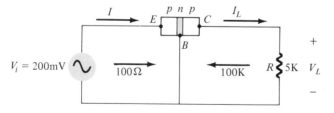

Figure 4.6 Basic voltage amplification action of the common-base configuration.

vary from 20 to 200 Ω, while the output resistance may vary from 100 K to 1 M. The difference in resistance is due to the forward-biased junction at the input (base to emitter) and the reverse-biased junction at the output (base to collector).

Using effective values and an average value of 100 Ω for the input resistance, we find

$$I = \frac{200 \times 10^{-3}}{100} = 2 \text{ mA}$$

If we assume for the moment that $\alpha = 1$ ($I_C = I_E$),

$$I_L = I = 2 \text{ mA}$$

and
$$V_L = I_L R$$
$$= (2 \times 10^{-3})(5 \times 10^{+3})$$
$$V_L = 10 \text{ V}$$

The voltage amplification is

$$A_v = \frac{V_L}{V_i} = \frac{10}{200 \times 10^{-3}} = 50$$

Typical values of voltage amplification for the common-base configuration vary from 20 to 100. The current amplification (I_C/I_E) is always less than 1 for the common-base configuration. This latter characteristic should be obvious since $I_C = \alpha I_E$ and α is always less than 1.

The basic amplifying action was produced by *transferring* a current I from a low- to a high-*resistance* circuit. The combination of the two terms in italics results in the label transistor; that is,

$$\textit{trans}\text{fer} + \text{re}\textit{sistor} \rightarrow \textit{transistor}$$

4.5 COMMON-BASE CONFIGURATION

The notation and symbols used in conjunction with the transistor in the majority of texts and manuals published today are indicated in Fig. 4.7 for the common-

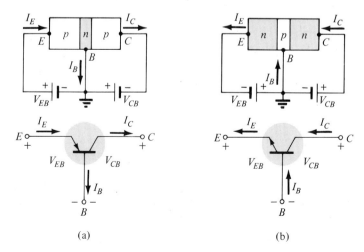

Figure 4.7 Notation and symbols used with the common-base configuration: (a) *pnp* transistor; (b) *npn* transistor.

base configuration with *pnp* and *npn* transistors. Throughout this text all current directions will refer to the conventional (hole flow) rather than the electron flow. This choice was based primarily on the fact that the vast majority of past and present publications in electrical engineering use conventional current flow.

Some texts prefer to show all the currents entering in Fig. 4.7 when they describe the basic operation of the transistor and simply include negative signs when appropriate. In other words, if the actual conventional flow direction is the opposite direction, a negative sign is included along with the magnitude. For clarity, all currents, as indicated in Fig. 4.7, will indicate the actual flow direction for the active region. *Note that the arrow in the symbol is the same as the direction of I_E (only true for conventional flow).* On specification sheets negative signs indicate that all currents are entering.

For the common-base configuration the applied potentials are written with respect to the base potential resulting in V_{EB} and V_{CB}. In other words, the second subscript will always indicate the transistor configuration. In all cases the first subscript is defined to be the point of higher potential, as shown in Fig. 4.7. For the *pnp* transistor, therefore, V_{EB} is positive and V_{CB} is negative (since the battery V_{CB} sets the collector or the lower potential), as indicated on the characteristics of Fig. 4.8. For the *npn* transistor V_{EB} is negative and V_{CB} is positive.

In addition, note that two sets of characteristics are necessary to represent the behavior of the *pnp* common-base transistor of Fig. 4.7: the *driving point* (or input) and the *output* set.

The output or collector characteristics of Fig. 4.8a relate the collector current to the collector-to-base voltage and emitter current. The collector characteristics have three basic regions of interest, as indicated in Fig. 4.8a: the *active, cutoff,* and *saturation* regions.

In the active region the collector junction is reverse-biased, while the emitter junction is forward-biased. These conditions refer to the situation of Fig. 4.5. The active region is the only region employed for the amplification of signals with minimum distortion. When the emitter current (I_E) is zero, the collector current is simply that due to the reverse saturation current I_{CO}, as indicated in Fig. 4.8a. The current I_{CO} is so small (microamperes) in magnitude compared to the vertical scale of I_C (milliamperes) that it appears on virtually the same horizontal line as $I_C = 0$. The circuit conditions that exist when $I_E = 0$ for the common-base configuration are shown in Fig. 4.9. The notation most frequently used for I_{CO} on data and specification sheets is, as indicated in Fig. 4.9, I_{CBO}. Because of improved construction techniques, the level of I_{CBO} for general-purpose transistors (especially silicon) in the low- and mid-power ranges is usually so low that its effect can be ignored. However, for higher power units I_{CBO} will still appear in the microampere range. In addition, keep in mind that I_{CBO} like I_S for the diode (both reverse leakage currents) is temperature-sensitive. At higher temperatures the effect of I_{CBO} for any power level unit may become an important factor since it increases so rapidly with temperature.

Note in Fig. 4.8a that as the emitter current increases above zero, the collector current increases to a magnitude slightly less ($\alpha < 1$) than that of the emitter current as determined by the basic transistor-current relations. Note also the almost negligible effect of V_{CB} on the collector current for the active region. The curves

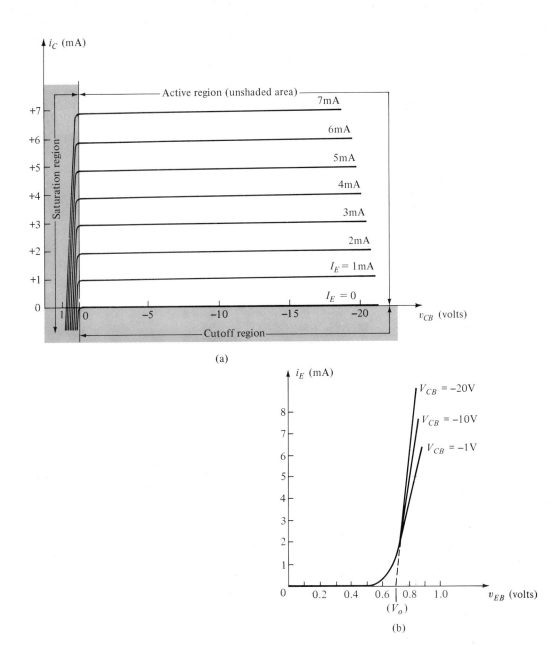

Figure 4.8 Characteristics of a *pnp* transistor in the common-base configuration: (a) collector characteristics; (b) emitter characteristics.

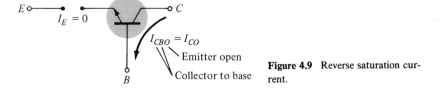

Figure 4.9 Reverse saturation current.

135

clearly indicate that *a first approximation to the relationship between I_E and I_C in the active region is given by $I_E = I_C$.*

In the cutoff region the collector and emitter junctions are both reverse-biased, resulting in negligible collector current as demonstrated in Fig. 4.8a.

The horizontal scale for V_{CB} has been expanded to the left of 0 V to represent clearly the characteristics in this region. *In the region, called the* saturation region, *the collector and emitter junctions are forward-biased,* resulting in the exponential change in collector current with small changes in collector-to-base potential.

The input or emitter characteristics have only one region of interest, as illustrated by Fig. 4.8b. For fixed values of collector voltage (V_{CB}), as the emitter-to-base potential increases, the emitter current increases, as shown. Increasing levels of V_{CB} result in a reduced level of V_{EB} to establish the same current.

Note the tight grouping of the curves for the wide range of values for V_{CB}. In addition, consider how closely the average value of the curves appears to begin its rise or about $V_0 = 0.7$ V for the silicon transistor. As with the semiconductor silicon diode, *a first approximation for the forward-biased base–emitter junction in the dc mode would be that $V_{EB} \cong 0.7$ V for all levels of V_{CB}.*

EXAMPLE 4.1   Using the characteristics of Fig. 4.8,
(a) Find the resulting collector current if $I_E = 3$mA and $V_{CB} = -10$ V.
(b) Find the resulting collector current if $V_{EB} = 750$ mV and $V_{CB} = -10$ V.
(c) Find V_{EB} for the conditions $I_C = 5$ mA and $V_{CB} = -1$ V.

Solution:

(a) $$I_C \cong I_E = \mathbf{3 \text{ mA}}$$

(b) On the input characteristics $I_E = 3.5$ mA at the intersection of $V_{EB} = 750$ mV, $V_{CB} = -10$ V, and $I_C \cong I_E = \mathbf{3.5 \text{ mA}}$.

(c) $$I_E \cong I_C = 5 \text{ mA}$$

On the input characteristics the intersection of $I_E = 5$ mA and $V_{CB} = -1$ V results in $V_{EB} \cong 800$ mV = **0.8 V.**

4.6 COMMON-EMITTER CONFIGURATION

The most frequently encountered transistor configuration is shown in Fig. 4.10 for the *pnp* and *npn* transistors. It is called the *common-emitter configuration* since the emitter is common to both the base and collector terminals. Two sets of characteristics are again necessary to fully describe the behavior of the common-emitter configuration: one for the input or base circuit and one for the output or collector circuit. Both are shown in Fig. 4.11.

The emitter, collector, and base currents are shown in their actual conventional current flow direction, while the potentials have the capital letter E as the second subscript to indicate the configuration. Even though the transistor configuration has changed, the current relations developed earlier for the common-base configuration are still applicable.

For the common-emitter configuration the output characteristics will be a plot

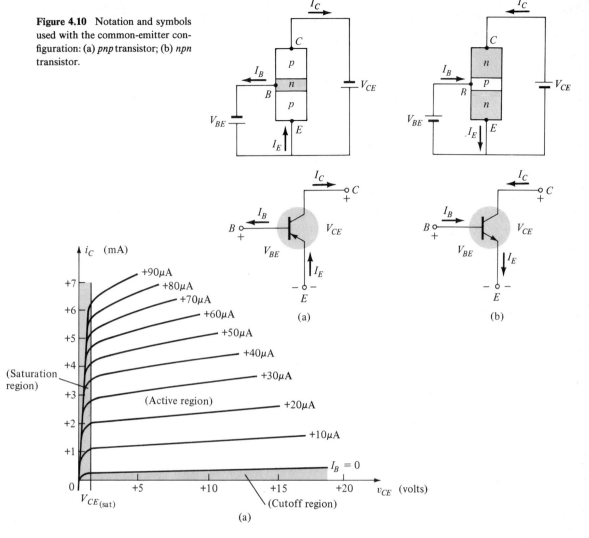

Figure 4.10 Notation and symbols used with the common-emitter configuration: (a) *pnp* transistor; (b) *npn* transistor.

Figure 4.11 Characteristics of a *npn* transistor in the common-emitter configuration: (a) collector characteristics; (b) base characteristics.

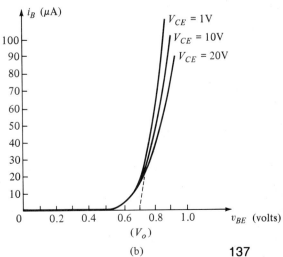

137

of the input current *(I_B)* versus the output voltage *(V_CE)* and output current *(I_C)*. The input characteristics are a plot of the output voltage *(V_CE)* versus the input voltage *(V_BE)* and input current *(I_B)*.

Note that on the characteristics of Fig. 4.11 the magnitude of I_B is in microamperes as compared to milliamperes for I_C. Consider also that the curves of I_B are not as horizontal as those obtained for I_E in the common-base configuration, indicating that the collector-to-emitter voltage will influence the magnitude of the collector current.

The active region for the common-emitter configuration is that portion of the upper-right quadrant that has the greatest linearity, that is, that region in which the curves for I_B are nearly straight and equally spaced. In Fig. 4.11a this region exists to the right of the vertical dashed line at $V_{CE_{sat}}$ and above the curve for I_B equal to zero. The region to the left of $V_{CE_{sat}}$ is called the saturation region. *In the active region the collector junction is reverse-biased, while the base junction is forward-biased.* You will recall that these were the same conditions that existed in the active region of the common-base configuration. The active region of the common-emitter configuration can be employed for voltage, current, or power amplification.

The cutoff region for the common-emitter configuration is not as well defined as for the common-base configuration. Note on the collector characteristics of Fig. 4.11 that I_C is not equal to zero when I_B is zero. For the common-base configuration, when the input current I_E was equal to zero, the collector current was equal only to the reverse saturation current I_{CO}, so that the curve $I_E = 0$ and the voltage axis were, for all practical purposes, one.

The reason for this difference in collector characteristics can be derived through the proper manipulation of Eqs. (4.1) and (4.5). That is,

$$I_C = \alpha I_E + I_{CO} \quad \text{[Eq. (4.5)]}$$

but

$$I_E = I_C + I_B \quad \text{[Eq. (4.1)]}$$

Therefore,

$$I_C = \alpha(I_C + I_B) + I_{CO}$$

and

$$I_C = \alpha I_C + \alpha I_B + I_{CO}$$

or

$$I_C(1 - \alpha) = \alpha I_B + I_{CO}$$

with

$$\boxed{I_C = \frac{\alpha I_B}{1 - \alpha} + \frac{I_{CO}}{1 - \alpha}} \quad (4.6)$$

If we consider the case discussed earlier, where $I_B = 0$, and substitute this value into Eq. (4.6), then

$$\boxed{I_C = \frac{I_{CO}}{1 - \alpha}\bigg|_{I_B = 0}} \quad (4.7)$$

For $\alpha = 0.996$,

$$I_C = \frac{I_{CO}}{1 - 0.996} = \frac{I_{CO}}{0.004}$$

and
$$I_C = 250 I_{CO}|_{I_B=0}$$

which accounts for the vertical shift in the $I_B = 0$ curve from the horizontal voltage axis.

For future reference, the collector current defined by Eq. (4.7) will be assigned the notation indicated by Eq. (4.8).

$$\boxed{I_{CEO} = \frac{I_{CO}}{1-\alpha}\bigg|_{I_B=0}} \qquad (4.8)$$

In Fig. 4.12 the conditions surrounding this newly defined current are demonstrated with its assigned reference direction.

The magnitude of I_{CEO} is typically much smaller for silicon materials than for germanium materials. For transistors with similar ratings I_{CEO} would typically be a few microamperes for silicon but perhaps a few hundred microamperes for germanium.

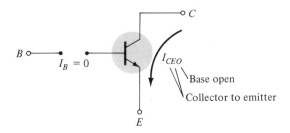

Figure 4.12 Circuit conditions related to I_{CEO}.

For linear (least distortion) amplification purposes, cutoff for the common-emitter configuration will be (for this text) determined by $I_C = I_{CEO}$. In other words, the region below $I_B = 0$ is to be avoided if an undistorted output signal is required.

When employed as a switch in the logic circuitry of a computer, a transistor will have two points of operation of interest: one in the cutoff and one in the saturation region. The cutoff condition should ideally be $I_C = 0$ for the chosen V_{CE} voltage. Since I_{CEO} is typically low in magnitude for silicon materials, *cutoff will exist for switching purposes when $I_B = 0$ or $I_C = I_{CEO}$ for silicon transistors only.* For germanium transistors, however, cutoff for switching purposes will be defined as those conditions that exist when $I_C = I_{CBO} = I_{CO}$. This condition can normally be obtained for germanium transistors by reverse-biasing the normally forward-biased base-to-emitter junction a few tenths of a volt.

EXAMPLE 4.2 Using the characteristics of Fig. 4.11,
(a) Find the value of I_C corresponding to $V_{BE} = +800$ mV and $V_{CE} = +10$ V.
(b) Find the value of V_{CE} and V_{BE} corresponding to $I_C = +4$ mA and $I_B = +40$ µA.

Solution: (a) On the input characteristics the intersection of $V_{BE} = +800$ mV and $V_{CE} = +10$ V results in

$$I_B \cong 50 \text{ µA}$$

On the output characteristics the intersection of $I_B = 50$ μA and $V_{CE} = 10$ V results in

$$I_C \cong 5.1 \text{ mA}$$

(b) On the output characteristics the intersection of $I_C = +4$ mA and $I_B = +40$ μA results in

$$V_{CE} = +6.2 \text{ V}$$

On the input characteristics the intersection of $I_B = +40$ μA and $V_{CE} = +6.2$ V results in

$$V_{BE} \cong 770 \text{ mV}$$

In Section 4.3 the symbol alpha (α) was assigned to the forward current transfer ratio of the common-base configuration. For the common-emitter configuration, the ratio of a small change in collector current to the corresponding change in base current at a fixed collector-to-emitter voltage (V_{CE}) is assigned the Greek letter beta (β) and is commonly called the *common-emitter forward current amplification factor*. In equation form, the magnitude of β is given by

$$\beta = \left.\frac{\Delta I_C}{\Delta I_B}\right|_{V_{CE} = \text{constant}} \tag{4.9}$$

As a first, but close, approximation, the magnitude of beta (β) can be determined by the following equation:

$$\beta \cong \frac{I_C}{I_B} \tag{4.10}$$

where I_C and I_B are collector and base currents of a particular operating point in the linear region (i.e., where the horizontal base current lines of the common-emitter characteristics are closest to being parallel and equally spaced). Since I_C and I_B in Eq. (4.10) are fixed or dc values, the value obtained for β from Eq. (4.10) is frequently called the *dc beta*, while that obtained by Eq. (4.9) is called the *ac* or *dynamic* value. Typical values of β vary from 20 to 600. Through the following manipulations of Eqs. (4.1), (4.4), and (4.10):

[Eq. (4.10)] $\quad\quad\quad\quad \beta = \dfrac{I_C}{I_B}; \quad I_B = \dfrac{I_C}{\beta}$

[Eq. (4.4)] $\quad\quad\quad\quad \alpha = \dfrac{I_C}{I_E}; \quad I_E = \dfrac{I_C}{\alpha}$

[Eq. (4.1)] $\quad\quad\quad\quad I_E = I_C + I_B$

$\quad\quad\quad\quad\quad\quad\quad\quad\quad\quad\quad \dfrac{I_C}{\alpha} = I_C + \dfrac{I_C}{\beta}$

and dividing by I_C:

$$\frac{1}{\alpha} = 1 + \frac{1}{\beta}$$

and
$$\beta = \alpha\beta + \alpha$$

or
$$\beta(1 - \alpha) = \alpha$$

we obtain
$$\boxed{\beta = \frac{\alpha}{1 - \alpha}} \quad (4.11)$$

or
$$\boxed{\alpha = \frac{\beta}{\beta + 1}} \quad (4.12)$$

In addition, since
$$I_{CEO} = \frac{I_{CO}}{1 - \alpha} = \frac{I_{CBO}}{1 - \alpha}$$

then
$$\boxed{I_{CEO} = (\beta + 1)I_{CBO} \cong \beta I_{CBO}} \quad (4.13)$$

EXAMPLE 4.3

(a) Find the dc beta at an operating point of $V_{CE} = +10$ V and $I_C = +3$ mA on the characteristics of Fig. 4.11.
(b) Find the value of α corresponding with this operating point.
(c) At $V_{CE} = +10$ V find the corresponding value of I_{CEO}.
(d) Calculate the approximate value of I_{CBO} using the β_{dc} obtained in part (a).

Solution

(a) At the intersection of $V_{CE} = +10$ V and $I_C = +3$ mA, $I_B = +25$ μA,

so that
$$\beta_{dc} = \frac{I_C}{I_B} = \frac{3 \times 10^{-3}}{25 \times 10^{-6}} = 120$$

(b)
$$\alpha = \frac{\beta}{\beta + 1} = \frac{120}{121} \cong 0.992$$

(c)
$$I_{CEO} = 300 \; \mu A$$

(d)
$$I_{CBO} \cong \frac{I_{CEO}}{\beta} = \frac{300 \; \mu A}{120} = 2.5 \; \mu A$$

The input characteristics for the common-emitter configuration are very similar to those obtained for the common-base configuration (Fig. 4.11). In both cases, the increase in input current is due to an increase in majority carriers crossing the base-to-emitter junction with increasing forward-bias potential. Note also that the variation in output voltage (V_{CE} for the CE configuration and V_{CB} for the CB configuration) does not result in a large relocation of the characteristics. In fact, for the dc voltage levels commonly encountered the variation in base-to-emitter voltage with change in output terminal voltage can, as a first approximation, be ignored. On this basis, if we use an average value, the curve of Fig. 4.13 for

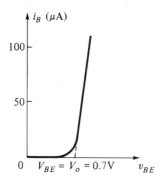

Figure 4.13 Reproduction of Fig. 4.11b ignoring the effects of V_{CE}.

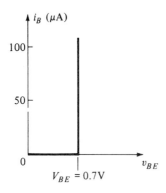

Figure 4.14 Approximate reproduction of Fig. 4.13 for dc analysis.

the CE configuration will result. Note the similarities with the silicon diode characteristics. Recall also from the description of the semiconductor diode that for dc analysis we approximated the curve of Fig. 4.13 with that indicated in Fig. 4.14. Essentially, therefore, for dc analysis a first approximation to the base-to-emitter voltage of a transistor configuration is to assume that $V_{BE} \cong 0.7$ V for silicon and 0.3 V for germanium. If insufficient voltage is present to provide the 0.7 V bias (for silicon transistors) with the proper polarity, the transistor cannot be in the active region. A number of applications of this approximation will appear in Chapter 5. Since the CB characteristics had a similar set of input characteristics [also true for the common-collector (CC) configuration to be discussed], we can conclude *as a first approximation for dc analysis that the base-to-emitter voltage of a BJT is assumed to be V_O when biased in the active region of the characteristics.*

Further, we found for the output characteristics of the CB configuration that $I_C = I_E$. For the CE configuration $I_C = \beta I_B$, where β is determined by the operating conditions.

In manuals, data sheets, and other transistor publications the common-emitter characteristics are the most frequently presented. The common-base characteristics can be obtained directly from the common-emitter characteristics using the basic current relations derived in the past few sections. In other words, for each point on the characteristics of the common-emitter configuration a sufficient number of variables can be obtained to substitute into the equations derived to come up with a point on the common-base characteristics. This process is, of course, time consuming, but it will result in the desired characteristics.

4.7 COMMON-COLLECTOR CONFIGURATION

The third and final transistor configuration is the *common-collector configuration*, shown in Fig. 4.15 with the proper current directions and voltage notation.

The common-collector configuration is used primarily for impedance matching purposes since it has a high input impedance and low output impedance, opposite to that which is true of the common-base and common-emitter configurations.

The common-collector circuit configuration is generally as shown in Fig. 4.16 with the load resistor from emitter to ground. Note that the collector is tied to ground even though the transistor is connected in a manner similar to the common-

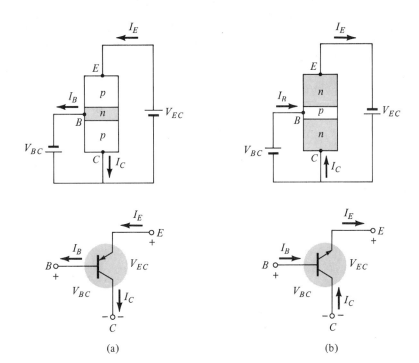

Figure 4.15 Notation and symbols used with the common-collector configuration: (a) *pnp* transistor; (b) *npn* transistor.

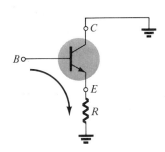

Figure 4.16 Common-collector configuration used for impedance matching purposes.

emitter configuration. From a design viewpoint, there is no need for a set of common-collector characteristics to choose the parameters of the circuit of Fig. 4.16. It can be designed using the common-emitter characteristics of Section 4.6. For all practical purposes, the output characteristics of the common-collector configuration are the same as for the common-emitter configuration. For the common-collector configuration the output characteristics are a plot of the input current I_B versus V_{EC} and I_E. The input current, therefore, is the same for both the common-emitter and common-collector characteristics. The horizontal voltage axis for the common-collector configuration is obtained by simply changing the sign of the collector-to-emitter voltage of the common-emitter characteristics since $V_{EC} = -V_{CE}$. Finally, there is an almost unnoticeable change in the vertical scale of I_C of the common-emitter characteristics if I_C is replaced by I_E for the common-collector characteristics (since $\alpha \cong 1$). For the input circuit of the common-collector configuration the common-emitter base characteristics are sufficient for obtaining any required information by simply writing Kirchhoff's voltage law around the loop indicated in Fig. 4.16 and performing the proper mathematical manipulations.

SEC. 4.7 COMMON-COLLECTOR CONFIGURATION

4.8 TRANSISTOR MAXIMUM RATINGS

The standard transistor data sheet will include at least three maximum ratings: *collector dissipation, collector voltage, and collector current.*

For the transistor whose characteristics were presented in Fig. 4.11, the following maximum ratings were indicated:

$$P_{C_{max}} = 30 \text{ mW}$$
$$I_{C_{max}} = 6 \text{ mA}$$
$$V_{CE_{max}} = 20 \text{ V}$$

The power or dissipation rating is the product of the collector voltage and current. For the common-emitter configuration,

$$\boxed{P_{C_{max}} = V_{CE} I_C} \qquad (4.14)$$

The nonlinear curve determined by this equation is indicated in Fig. 4.17. The curve was obtained by choosing various values of V_{CE} or I_C and finding the other variable using Eq. (4.14). For example, at $V_{CE} = 10$ V,

$$I_C = \frac{P_{C_{max}}}{V_{CE}} = \frac{30 \times 10^{-3}}{10} = 3 \text{ mA}$$

as indicated in Fig. 4.17. The region above this curve must be avoided in the design of systems using this particular transistor if the maximum power rating is not to be exceeded. The maximum collector voltage, in this case V_{CE}, is indicated as a vertical line in Fig. 4.17. The maximum collector current is also indicated as a horizontal line.

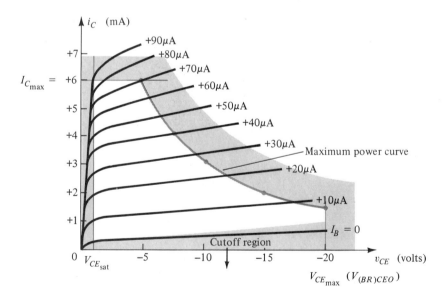

Figure 4.17 Region of operation for amplification purposes.

For the common-base configuration the collector dissipation is determined by the following equation. The maximum collector voltage would refer to V_{CB}.

$$\boxed{P_{C_{\max}} = V_{CB} I_C} \qquad (4.15)$$

For amplification purposes the nonlinear characteristics of the saturation and the cutoff regions are also avoided. The saturation region has been indicated by the vertical line at $V_{CE_{sat}}$ and the cutoff region by $I_B = 0$ in Fig. 4.17. The unshaded region remaining is the region employed for amplification purposes. Although it appears as though the area of operation has been drastically reduced, we must keep in mind that many signals are in the microvolt or millivolt range, while the horizontal axis of the characteristics is measured in volts. In addition to maximum ratings, data and specification sheets on transistors also include other important information about their operation. The discussion of this additional data will not be considered until each parameter is fully defined.

4.9 TRANSISTOR SPECIFICATION SHEET

The information provided in the RCA power transistors data book for the 2N1711 transistor appears in Figs. 4.18 through 4.25. As noted, this is a general-purpose small signal/medium power device.

The letter o at the end of a parameter indicates that the terminal not listed is left open. Note in Fig. 4.18 that I_{CBO} is only 0.01 μA at 25°C but 10 μA at 150°C. The quantity h_{FE}, which is synonymous with $\beta_{dc} = I_C/I_B$, has a minimum value of 20. Priorities do not permit an introduction to all the quantities listed in Fig. 4.18. However, most companies carefully define these quantities in the introductory pages of their catalogs. A number of the remaining quantities in the figure will be introduced in a later chapter. Certainly, V_{CE}(sat.), the capacitance levels and thermal resistance quantities are familiar. The *hybrid* parameters h_{fe}, h_{ib}, h_{rb}, and h_{ob} will be introduced in Chapter 5.

The output characteristics for the device appear in Fig. 4.20. Note the resulting distortion at high levels of voltage and current—a region that must be avoided for linear operation. Linear operation suggests that the output waveform has the same appearance as the input (but amplified) and is not distorted by the amplifying device.

Note the shift in the V_{BE} versus I_C curve for increasing levels of base current. Consider also that I_B is measured in milliamperes since it is a power device. It would appear that our use of $V_o = 0.7$ V is a good average value since base currents will probably not exceed 25 mA, as shown in Fig. 4.26.

The familiar power derating curve is provided in Fig. 4.22. Note that a curve has been provided for the case and free-air temperature.

The variation in h_{FE} is provided versus I_C for a range of temperatures. Note that the ratio is less at each temperature if the collector current is too high.

In Fig. 4.24 the level of I_{CBO} is provided versus junction temperature for different levels of collector-to-base voltage. It appears the I_{CBO} will not reach 1 μA until

Figure 4.18 Electrical characteristics of the RCA 2N1711 power transistor. (Courtesy RCA Solid State Division.)

Characteristic	Symbol	Case Temperature °C	Frequency kHz	Test Conditions dc Collector-to-Base Voltage V V_{CB}	dc Collector-to-Emitter Voltage V V_{CE}	dc Emitter-to-Base Voltage V V_{EB}	dc Collector Current mA I_C	dc Emitter Current mA I_E	dc Base Current mA I_B	Limits Min.	Limits Max.	Units
Collector-cutoff current	I_{CBO}	25		60			0	0		—	0.01	μA
		150		60			0	0		—	10	
Emitter-cutoff current	I_{EBO}	25				5				—	0.005	μA
dc-pulse forward-current transfer ratio[a]	h_{FE}	25			10		10			75	—	
		25			10		150			100	300	
		25			10		500			40	—	
dc forward-current transfer ratio	h_{FE}	25			10		0.01			20	—	
		25			10		0.1			35	—	
		−55			10		10			35	—	
Collector-to-base breakdown voltage	$V_{(BR)CBO}$	25					0.1	0		75	—	V
Emitter-to-base breakdown voltage	$V_{(BR)EBO}$	25					0	0.1		7	—	V
Collector-to-emitter reach-through voltage	V_{RT}	25				1.5[b]	0.1			75	—	V

Parameter	Symbol								Unit
Collector-to-emitter sustaining voltage with external base-to-emitter resistance = 10 ohms	$V_{CER}(\text{sus})$	25			100 (pulsed)		50	—	V
Collector-to-emitter saturation voltage	$V_{CE}(\text{sat})$	25			150		15	1.5	V
Base-to-emitter saturation voltage	$V_{BE}(\text{sat})$	25			150		15	1.3	V
Small-signal forward-current transfer ratio	h_{fe}	25	1	5	1			200	
		25	1	10	5			300	
		25	20 MHz	10	50			—	
Noise figure: Generator resistance (R_G) = 510 ohms, circuit bandwidth (BW) = 1 cycle	NF	25	1		0.3			8	dB
Output capacitance	C_{ob}	25		10		0		15	pF
Input capacitance	C_{ib}	25	1		0	0.5		80	pF
Input resistance	h_{ib}	25	1	5	1		24	34	Ω
		25	1	10	5		4	8	
Voltage-feedback ratio	h_{rb}	25	1	5	1			5×10^{-4}	
		25	1	10	5			5×10^{-4}	
Output conductance	h_{ob}	25	1	5	1		0.1	0.5	μmho
		25	1	10	5		0.1	1	
Thermal resistance: Junction-to-case	$R_{\theta JC}$	—					—	58.3	°C/W
Junction-to-free air	$R_{\theta JA}$	—					—	219	

[a] Pulse duration = 300 μs; duty factor ≤2%.
[b] V_{EBF} = Emitter-to-base floating potential.

Power Transistors

2N1711

Silicon N-P-N Planar Transistors

General-Purpose Type for Small-Signal, Medium-Power Applications

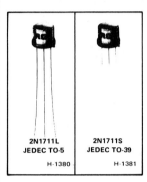

2N1711L JEDEC TO-5
H-1380

2N1711S JEDEC TO-39
H-1381

Features:

- Minimum gain-bandwidth product = 70 MHz; useful in applications from dc to 25 MHz
- Operation at high junction temperatures
- Planar construction for low-noise and low-leakage characteristics
- Low output capacitance

> These devices are available with either 1½-inch leads (TO-5 package) or ½-inch leads (TO-39 package). The longer-lead versions are specified by suffix "L" after the type number; the shorter-lead versions are specified by suffix "S" after the type number.

RCA-2N1711 is a silicon n-p-n planar transistor intended for a wide variety of small-signal and medium-power applications in military and industrial equipment. It features exceptionally low noise and leakage characteristics, high pulse beta (h_{FE}), high breakdown-voltage ratings, low saturation voltages, high sustaining voltages, and low output capacitance.

MAXIMUM RATINGS, *Absolute-Maximum Values:*

COLLECTOR-TO-BASE VOLTAGE	V_{CBO}	75	V
COLLECTOR-TO-EMITTER VOLTAGE:			
With external base-to-emitter resistance (R_{BE}) ≤ 10 Ω	V_{CER}	50	V
EMITTER-TO-BASE VOLTAGE	V_{EBO}	7	V
COLLECTOR CURRENT	I_C	1	A
TRANSISTOR DISSIPATION:	P_T		
At case temperatures up to 25°C		3	W
At case temperatures above 25°C		See Fig. 4.22	
At free-air temperatures up to 25°C		0.8	W
At free-air temperatures above 25°C		See Fig. 4.22	
TEMPERATURE RANGE:			
Storage and Operating (Junction)		−65 to +200	°C
LEAD TEMPERATURE (During soldering):			
At distance ≥ 1/16 in. (1.58 mm) from seating plane for 10 s max.		230	°C

Figure 4.19 RCA 2N1711 power transistor. (Courtesy RCA Solid State Division.)

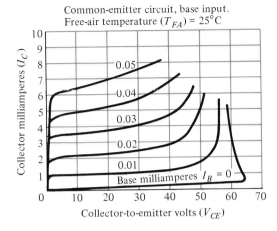

Figure 4.20 RCA 2N1711 output characteristics. (Courtesy RCA Solid State Division.)

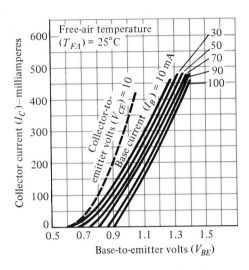

Figure 4.21 RCA 2N1711 transfer characteristics. (Courtesy RCA Solid State Division.)

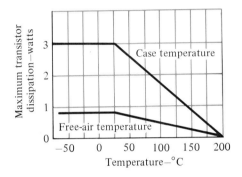

Figure 4.22 RCA 2N1711 derating curves. (Courtesy RCA Solid State Division.)

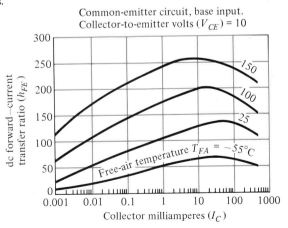

Figure 4.23 RCA 2N1711 dc beta characteristics. (Courtesy RCA Solid State Division.)

149

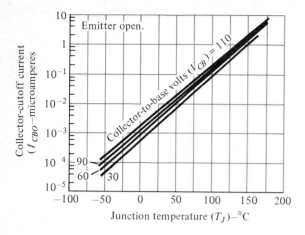

Figure 4.24 RCA 2N1711 collector-cutoff-current characteristics. (Courtesy RCA Solid State Division.)

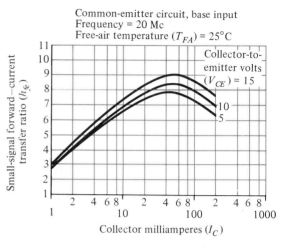

Figure 4.25 RCA 2N1711 small-signal beta characteristics. (Courtesy RCA Solid State Division.)

the junction temperature approaches 135°C. Even with a β of 100, I_{CEO} is still limited to 100 μA = 0.1 mA at this temperature. For a medium-power device, this is a fairly low level for this undesirable effect.

The small-signal forward current transfer ratio (h_{fe}) will be defined in Chapter 5. Briefly, it is a measure of the small-signal ac gain of the device (the increase in the peak-to-peak level of a sinusoidal signal).

Figure 4.26 demonstrates the change in collector characteristics at the high levels of current flow. The nice even spacing between curves in Fig. 4.20 is no longer present, but the increasing density of the lines follows a fairly linear pattern.

We will refer to these figures as we introduce new quantities of importance in succeeding chapters.

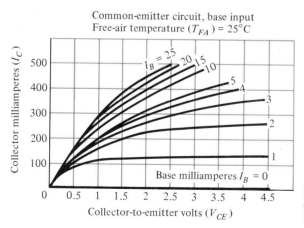

Figure 4.26 RCA 2N1711 output characteristics (high I_C). (Courtesy RCA Solid State Division.)

4.10 TRANSISTOR FABRICATION

The majority of the methods used to fabricate transistors are simply extensions of the methods used to manufacture semiconductor diodes. The methods most frequently employed today include *point-contact, alloy junction, grown junction, and diffusion*. The following discussion of each method will be brief, but the fundamental steps included in each will be presented. A detailed discussion of each method would require a text in itself.

Point-Contact

The point-contact transistor is manufactured in a manner very similar to that used for point-contact semiconductor diodes. In this case two wires are placed next to an *n*-type wafer as shown in Fig. 4.27. Electrical pulses are then applied to each wire resulting in a *p-n* junction at the boundary of each wire and the semiconductor wafer. The result is a *pnp* transistor as shown in Fig. 4.27. This method of fabrication is today limited to high frequency/low power devices. It was the method used in the fabrication of the first transistor shown in Fig. 4.1.

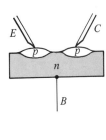

Figure 4.27 Point-contact transistor.

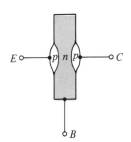

Figure 4.28 Alloy junction transistor.

Alloy Junction

The alloy junction technique is also an extension of the alloy method of manufacturing semiconductor diodes. For a transistor, however, two dots of the same impurity are deposited on each side of a semiconductor wafer having the opposite impurity as shown in Fig. 4.28. The entire structure is then heated until melting occurs and each dot is alloyed to the base wafer resulting in the *p-n* junctions indicated in Fig. 4.28 as described for semiconductor diodes.

The collector dot and resulting junction are larger, to withstand the heavy current and power dissipation at the collector-base junction. This method is not employed as much as the diffusion technique to be described shortly, but it is still used extensively in the manufacture of high-power diodes.

Grown Junction

The Czochralski technique (Section 1.6) is used to form the two *p-n* junctions of a grown junction transistor. The process, as depicted in Fig. 4.29, requires that the impurity control and withdrawal rate be such as to ensure the proper base width and doping levels of the *n*- and *p*-type materials. Transistors of this type are, in general, limited to less than $\frac{1}{4}$-W rating.

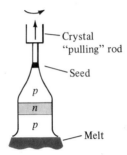

Figure 4.29 Grown junction transistor.

Diffusion

The most frequently employed method of manufacturing transistors today is the diffusion technique. The basic process was introduced in the discussion of semiconductor diode fabrication. The diffusion technique is employed in the production of *mesa* and *planar* transistors, each of which can be of the *diffused* or *epitaxial* type.

In the *pnp*, diffusion-type mesa transistor the first process is an *n*-type diffusion into a *p*-type wafer, as shown in Fig. 4.30, to form the base region. Next, the *p*-type emitter is diffused or alloyed to the *n*-type base as shown in the figure. Etching is done to reduce the capacitance of the collector junction. The term "mesa" is derived from its similarities with the geographical formation. As mentioned earlier in the discussion of diode fabrication, the diffusion technique permits very tight control of the doping levels and thicknesses of the various regions.

The major difference between the epitaxial mesa transistor and the mesa transis-

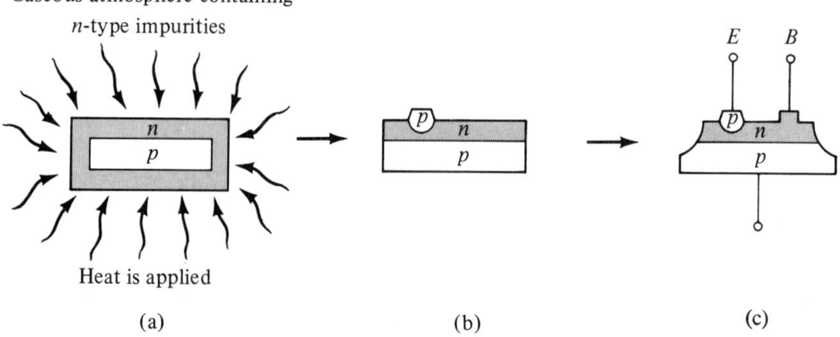

Figure 4.30 Mesa transistor: (a) diffusion process; (b) alloy process; (c) etching process.

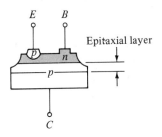

Figure 4.31 Epitaxial mesa transistor.

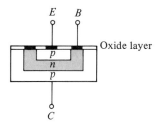

Figure 4.32 Planar transistor.

tor is the addition of an epitaxial layer on the original collector substrate. The term epitaxial is derived from the Greek words *epi*—upon, and *taxi*—arrange, which describe the process involved in forming this additional layer. The original *p*-type substrate (collector of Fig. 4.31) is placed in a closed container having a vapor of the same impurity. Through proper temperature control, the atoms of the vapor will *fall upon* and *arrange* themselves on the original *p*-type substrate resulting in the epitaxial layer indicated in Fig. 4.31. Once this layer is established, the process continues, as described above for the mesa transistor, to form the base and emitter regions. The original *p*-type substrate will have a higher doping level and correspondingly less resistance than the epitaxial layer. The result is a low-resistance connection to the collector lead that will reduce the dissipation losses of the transistor.

The planar and epitaxial planar transistors are fabricated using two diffusion processes to form the base and emitter regions. The planar transistor, as shown in Fig. 4.32, has a flat surface, which accounts for the term planar. An oxide layer is added as shown in Fig. 4.32 to eliminate exposed junctions, which will reduce substantially the surface leakage loss (leakage currents that flow on the surface rather than through the junction).

4.11 TRANSISTOR CASING AND TERMINAL IDENTIFICATION

After the transistor has been manufactured using one of the techniques indicated in Section 4.10, leads of, typically, gold, aluminum, or nickel are then attached and the entire structure is encapsulated in a container such as that shown in Fig. 4.33. Those with the studs and heat sinks are high-power devices, while those with the small can (top hat) or plastic body are low- to medium-power devices.

Whenever possible, the transistor casing will have some marking to indicate which leads are connected to the emitter collector or base of a transistor. A few of the methods commonly used are indicated in Fig. 4.34.

The internal construction of a TO-92 package in the Fairchild line appears in Fig. 4.35. Note the very small size of the actual semiconductor device. There are gold bond wires, a copper frame, and an epoxy encapsulation.

Four (quad) individual *pnp* silicon transistors can be housed in the 14-pin plastic dual-in-line package appearing in Fig. 4.36a. The internal pin connections appear in Fig. 4.36b. As with the diode IC package, the indentation in the top surface reveals the number 1 and 14 pins.

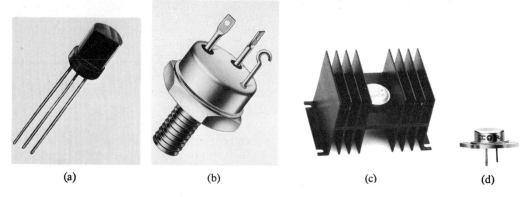

Figure 4.33 Various types of transistors. (*a* and *b*, courtesy General Electric Company; *c* and *d*, courtesy International Rectifier Corporation.)

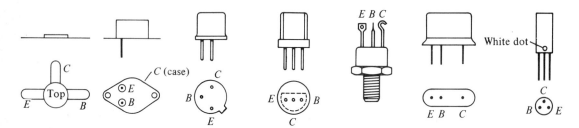

Figure 4.34 Transistor terminal identification.

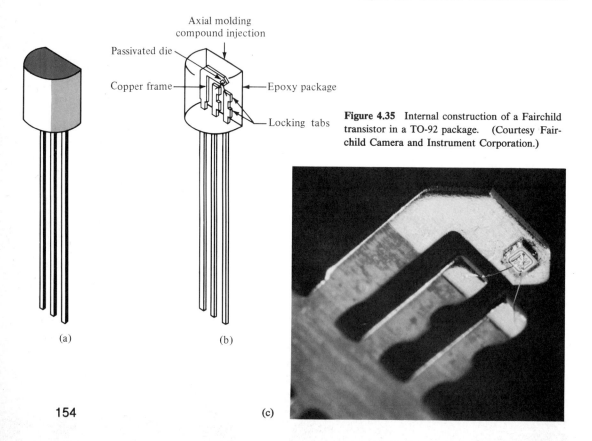

Figure 4.35 Internal construction of a Fairchild transistor in a TO-92 package. (Courtesy Fairchild Camera and Instrument Corporation.)

154

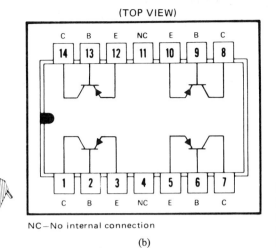

Figure 4.36 Type Q2T2905 Texas Instruments quad *pnp* silicon transistor: (a) appearance; (b) pin connections. (Courtesy Texas Instruments Incorporated.)

4.12 TRANSISTOR TESTING

Using only the ohmmeter section of a VOM, we can apply a number of tests to a transistor. The first to be described will determine whether a short-circuit or open-circuit state exists within the device.

Short-Circuit or Open-Circuit Test

If the terminal identification and type (*npn* or *pnp*) are known, the tests indicated in Fig. 4.37 will determine whether the device is in the defective short-circuit or open-circuit state. For the *pnp* transistor appearing in Fig. 4.37 the expected readings for the tests applied are indicated. If a low or high reading results between the base and either terminal for each polarity applied, the device is in the short- or open-circuit state, respectively. For an *npn* transistor the readings should be the direct opposite.

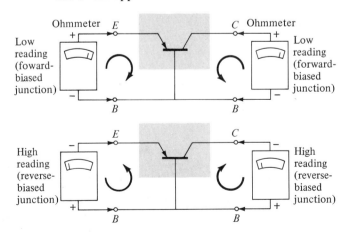

Figure 4.37 Short-circuit or open-circuit test for *pnp* transistor.

Determining the Base Lead

For each set of two terminals (there are three possibilities) measure the resistance between each set with each polarity applied. That set for which the readings remains high for each polarity *does not* include the base terminal.

Determining Whether It Is *pnp* or *npn*

Place the black (or negative) lead on the base terminal and the red (or positive) lead on either of the two remaining terminals. A low reading (as indicated in Fig. 4.37a indicates a *pnp* type, while a high reading indicates an *npn* type.

Determining Which Is the Collector Terminal and Which is the Emitter Terminal

For the *pnp* transistor the meter is connected as indicated in Fig. 4.37a to the collector and emitter terminals. The measurement that results in the higher reading resistance indicates the emitter terminal. For the *npn* transistor the meter is reversed at each terminal (collector and emitter), but the higher reading again indicates the emitter terminal.

PROBLEMS

§ 4.2

1. What names are applied to the two types of transistors? Sketch each transistor and indicate the type of majority and minority carrier in each layer. Is any of this information altered by changing from silicon to germanium material?

2. What is the major difference between a bipolar and a unipolar device?

§ 4.3

3. How must the two transistor junctions be biased for proper transistor operation?
4. What is the source of the leakage current in a transistor?
5. Sketch a figure similar to Fig. 4.3 for the forward-biased junction of an *npn* transistor. Describe the resulting carrier motion.
6. Sketch a figure similar to Fig. 4.4 for the reverse-biased junction of an *npn* transistor. Describe the resulting carrier motion.
7. Sketch a figure similar to Fig. 4.5 for the majority- and minority-carrier flow of an *npn* transistor. Describe the resulting carrier motion.
8. Determine the resulting change in emitter current for a change in collector current of 2 mA and an alpha of 0.98.
9. A transistor has an emitter current of 8 mA and an α of 0.99. How large is the collector current?

§ 4.4

10. Calculate the voltage gain $(A_v = V_o/V_i)$ for the circuit of Fig. 4.6 if $V_i = 500$ mV and $R = 1$ K. (The other circuit values remain the same.)

11. Calculate the voltage gain $(A_v = V_o/V_i)$ for the circuit of Fig. 4.6 if the source has an internal resistance of 100 Ω in series with V_i.

§ 4.5

12. From memory, and memory only, sketch the common-base transistor configuration (for *npn* and *pnp*) and indicate the polarity of the applied bias and resulting current directions.

13. Using the characteristics of Fig. 4.8:
    (a) Find the resulting collector current if $I_E = 5$ mA and $V_{CB} = -10$ V.
    (b) Find the resulting collector current if $V_{EB} = 750$ mV and $V_{CB} = -10$ V.
    (c) Find V_{EB} for the conditions $I_C = 4$ mA and $V_{CB} = -15$ V.

14. The characteristics of Fig. 4.8b are for a silicon transistor. How would you expect them to be different for a germanium transistor? What would be a first approximation for the base-to-emitter voltage of the forward-biased junction?

§ 4.6

15. Define I_{CO} and I_{CEO}. How are they different? How are they related? Are they typically close in magnitude?

16. Using the characteristics of Fig. 4.11:
    (a) Find the value of I_C corresponding to $V_{BE} = +750$ mV and $V_{CE} = +5$ V.
    (b) Find the value of V_{CE} and V_{BE} corresponding to $I_C = 3$ mA and $I_B = 30$ μA.

17. (a) For the common-emitter characteristics of Fig. 4.11, find the dc beta at an operating point of $V_{CE} = +8$ V and $I_C = 2$ mA.
    (b) Find the value of α corresponding to this operating point.
    (c) At $V_{CE} = +8$ V, find the corresponding value of I_{CEO}.
    (d) Calculate the approximate value of I_{CBO} using the dc beta value obtained in part (a).

§ 4.7

18. An input voltage of 2 V rms (measured from base to ground) is applied to the circuit of Fig. 4.16. Assuming that the emitter voltage follows the base voltage exactly and that V_{be} (rms) = 0.1 V, calculate the circuit voltage amplification $(A_v = V_o/V_i)$ and emitter current for $R_E = 1$ K.

19. For a transistor having the characteristics of Fig. 4.11, sketch the input and output characteristics of the common-collector configuration.

§ 4.8

20. Determine the region of operation for a transistor having the characteristics of Fig. 4.11 if $I_{C_{max}} = 5$ mA, $V_{CE_{max}} = 15$ V, and $P_{C_{max}} = 40$ mW.

21. Determine the region of operation for a transistor having the characteristics of Fig. 4.8 if $I_{C_{max}} = 6$ mA, $V_{CB_{max}} = -15$ V, and $P_{C_{max}} = 30$ mW.

§ 4.9

22. Referring to Fig. 4.19, determine the temperature range for the device in degrees Fahrenheit.

23. Determine the value of α when $I_C = 0.1$ mA from the provided value of dc forward-current transfer ratio (Fig. 4.18).

24. Using the results of Problem 23 and Eq. (4.7), calculate I_{CO} and note whether it falls within the limits of I_{CBO} in Fig. 4.18.

25. Using the information provided in Figs 4.18 and 4.19, sketch the boundaries of the maximum power dissipation region ($T_{case} = 25°C$).

26. Sketch the maximum power curve in the characteristics of Fig. 4.26 if the free-air temperature is 100°C.

27. Referring to Fig. 4.20, determine the following:
    (a) I_C if $V_{CE} = 30$ V, $I_B = 25$ μA.
    (b) V_{CE} if $I_C = 4$ mA, $I_B = 30$ μA.
    (c) I_B if $I_C = 3$ mA, $V_{CE} = 20$ V.

28. (a) At an operating point of $V_{CE} = 30$ V and $I_C = 7$ mA, determine the level of V_{BE} using Figs. 4.20 and 4.21. For base currents significantly less than 10 mA, use the curve of Fig. 4.21 denoted $V_{CE} = 10$ V.
    (b) At an operating point of $V_{CE} = 2$ V and $I_C = 300$ mA, determine the level of V_{BE} using Figs. 4.21 and 4.26.
    (c) What would be an appropriate level of V_o for the approximate equivalent circuit for each of the operating conditions of parts (a) and (b)?

29. (a) Calculate the power derating factor for each of the curves of Fig. 4.22.
    (b) Using the value obtained for the case temperature curve, find the power rating at a temperature of 100°C.
    (c) Compare the results of part (b) with the value obtained from the graph.

30. (a) Determine the value of β_{dc} at $I_C = 5$ mA and a temperature of 100°C, using Fig. 4.23.
    (b) What is the level of α at this point?
    (c) What is the average difference in β level between room temperature and 100°C for the range $I_C = 0.01$ mA to $I_C = 10$ mA? Is it something we should carefully consider in a design problem? Why?

31. (a) Determine I_{CBO} from Fig. 4.24 for $V_{CB} = 30$ V and a junction temperature of 50°C. Note the log scale.
    (b) If $\beta = 200$, what is the level of I_{CEO}?
    (c) What is the rate of change of I_{CBO} per degree change in temperature near 50°C for $V_{CB} = 30$ V?

32. (a) Using Fig. 4.25, determine the rate of change of h_{fe} with change in collector voltage for the range $I_C = 2$ to 4 mA. ($V_{CE} = 10$ V.)
    (b) Why would you assume that the magnitude of h_{fe} in this particular figure is so small on the vertical scale when typical values of h_{fe} approach 100 or more?

§ 4.10

33. (a) Describe the basic differences among the various techniques of transistor construction.

(b) Which would you categorize as acceptable for high-power applications?
(c) Define the *diffusion* process.

§ 4.11

34. (a) Find three transistors with different casings, identify the terminals, and sketch the device.
(b) Search through a manufacturer's data book for another IC structure limited totally to transistors. Sketch the internal schematic and identify the terminals.

§ 4.12

35. (a) Use the technique described in Section 4.12 to check out the three transistors mentioned in Problem 34. That is, check the identity of the terminals and determine whether they are *npn* or *pnp*.
(b) Also determine whether they have a defective short-circuit or open-circuit state.

GLOSSARY

Transistor   A three-terminal semiconductor device employed for both current and voltage amplification.

Bipolar Junction Transistor (BJT)   A term often applied to the basic transistor to reflect the fact that both holes and electrons participate in the injection process into the oppositely polarized material.

Leakage Current   The minority current component of the collector current.

Common-Base Short-Circuit Amplification Factor (α)   Defined by a change in collector current divided by a change in the emitter current with V_{EB} constant—on an approximate basis simply the ratio of I_C divided by I_E.

Common-Base Transistor Configuration   A transistor configuration in which the base is common to both the collector and emitter terminals.

Common-Emitter Transistor Configuration   A transistor configuration in which the emitter is common to both the collector and base terminals.

Cutoff Region   The region below $I_B = 0$ on the collector characteristics of the common-emitter configuration.

Common-Emitter Forward-Current Amplification Factor (β)   Defined by the ratio of the change in collector current to a change in base current with $V_{CE} = $ constant—defined approximately by the ratio of I_C to I_B at the particular operating point.

Common-Collector Transistor Configuration   The transistor configuration in which the collector is common to both the emitter and base terminals.

Collector Dissipation   The power dissipation at the collector determined by the product of the collector voltage and current.

Point-Contact Transistor   A transistor constructed by placing two wires on an *n*-type wafer and applying electrical impulses to establish the bond.

Alloy Junction Transistor   A transistor constructed by placing a dot of the same impurity on each side of a semiconductor wafer (having the opposite impurity) and heating to form the necessary bonds.

Grown Junction Transistor   A transistor manufactured through the pulling of a seed from a melt of alternating n- and p-type impurities using the Czochralski technique.

Planar Transistor   A flat-surface transistor fabricated through two diffusion processes.

Mesa Transistor   A transistor constructed through two different diffusion processes or the combination of a diffusion and/or alloy process—having an appearance similar to the geographical formation.

CHAPTER 5

Dc and Ac BJT Analysis

5.1 INTRODUCTION

The basic construction, appearance, and characteristics of the transistor were introduced in Chapter 4. We now briefly examine those considerations that must be carefully examined if the device is to function properly as an amplifier.

There are two states that must be considered for the complete analysis or design of an amplifier system. Fortunately, they can be isolated and considered separately. In this chapter we first examine the dc conditions and how they affect the operation of the device. We then introduce an ac model for the transistor and examine its sinusoidal response.

It would be virtually impossible to examine all the possible configurations within a single chapter. Those of primary importance will be examined, however, and it would be assumed that an analysis of those systems not included could be undertaken based on the content of this chapter.

5.2 OPERATING POINT

You will recall from your introductory dc analysis courses that a dc system has fixed values of current and voltage after the transient or turn-on period has passed. In a transistor network in which only dc voltages have been applied, the transistor voltages and currents will also be steady-state fixed values. The applied emfs and resistors present will result in a particular set of fixed voltages and currents that will define an *operating* point (also referred to as the *quiescent point*). The choice

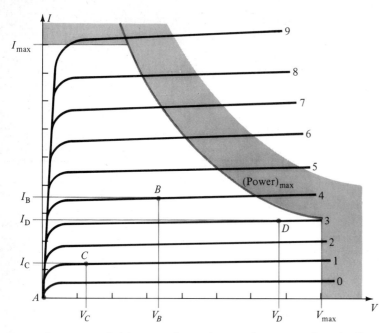

Figure 5.1 Various operating points on device static characteristics.

of this operating point can have a profound effect on the behavior of the system if not chosen properly.

Figure 5.1 shows a general device characteristic with four indicated operating points. The biasing circuit may be designed to set the device operation at any of these points or others within the *operating region.* The operating region is the area of current or voltage within the maximum limits for the particular device. These maximum ratings are indicated on the characteristic of Fig. 5.1 by a horizontal line for the maximum current, I_{max}, and a vertical line for the maximum voltage, V_{max}. An additional consideration of maximum power (product of voltage and current) must also be taken into consideration in defining the operating region of a particular device, as shown by the gray line marked P_{max} on Fig. 5.1.

It should be realized that the device could be biased to operate outside these maximum limit points but that the result of such operation would be either a considerable shortening of the lifetime of the device or destruction of the device. Confining ourselves to the safe operating region, we may select many different operating areas or points. The exact point or area often depends on the intended use of the circuit. Still, we can consider some differences between operation at the different points shown in Fig. 5.1 to present some basic ideas about the operating point and, thereby, the bias circuit.

If no bias were used, the device would initially be completely off, which would result in the current of point *A*—namely, zero current through the device (and zero voltage across it). It is necessary to bias the device so that it can respond or change in current and voltage for the entire range of the input signal. While point *A* would not be suitable, point *B* provides this desired operation. If a signal is applied to the circuit, *in addition to the bias level,* the device will vary in current and voltage from operating point *B,* allowing the device to react to (and possibly amplify) both the positive and negative part of the input signal. If, as could be the case, the input signal is small, the voltage and current of the device will vary but not enough to drive the device into *cutoff* or *saturation*. Cutoff is the condition in which the device no longer conducts. Saturation is the condition in which voltage across the device is as small as possible with the current flow in the device path reaching a limiting value depending on the external circuit. The usual amplifier action desired occurs within the operating region of the device, that is, between saturation and cutoff.

Point C would also allow some positive and negative variation with the device still operating, but the output could not vary too negatively (left of V_C) because bias point C is lower in voltage than point B. Point C is also in a region of operation in which the current level in the device is smaller and the device gain is *not* linear; that is, the spacing between curves is unequal. This nonlinearity shows that the amount of gain of the device is less in the lower regions on the characteristic and greater in the upper regions. It is preferable to operate where the gain of the device is most constant (or linear) so that the amount of amplification over the entire swing of input signal is the same. Point B is in a region of more linear spacing and, therefore, more linear operation, as shown in Fig. 5.1. Point D sets the device operating point near the maximum voltage level. The output voltage swing in the positive direction is thus limited if the maximum voltage is not to be exceeded. Point B, therefore, seems the best operating point in terms of linear gain or largest possible voltage and current swing. This is usually the desired condition for small-signal amplifiers but not necessarily for power amplifiers and logic circuits, as we see later.

One other very important biasing factor must be considered. Having selected and biased for a desired operating point, the effect of temperature must also be taken into account. Higher temperature results in more current flow in the device than at room temperature, thereby upsetting the operating condition set by the bias circuit. Because of this, the bias circuit must also provide a degree of *temperature stability* to the circuit so that temperature changes at the device produce minimum change in its operating point. This maintenance of operating point may be specified by a *stability factor, S,* indicating the amount of change in operating-point current due to temperature.

5.3 COMMON-BASE BIAS CIRCUIT

The common-base (CB) configuration provides a relatively straightforward and simple starting point in our dc bias considerations. Figure 5.2a shows a common-base circuit configuration. By CB we mean that the base is the reference point for measurement for both input (emitter, in this case) and output (collector).

The dc supply (battery) terminals are marked with a double-letter designation. V_{EE} is the dc voltage supply associated with the emitter section, and V_{CC} is the dc supply for the collector section of the circuit. Two separate voltage supplies may be required for the CB configuration. Resistor R_E is basically a current-limiting resistor for setting the emitter current I_E. Resistor R_C is the collector (output or load) resistor and the output ac signal is developed across it. It is also one of the components that is used to set a desired operating point.

Figure 5.2b shows the CB collector characteristic, which is the output characteristic for this circuit connection. The abscissa is the collector-to-base voltage V_{CB}, which is a negative voltage for the *pnp* circuit of Fig. 5.2a. The ordinate is the collector current, I_C. The family of curves for the characteristic are for various emitter currents.

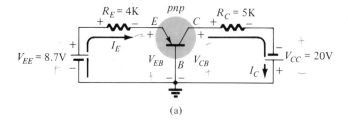

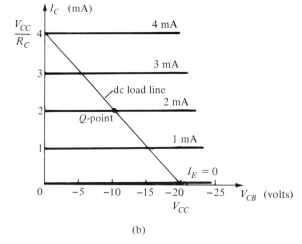

Figure 5.2 Common-base circuit and transistor characteristic: (a) common-base circuit; (b) dc load line on common-base characteristic.

The theory developed for either the *pnp* or *npn* transistor device can be equally applied to the other device by merely changing all current directions and all voltage polarities. We shall consider mostly *npn* transistors in developing concepts because it is the more popular unit at present.

It is possible to consider biasing the CB circuit by analyzing separately the input (base-emitter loop) and the output (base-collector loop) sections of the circuit. Although in reality there is some interaction between the operation of the base-collector section and that of the base-emitter section, this can be neglected with excellent practical results obtained.

Input Section

The input loop (see Fig. 5.3a) is composed of the battery, V_{EE}, the resistor, R_E, and the emitter-base junction of the transistor, V_{EB}. Writing the voltage loop equation (using Kirchhoff's voltage law) for the input loop,

$$+V_{EE} - I_E R_E - V_{EB} = 0$$

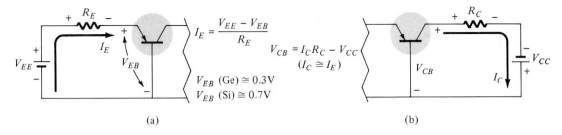

Figure 5.3 Input and output sections for common-base circuit: (a) input (emitter-base) section only; (b) output (collector-base) section only.

from which we obtain

$$I_{EE} = \frac{V_{EE} - V_{EB}}{R_E} \qquad (5.1a)$$

When forward-biased, the emitter-base voltage V_{EB} is small—on the order of 0.3 V for germanium transistors and 0.7 V for silicon. Although the actual emitter–base voltage is slightly affected by the collector-base voltage, this effect can be neglected for practical considerations. In fact, if the supply voltage V_{EE} is, say, 10 V or more, the emitter-base voltage could be neglected, leaving

$$I_E \cong \frac{V_{EE}}{R_E} \qquad (5.1b)$$

as a good approximation. Observe that the emitter current is set essentially by the emitter supply voltage and the emitter resistor. Since the supply voltage is usually fixed by voltage requirements in other parts of the electronic circuit, the emitter current is specifically determined by the emitter resistor R_E.

Output Section

The output loop (see Fig. 5.3b) consists of the battery, V_{CC}, the resistor, R_C, and the voltage across the collector-base junction of the transistor, V_{CB}. For operation as an amplifier the *collector-base* junction must be *reverse-biased* in addition to the *emitter-base* being *forward-biased*. This *reverse bias* is provided by the V_{CC} battery voltage connected in polarity so that the *p*ositive battery terminal connects to the *n*-material and the *n*egative battery terminal to the *p*-material. (Note the letter opposites for *p* and *n* for battery and transistor type.) The result of this consideration is that for *pnp* transistors the battery polarity should be positive terminal to the common-base point and negative terminal to the resistor connected to the collector terminal. (For *npn* transistors we have the opposite—the negative battery terminal connected to the CB terminal and the positive terminal of the battery connected to the resistor, which then connects to the collector terminal.)

Summing the voltage drops around the output or collector-base loop of the circuit of Fig. 5.3b, we get

$$+V_{CC} - I_C R_C + V_{CB} = 0$$

Solving for the collector-base voltage results in

$$V_{CB} = -V_{CC} + I_C R_C \quad (5.2)$$

The collector current I_C is approximately the same magnitude as the emitter current I_E [obtained from Eqs. (5.1a) or (5.1b)]. This is a very good approximation that will be true for any type of transistor connection used. For the purposes of calculating circuit bias values we may write the relation as

$$I_C \cong I_E \quad (5.3)$$

Actually, $I_C = \alpha I_E$, where α (alpha) is typically 0.9 to 0.998 in value.

The load line and Q-point appearing in Fig. 5.2b can be obtained quite quickly if we consider that in the output section if we set $I_C = 0$ in Eq. (5.2), then $V_{CB} = -V_{CC}$. In other words, by setting $I_C = 0$ we know we are somewhere along the horizontal line of $I_C = 0$ on the graph of Fig. 5.2. The result $V_{CB} = -V_{CC}$ therefore defines a point on the load line at the intersection of $I_C = 0$. By setting V_{CB} equal to zero in Eq. (5.2), we obtain $I_C = V_{CC}/R_C$ and define a second point at the intersection of $V_{CB} = 0$ V. Since two points define a straight line, the load line is sketched as shown in Fig. 5.2. The level of I_E is determined from the input section and the Q-point is determined by the intersection of the load line and resulting line for I_E.

EXAMPLE 5.1 Calculate the bias voltages V_{EB} and V_{CB} and currents I_E and I_C for the circuit of Fig. 5.4. The circuit contains a *pnp* silicon transistor with an alpha of 0.99.

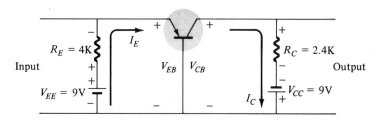

Figure 5.4 Common-base bias circuit for Example 5.1.

Solution: Using the step-by-step procedure outlined previously, we get

(a) $V_{EB} \cong 0.7$ (silicon).

(b) $I_E = \dfrac{V_{EE} - V_{EB}}{R_E} = \dfrac{9 - 0.7 \text{ V}}{4 \text{ K}} \cong 2.075$ mA.

(c) $I_C \cong I_E = 2.075$ mA.

(d) $V_{CB} = -V_{CC} + I_C R_C = -9 + (2.075 \text{ mA})(2.4 \text{ K}) = -4.02$ V.

5.4 COMMON-EMITTER CIRCUIT CONNECTION

A more popular amplifier connection applies the input signal to the base of the transistor with the emitter as the common terminal. The common-emitter (CE) circuit of Fig. 5.5 shows only one supply voltage. Recall that the CB circuit

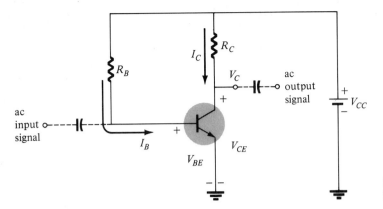

Figure 5.5 Common-emitter fixed-bias circuit.

used two supply voltages, one to forward-bias the base-emitter and the second to provide reverse bias for the base-collector. Both forward- and reverse-bias conditions are achieved in the CE connection using one voltage supply. In addition, we shall show a number of other important advantages of the CE circuit relating to input and output impedances, current and voltage gain, and so on, which apply to the ac operation of the circuit.

Note that the capacitors act as open circuits to the dc currents, resulting in the paths shown in Fig. 5.5.

Input Section

To provide for simple step-by-step analysis, consider only the base-emitter circuit loop shown in the partial circuit diagram of Fig. 5.6a. Writing the Kirchhoff

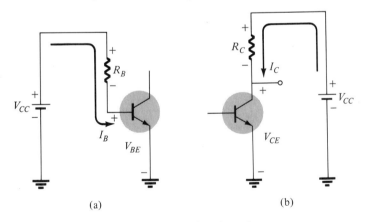

Figure 5.6 Separate input and output loops for fixed-bias circuit: (a) input base-emitter loop; (b) output collector-emitter loop.

voltage equation for the given loop, we get

$$+V_{CC} - I_B R_B - V_{BE} = 0$$

We can solve the equation above for the base current I_B:

$$\boxed{I_B = \frac{V_{CC} - V_{BE}}{R_B}} \quad (5.4a)$$

Since the supply voltage V_{CC} and the base-emitter voltage V_{BE} are fixed values of voltage, the selection of a base bias resistor fixes the value of the base current. As a good approximation we may even neglect the few tenths volt drop across the forward-biased base-emitter V_{BE}, obtaining the simplified form for calculating base current,

$$I_B \cong \frac{V_{CC}}{R_B} \tag{5.4b}$$

Output Section

The output section of the circuit (Fig. 5.6b) consists of the supply battery, the collector (load) resistor, and the transistor collector-emitter junctions. The currents in the collector and emitter are about the same since I_B is small in comparison to either. For linear amplifier operation the collector current is related to the base current by the transistor current gain, beta (β) or h_{FE}. Expressed mathematically,

$$I_C = \beta I_B \tag{5.5}$$

The base current is determined from the operation of the base-emitter section of the circuit as provided by Eq. (5.4a) or (5.4b). The collector current as shown by Eq. (5.5) is β times greater than the base current *and* not dependent on the resistance in the collector circuit. From the previous consideration of the common-base circuit we know that the collector current is controlled in the base-emitter section of the circuit and not in the collector-base (or collector-emitter, in this case) section of the circuit.

Calculating voltage drops in the output loop, we get

$$V_{CC} - I_C R_C - V_{CE} = 0$$

$$V_{CE} = V_{CC} - I_C R_C \tag{5.6}$$

Equation (5.6) shows that the sum of voltages across the collector-emitter and across the collector resistor is the supply voltage value.

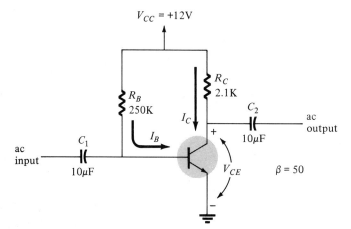

Figure 5.7 dc fixed-bias circuit for Example 5.2.

EXAMPLE 5.2 Compute the dc bias voltages and currents for the *npn* CE circuit of Fig. 5.7.

Solution:

(a) $I_R = \dfrac{V_{CC} - V_{BE}}{R_B} = (12 - 0.7)/250 \text{ K} = \mathbf{45.2\ \mu A}.$

(b) $I_C = \beta I_B = 50(45.2\ \mu A) = \mathbf{2.26\ mA}.$

(c) $V_{CE} = V_{CC} - I_C R_C = 12 - (2.26\text{ mA})(2.1\text{ K}) = 12 - 5 = \mathbf{7.254\ V}.$

5.5 DC BIAS CIRCUIT WITH EMITTER RESISTOR

The dc bias circuit of Fig. 5.8 contains an emitter resistor to provide better bias stability (less temperature sensitive) than the fixed-bias circuit considered in Section 5.4. For the analysis of the circuit operation we shall deal separately with the base-emitter loop of the circuit and the collector-emitter loop of Fig. 5.8.

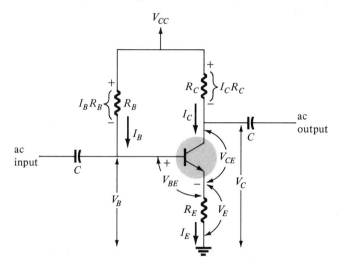

Figure 5.8 dc bias circuit with emitter stabilization resistor.

Input Section (Base-Emitter Loop)

A partial circuit diagram of the base-emitter loop is shown in Fig. 5.9a. Writing Kirchhoff's voltage equation for the loop, we get

$$V_{CC} - I_B R_B - V_{BE} - I_E R_E = 0$$

We can replace $I_E = I_C$ with $(\beta + 1) I_B$, and the equation will appear as

$$V_{CC} - I_B R_B - V_{BE} - (\beta + 1) I_B R_E = 0$$

Solving for the base current:

$$I_B = \dfrac{V_{CC} - V_{BE}}{R_B + (\beta + 1) R_E} \cong \dfrac{V_{CC}}{R_B + \beta R_E} \qquad (5.7)$$

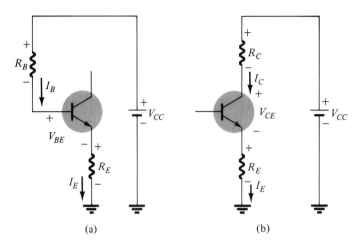

Figure 5.9 Input and output loops of the circuit of Fig. 5.8: (a) input loop; (b) output loop.

Note that the difference between the fixed-bias current calculation [Eq. (5.4a)] and Eq. (5.7) is the additional term of $(\beta + 1)R_E$ in the denominator.

Output Section (Collector-Emitter Loop)

The collector-emitter loop is shown in Fig. 5.9b. Writing the Kirchhoff voltage equation for this loop, we obtain

$$V_{CC} - I_C R_C - V_{CE} - I_E R_E = 0$$

Using the relationship

$$I_C \cong I_E \qquad I_c = \alpha I_E$$

we can solve for the voltage across the collector-emitter:

$$\boxed{V_{CE} \cong V_{CC} - I_C(R_C + R_E)} \qquad (5.8)$$

$V_{CE} = V_{CC} - I_C(R_C + \frac{1}{\alpha}R_E)$

The voltage measured from emitter to ground is

$$V_E = I_E R_E \cong I_C R_E$$

$\frac{I_c}{\alpha} R_E$

and the voltage measured from collector to ground is

$$V_C = V_{CC} - I_C R_C$$

The voltage at which the transistor is biased is measured from collector to emitter, V_{CE}, which is given by Eq. (5.8) and may also be calculated as

$$V_{CE} = V_C - V_E$$

EXAMPLE 5.3 Calculate all dc bias voltages and currents in the circuit of Fig. 5.10.

Solution:

(a) $I_B \cong \dfrac{V_{CC}}{R_B + \beta R_E} = \dfrac{20\text{ V}}{400\text{ K} + 100(1\text{ K})} = 20\text{ V}/500\text{ K} = \mathbf{40\ \mu A}.$

(b) $I_C = \beta I_B = 100(40\ \mu A) = \mathbf{4\ mA} \cong I_E$.
(c) $V_{CE} = V_{CC} - I_C R_C - I_E R_E = 20 - (4\ \text{mA})(2\ \text{K}) - (4\ \text{mA})(1\ \text{K})$.
   $= 20 - 8 - 4 = \mathbf{8\ V}$.

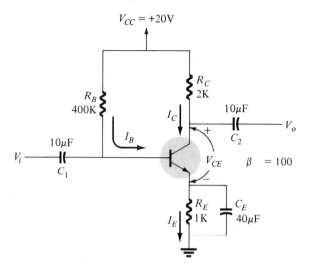

Figure 5.10 Emitter-stabilized bias circuit for Example 5.3.

5.6 DC BIAS CIRCUIT INDEPENDENT OF BETA

In the previous dc bias circuits the values of the bias current and voltage of the collector depended on the current gain (β) of the transistor. Since the value of beta is temperature-sensitive (especially for silicon transistors) and the nominal value of beta is not well defined, it would be desirable for these as well as other reasons (transistor replacement and stability) to provide a dc bias circuit that is *independent* of the transistor beta. The circuit of Fig. 5.11 meets these conditions and is thus a very popular bias circuit.

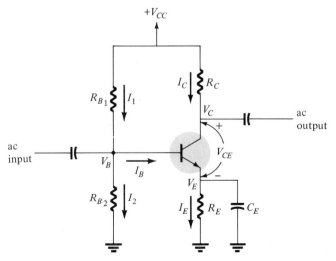

Figure 5.11 Beta-independent dc bias circuit.

Let us first analyze the base-emitter input circuit. A basic assumption (which will be proved later) is that the resistance seen looking into the base (see Fig. 5.12) is much larger than that of resistor R_{B2}. If this is so, then the current through R_{B1} flows almost completely into R_{B2} and the two resistors may be considered effectively in series. The voltage at the junction of the resistors, which is

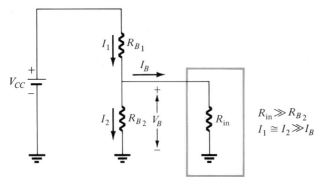

Figure 5.12 Partial bias circuit for calculating base voltage V_B.

also the voltage of the base of the transistor, is then determined simply by the voltage divider network of R_{B1} and R_{B2} and the supply voltage. Calculating the voltage at the transistor base due to the voltage divider network of resistors R_{B1} and R_{B2}, we obtain

$$V_B \cong \frac{R_{B2}}{R_{B1} + R_{B2}} V_{CC} \tag{5.9}$$

where V_B is the voltage measured from base to ground.

We can calculate the voltage at the emitter from

$$V_E = V_B - V_{BE} \tag{5.10}$$

The current in the emitter may then be calculated from

$$I_E = \frac{V_E}{R_E} \tag{5.11a}$$

and the collector current is then

$$I_C \cong I_E \tag{5.11b}$$

The voltage drop across the collector resistor is

$$V_{RC} = I_C R_C$$

The voltage at the collector (measured with respect to ground) can then be obtained:

$$V_C = V_{CC} - V_{RC} = V_{CC} - I_C R_C \tag{5.12}$$

and, finally, the voltage from collector to emitter is calculated from

$$V_{CE} = V_C - V_E$$

$$V_{CE} = V_{CC} - I_C(R_C + R_E) \qquad (5.13)$$

Look back at the procedure just outlined and notice that the value of beta was never used in Eqs. (5.9)–(5.13). The base voltage is set by resistors R_{B1} and R_{B2} and the supply voltage. The emitter voltage is fixed at approximately the same voltage value as the base. Resistor R_E then determines emitter and collector currents. Finally, R_C determines the collector voltage and, thereby, the collector-emitter bias voltage.

The base voltage V_B is best adjusted using resistor R_{B2}, the collector current by resistor R_E, and the collector-emitter voltage by resistor R_C. Varying other components will have less effect on the dc bias adjustments. The capacitor components are part of the ac amplifier operation but have no effect on the dc bias and will not be discussed at this time.

EXAMPLE 5.4 Calculate the dc bias voltages and currents for the circuit of Fig. 5.13.

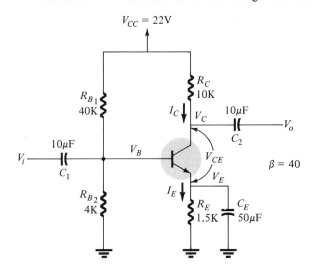

Figure 5.13 Beta-stabilized bias circuit for Example 5.4.

Solution:

(a) $V_B = \dfrac{R_{B2}}{R_{B1} + R_{B2}}(V_{CC}) = \dfrac{4}{40+4}(22) = \mathbf{2\ V.}$

(b) $V_E = V_B - V_{BE} = 2 - 0.7 = \mathbf{1.3\ V.}$

(c) $I_E = \dfrac{V_E}{R_E} \cong I_C = \dfrac{1.3\ V}{1.5\ K} = \mathbf{0.867\ mA.}$

(d) $V_C = V_{CC} - I_C R_C = 22 - (0.867\ \text{mA})(10\ K) = \mathbf{13.33\ V.}$

(e) $V_{CE} = V_C - V_E = 13.33 - 1.30 = \mathbf{12.03\ V.}$

5.7 COMMON-COLLECTOR (EMITTER-FOLLOWER) DC BIAS CIRCUIT

A third connection for the transistor provides input to the base circuit and output from the emitter circuit, with the collector common (CC) to the ac input and output signals. A simple CC circuit (usually referred to as emitter-follower) is

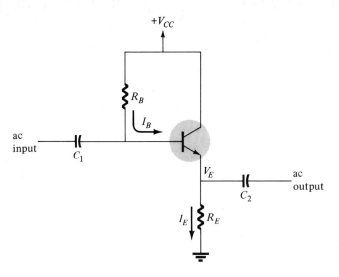

Figure 5.14 Emitter-follower dc bias circuit.

shown in Fig. 5.14. The collector voltage is fixed at the positive supply voltage value. For V_{CE} to be approximately one-half the voltage of V_{CC}, allowing the widest voltage swing in the output before distortion occurs, the emitter voltage should be set at a voltage of about one-half V_{CC}.

Input Section

For the input section of the circuit the voltages summed around the base-emitter loop give

$$+V_{CC} - I_B R_B - V_{BE} - I_E R_E = 0$$

Using the current relation, we obtain

$$I_E = (\beta + 1) I_B \cong \beta I_B$$

We can then solve for the base current:

$$I_B = \frac{V_{CC} - V_{BE}}{R_B + (\beta + 1) R_E} \cong \frac{V_{CC}}{R_B + \beta R_E} \tag{5.14}$$

Output Section

The voltage from emitter to ground is

$$V_E = I_E R_E \tag{5.15}$$

and the collector-emitter voltage is

$$V_{CE} = V_{CC} - V_E = V_{CC} - I_E R_E \tag{5.16}$$

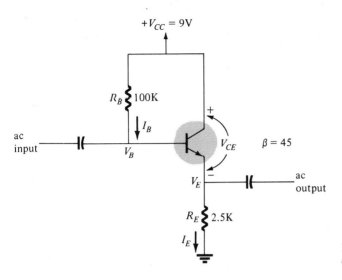

Figure 5.15 Emitter-follower bias circuit for Example 5.5.

EXAMPLE 5.5 Calculate all dc bias currents and voltages for the circuit of Fig. 5.15.

Solution:

(a) $I_B \cong \dfrac{V_{CC}}{R_B + \beta R_E} = \dfrac{9 \text{ V}}{100 + (45)(2.5)} = \mathbf{42.35\ \mu A.}$

(b) $I_E = (\beta + 1) I_B = 46(42.35\ \mu A) = \mathbf{1.95\ mA.}$

(c) $V_{CE} = V_{CC} - I_C R_C = 9 - (1.95\text{ mA})(2.5\text{ K}) = \mathbf{4.125\ V.}$

(d) $V_E = I_E R_E = (1.95\text{ mA})(2.5\text{ K}) = \mathbf{4.875\ V.}$

A second emitter-follower dc bias circuit is shown in Fig. 5.16. Like the similar CE dc bias circuit of Fig. 5.11, this circuit provides a bias operating condition that depends not on the current gain (β) of the transistor, but only on the resistor components and the supply voltage.

EXAMPLE 5.6 Calculate the dc bias currents and voltages for the CC circuit of Fig. 5.16.

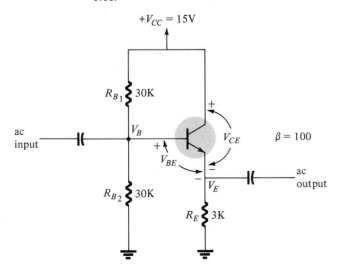

Figure 5.16 Beta-stabilized emitter-follower bias circuit for Example 5.5.

175

Solution:

(a) $V_B = \dfrac{R_{B2}}{R_{B1} + R_{B2}}(V_{CC}) = \dfrac{30}{30+30}(15) = 7.5$ V.

(b) $V_E = V_B - V_{BE} = 7.5 - 0.7 = 6.8$ V.

(c) $I_E = \dfrac{V_E}{R_E} = \dfrac{6.8\text{ V}}{3\text{ K}} = 2.267$ mA.

(d) $V_{CE} = V_{CC} - I_E R_E = 15 - (2.267\text{ mA})(3\text{ K}) = 8.2$ V.

5.8 COMMON-EMITTER CONFIGURATION— GRAPHICAL ANALYSIS

The graphical analysis of the common-emitter configuration is simply an extension of the approach applied to the common-base bias circuit in Section 5.3. Conditions are set on both the input and output networks to determine the load line and

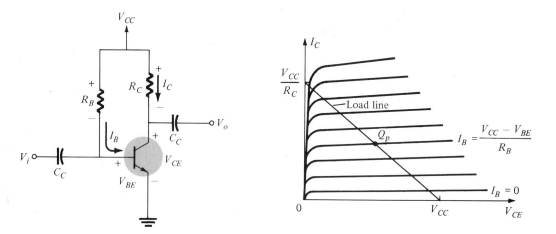

Figure 5.17 Graphical analysis of fixed-bias transistor network.

Q-point. Consider the fixed-bias network of Fig. 5.17, for example. For the output collector network

$$V_{CC} = I_C R_C + V_{CE}$$

If we set $I_C = 0$, we find that

$$V_{CE} = V_{CC}$$

and we have the load-line intersection as $I_C = 0$ in the collector characteristics. If we set $V_{CE} = 0$, we find that

$$I_C = \dfrac{V_{CC}}{R_C}$$

and we have the load-line intersection at $V_{CE} = 0$ on the collector characteristics. These two points then determine the load line.

For the input base network,

$$V_{CC} = I_B R_B + V_{BE}$$

or

$$I_B = \frac{V_{CC} - V_{BE}}{R_B}$$

Assuming that $V_{BE} = 0.7$ V (silicon) or 0.3 V (germanium), we can calculate I_B immediately and the Q-point is determined.

EXAMPLE 5.7 Given the circuit and the transistor collector characteristics of Fig. 5.18:
(a) Plot the dc load line and obtain the Q-point.
(b) Find V_{CE} and I_C from the graph.

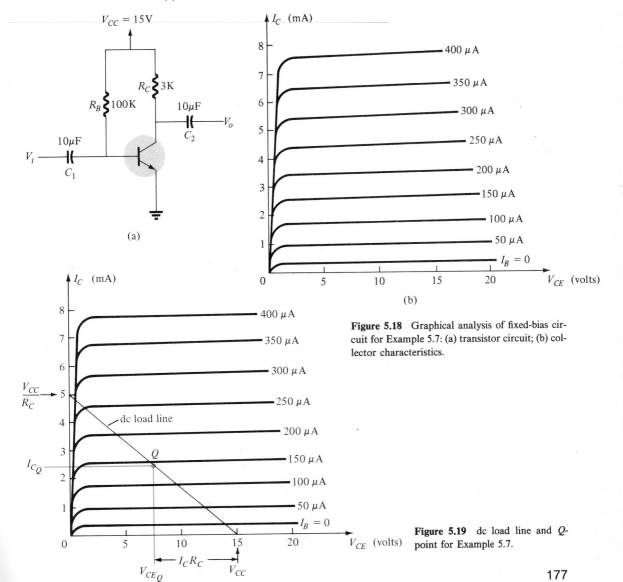

Figure 5.18 Graphical analysis of fixed-bias circuit for Example 5.7: (a) transistor circuit; (b) collector characteristics.

Figure 5.19 dc load line and Q-point for Example 5.7.

177

Solution:
(a) Draw the dc load line as shown in Fig. 5.19. The two points for the load line are
   a. at $I_C = 0$, $V_{CE} = V_{CC} = 15$ V.
   b. at $V_{CE} = 0$, $I_C = V_{CC}/R_C = 15$ V/3 K = 5 mA.

Calculate the base current using

$$I_B = \frac{V_{CC} - V_{BE}}{R_B} = \frac{15 - 0.7}{100 \text{ K}} = 143 \text{ } \mu A$$

The intersection of the load line and the transistor curve for $I_B = 143$ μA defines the quiescent operating point.

(b) At the Q-point, we find $V_{CE_Q} = $ **7.2 V** and $I_{C_Q} = $ **2.4 mA**.

5.9 PROPER BIASING INSURANCE

One technique for ensuring that a transistor is properly biased will now be described using the fact that the arrow in the transistor symbol points in the actual emitter conventional flow direction. As an example, consider the *npn* transistor of Fig. 5.20a in the CE configuration. In Fig. 5.20b, the current directions for I_B and I_C are included as defined by the fact that the emitter current is the sum of the base and collector currents. That is, if the emitter current leaves the device, both the base and collector currents must enter their respective terminals. Now we have to ensure that the battery inserted between the base and emitter terminals will establish the indicated base current direction (conventional flow through the battery from the negative to positive terminal of the battery). This is indicated in Fig. 5.20c for the base and collector circuits. The transistor junctions are now

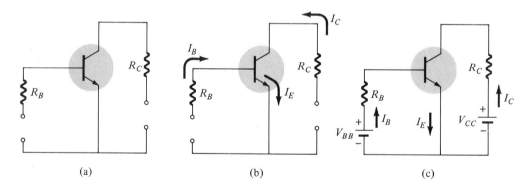

Figure 5.20 Steps leading to the proper biasing of a transistor: (a) unbiased network; (b) conventional current directions inserted; (c) proper biasing inserted as determined by current directions.

properly biased. This technique of starting with the necessary current directions as defined by I_E can be applied to any configuration whether it be *pnp* or *npn*.

One easy way to remember whether the arrow of a *pnp* or *npn* transistor symbol points in or out is to associate *p*ointing in with *pnp* and *n*ot *p*ointing in with *npn*.

5.10 SMALL-SIGNAL (AC) ANALYSIS

It was noted earlier that the transistor is an amplifying device. That is, the output sinusoidal signal is greater than the input signal or, stated another way, the output ac power is greater than the input ac power. The question often arises as to where this additional ac power has been generated. The sole purpose of the next few paragraphs will be to answer that question, since it is fundamental to the clear understanding of a number of important efficiency criteria to be defined later in the text.

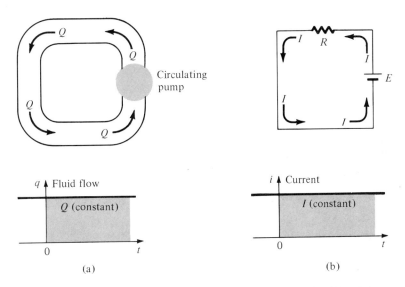

Figure 5.21 Fluid flow analogy of a series electrical circuit with a dc input: (a) fluidic system; (b) electrical system.

Analogies are seldom perfect, but the following will be useful in describing the events leading to the foregoing conclusions. In Fig. 5.21a a steady heavy flow of a liquid has been established by the pump. The electrical analogy of this system appears in Fig. 5.21b. In each case there is some resistance to the flow, with the result that the magnitude of that flow is determined by an Ohm's law relationship. A graph of the flow versus time appears in each figure.

Let us now install a control mechanism in each system, as shown in Fig. 5.22. A small signal at the input to each of these control elements can have a marked effect on the established steady-state (dc) flow of each system. For the fluid system it could be the oscillatory partial closing of the passage to limit the flow through the pipe. For the electrical system a mechanism is established for the control of the current i through the system. Recall that very small variations in I_B can have a pronounced effect on the collector current for the common-emitter transistor configuration.

In other words, a small input signal can have a pronounced effect on the steady-state flow of the system. Consider that the resulting output flow for the two systems may be as shown in Fig. 5.22. The sinusoidal swing of the output

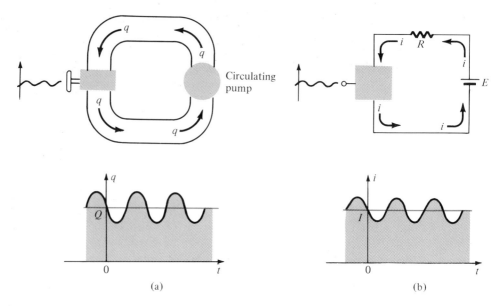

Figure 5.22 The effect of a control element on the steady-state flow of a: (a) fluid system; (b) electrical system.

flow is certainly greater than the applied input—*amplification in the ac domain is therefore a reality!*

We can conclude, therefore, that amplifiers are simply devices having a control point or terminal that can establish a heavy variation in flow between the other two terminals (normally part of the output circuit). The dc biasing circuits are necessary to establish the heavy flow of charge that will be very sensitive to the magnitude of the input signal. The increased ac power is only the result of the conversion of some of the dc power to the sinusoidal domain. The efficiency of an electronic amplifier is typically the ratio of the ac power out to the dc power in.

In ac (sinusoidal) analysis our first concern is the magnitude of the input signal. It will determine whether *small-signal* or *large-signal* techniques must be used. There is no set dividing line between the two, but the application, and the magnitude of the variables of interest (i, v) relative to the scales of the device characteristics, will usually make it clear which is the case in point. The small-signal technique will be discussed in this chapter; large-signal applications will be considered in a later chapter.

The key to the small-signal approach is the use of equivalent circuits to be derived in this chapter. It is that combination of circuit elements, properly chosen, that will best approximate the actual semiconductor device in a particular operating region. Once the ac equivalent circuit has been determined, the graphic symbol of the device can be replaced in the schematic by this circuit and the basic methods of ac circuit analysis (branch-current analysis, mesh analysis, nodal analysis, and Thévenin's theorem) can be applied to determine the response of the circuit.

There are two schools of thought in prominence today regarding the equivalent circuit to be substituted for the transistor. For many years the industrial and educational institutions relied heavily on the *hybrid parameters* (to be introduced shortly). The hybrid parameter equivalent continues to be very popular, although it must now share the spotlight with an equivalent circuit derived directly from the operating conditions of the transistor. Manufacturers continue to specify the hybrid parameters for a particular operating region on their specification sheets. The parameters (or components) of the other equivalent circuit can be derived directly from the hybrid parameters in this region. However, the hybrid equivalent circuit suffers from being limited to a particular set of operating conditions if it is to be considered accurate. The parameters of the other equivalent circuit can be determined for any region of operation within the active region and are not limited by the single set of parameters provided by the specification sheet. For the purposes of this text, if the operating region corresponds with that indicated on the specification sheet, then either equivalent will be used. If not specified, the equivalent circuit derived from the operating conditions will be used. Be encouraged by the fact that both equivalent circuits are very similar in appearance and application. Developed skills with one will result in a measure of ability with the other.

In an effort to demonstrate the effect that the ac equivalent circuit will have on the analysis to follow, consider the circuit of Fig. 5.23. Let us assume for

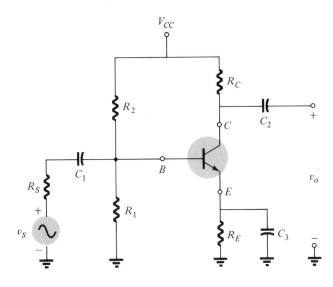

Figure 5.23 Transistor circuit under examination in this introductory discussion.

the moment that the small-signal ac equivalent circuit for the transistor has already been determined. Since we are interested only in the ac response of the circuit, all the dc supplies can be replaced by a zero potential equivalent (short circuit) since they determine only the dc or quiescent level of the output voltage and not the magnitude of the swing of the ac output. This is clearly demonstrated by Fig. 5.24. The dc levels were simply important for determining the proper Q-point of operation. Once determined, the dc levels can be ignored for the ac analysis of the network. In addition, the coupling capacitors C_1 and C_2 and bypass capacitor C_3 were chosen to have a very small reactance at the frequency of application. Therefore, they too may for all practical purposes be replaced by a low-

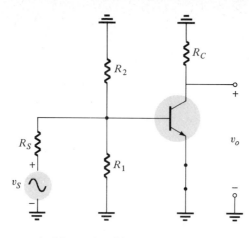

Figure 5.24 The network of Fig. 5.23 following the removal of the dc supply and inserting the short-circuit equivalent for the capacitors.

resistance path (short circuit). Note that this will result in the "shorting out" of the dc biasing resistor R_E. Connecting common grounds will result in a parallel combination for resistors R_1 and R_2, and R_C will appear from collector to emitter as shown in Fig. 5.25. Since the components of the transistor equivalent circuit inserted in Fig. 5.25 are those we are already familiar with (resistors, controlled sources, etc.), analysis techniques such as superposition, Thévenin's theorem, and so on, can be applied to determine the desired quantities.

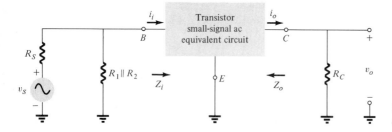

Figure 5.25 Circuit of Fig. 5.24 redrawn for small-signal ac analysis.

Let us further examine Fig. 5.25 and identify the important quantities to be determined for the system. Certainly, we would like to know the input and output impedance Z_i and Z_o as shown in Fig. 5.25. Since we know that the transistor is an amplifying device, we would expect some indication of how the output current i_o is related to the input current—the *current gain*. Note in this case that $i_o = i_C$ and $i_i = i_B$. The ratio of these two quantities certainly relates directly to the β of the transistor. In Chapter 4 we found that the collector-to-emitter voltage did have some effect (if even slight) on the input relationship between i_B and v_{BE}. We might, therefore, expect some "feedback" from the output to input circuit in the equivalent circuit. The following section, through its brief introduction to *two-port theory*, will introduce the hybrid equivalent circuit, which will have parameters that will permit a determination of each of the quantities discussed above.

5.11 TRANSISTOR HYBRID EQUIVALENT CIRCUIT

The development that follows is an introduction to a subject called *two-port* theory. For the basic three-terminal device it is obvious, from Fig. 5.26, that there are two ports (pairs of terminals) of interest. For our purposes, the set at the left

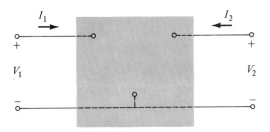

Figure 5.26 Two-port system.

will represent the input terminals, and the set at the right, the output terminals. Note that, for each set of terminals, there are two variables of interest.

The following set of equations, (5.17), is only one of a number of ways in which the four variables can be related. It is the most frequently employed in transistor circuit analysis, however, and therefore will be discussed in detail in this chapter.

$$V_1 = h_{11}I_1 + h_{12}V_2 \tag{5.17a}$$

$$I_2 = h_{21}I_1 + h_{22}V_2 \tag{5.17b}$$

The parameters relating the four variables are called *h-parameters* from the word "hybrid." The term "hybrid" was chosen because the mixture of variables (*v* and *i*) in each equation results in a "hybrid" set of units of measurement for the *h*-parameters.

A clearer understanding of what the various *h*-parameters represent and how we can expect to treat them later can be developed by isolating each and examining the resulting relationship.

If we arbitrarily set $V_2 = 0$ (short-circuit the output terminals), and solve for h_{11} in Eq. (5.17a), the following will result:

$$h_{11} = \frac{V_1}{I_1}\bigg|_{V_2=0} \quad \text{(ohms)} \tag{5.18}$$

The ratio indicates that the parameter h_{11} is an impedance parameter to be measured in ohms. Since it is the ratio of the *input* voltage to the *input* current with the output terminals *shorted*, it is called the *short-circuit input impedance parameter*.

If I_1 is set equal to zero by opening the input leads, the following will result for h_{12}:

$$h_{12} = \frac{V_1}{V_2}\bigg|_{I_1=0} \tag{5.19}$$

The parameter h_{12}, therefore, is the ratio of the input voltage to the output voltage with the input current equal to zero. It has no units since it is a ratio of voltage levels. It is called the *open-circuit reverse transfer voltage ratio parameter*. The term "reverse" is included to indicate that the voltage ratio is an input quantity over an output quantity rather than the reverse, which is usually the ratio of interest.

If in Eq. (5.17b), V_2 is set equal to zero by again shorting the output terminals, the following will result for h_{21}:

$$h_{21} = \left.\frac{I_2}{I_1}\right|_{V_2=0} \tag{5.20}$$

Note that we now have the ratio of an output quantity to an input quantity. The term *forward* will now be used rather than *reverse* as indicated for h_{12}. The parameter h_{21} is the ratio of the output current divided by the input current with the output terminals shorted. It is, for most applications, the parameter of greatest interest. This parameter, like h_{12}, has no units since it is the ratio of current levels. It is formally called the *short-circuit forward transfer current ratio parameter*.

The last parameter, h_{22}, can be found by again opening the input leads to set $I_1 = 0$ and solving for h_{22} in Eq. (5.17b).

$$h_{22} = \left.\frac{I_2}{V_2}\right|_{I_1=0} \quad \text{(siemens)} \tag{5.21}$$

Since it is the ratio of the output current to the output voltage, it is the output conductance parameter and is measured in *siemens*. It is called the *open-circuit output conductance parameter*.

Since each term of Eq. (5.17a) has the units of volts, let us apply Kirchhoff's voltage law in reverse to find a circuit that "fits" the equation. Performing this operation will result in the circuit of Fig. 5.27. Since the parameter h_{11} has the units of ohms, it is represented as an impedance which for the transistor becomes a resistor in Fig. 5.27, h_{12} is a dimensionless quantity. Note that it is a "feedback" of the output voltage to the input circuit.

Each term of Eq. (5.17b) has the units of current. Let us now apply Kirchhoff's current law in reverse to obtain the circuit of Fig. 5.28.

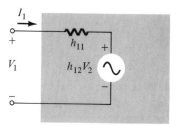

Figure 5.27 Hybrid input equivalent circuit.

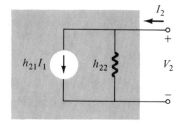

Figure 5.28 Hybrid output equivalent circuit.

Since h_{22} has the units of conductance, it is represented by the resistor symbol. Keep in mind, however, that the resistance in ohms of this resistor is equal to the reciprocal of conductance ($1/h_{22}$).

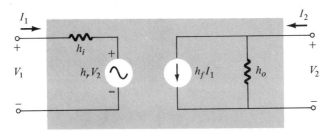

Figure 5.29 Complete hybrid equivalent circuit.

The complete "ac" equivalent circuit for the basic three-terminal linear device is indicated in Fig. 5.29 with a new set of subscripts for the *h*-parameters.

The notation of Fig. 5.29 is of a more practical nature since it relates the *h*-parameters to the resulting ratio obtained in the last few paragraphs. The choice of letters is obvious from the following listing:

$$h_{11} \rightarrow \textit{i}\text{nput resistance} \rightarrow h_i$$
$$h_{12} \rightarrow \textit{r}\text{everse transfer voltage ratio} \rightarrow h_r$$
$$h_{21} \rightarrow \textit{f}\text{orward transfer current ratio} \rightarrow h_f$$
$$h_{22} \rightarrow \textit{o}\text{utput conductance} \rightarrow h_o$$

The circuit of Fig. 5.29 is applicable to any linear three-terminal device with no internal independent sources. For the transistor, therefore, even though it has three basic configurations, *they are all three-terminal configurations,* so that the resulting equivalent circuit will have the same format as shown in Fig. 5.29. The *h*-parameters, however, will change with each configuration. To distinguish which parameter has been used or which is available, a second subscript has been added to the *h*-parameter notation. For the common-base configuration the lower-case letter *b* was added, while for the common-emitter and common-collector configurations the letters *e* and *c* were added, respectively. The hybrid equivalent circuit for the common-base and common-emitter configurations with the standard notation is presented in Fig. 5.30. The circuits of Fig. 5.30 are applicable for *pnp* or *npn* transistors.

The hybrid equivalent circuit of Fig. 5.29 is an extremely important one in the area of electronics today. It will appear over and over again in the analysis to follow. It would be time well spent, at this point, for the reader to memorize and draw from memory its basic construction and define the significance of the various parameters. The fact that both a Thévenin and a Norton circuit appear in the circuit of Fig. 5.29 was further impetus for calling the resultant circuit a *hybrid* equivalent circuit. Two additional transistor equivalent circuits, not to be discussed in this text, called the *Z*-parameter and *Y*-parameter equivalent circuits, use either the voltage source or the current source but not both in the same equivalent circuit.

The hybrid parameters can be determined from the characteristics of the device, but since it is seldom done in practice the technique will not be covered here. As noted earlier, the hybrid parameters are normally provided on specification sheets or in data books. Review the values provided in Section 4.9 for the RCA 2N1711 transistor. Typical values for each parameter for the broad range of transistors available today in each of its three configurations are provided in Table 5.1.

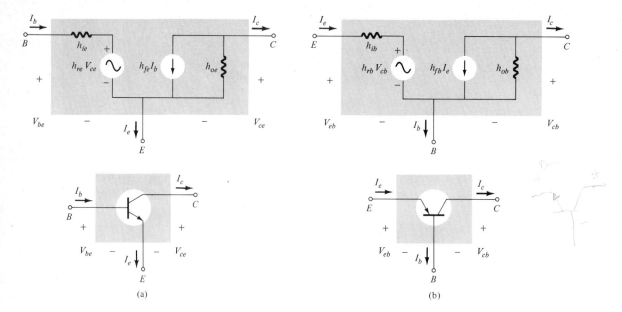

Figure 5.30 Complete hybrid equivalent circuits: (a) common-emitter configuration; (b) common-base configuration.

The minus sign indicates that as one quantity of the defining equation increased in magnitude, the other decreased in magnitude.

TABLE 5.1 Table Parameter Values for the CE, CC, and CB Transistor Configurations

Parameter	CE	CC	CB
h_i	1 K	1 K	20 Ω
h_r	2.5×10^{-4}	$\cong 1$	3.0×10^{-4}
h_f	50	−50	−0.98
h_o	25 μA/V	25 μA/V	0.5 μA/V
$1/h_o$	40 K	40 K	2 M

Note in retrospect (Section 4.4) that the input resistance of the common-base configuration is low, while the output impedance is high. Consider also that the short-circuit gain is very close to 1. For the common-emitter and common-collector configurations note that the input impedance is much higher than that of the common-base configuration and that the ratio of output to input resistance is about 40:1. Consider also for the common-emitter and common-base configuration that h_r is very small in magnitude. Transistors are available today with values of h_{fe} that vary from 20 to 600. For any transistor the region of operation and conditions under which it is being used will have an effect on the various h-parameters. The effect of temperature and collector current and voltage on the h-parameters will be introduced in Section 5.12.

5.12 VARIATIONS OF TRANSISTOR PARAMETERS

There is a wide variety of curves that can be drawn to show the variations of the *h*-parameters with temperature, frequency, voltage, and current. The most interesting and useful at this stage of the development include the *h*-parameter variations with junction temperature and collector voltage and current.

In Fig. 5.31 the effect of the collector current on the *h*-parameter has been indicated. Take careful note of the logarithmic scale on the vertical and horizontal axes. The parameters have all been normalized to unity so that the relative change in magnitude with collector current can easily be determined. On every set of curves, such as in Fig. 5.31, the operating point at which the parameters were

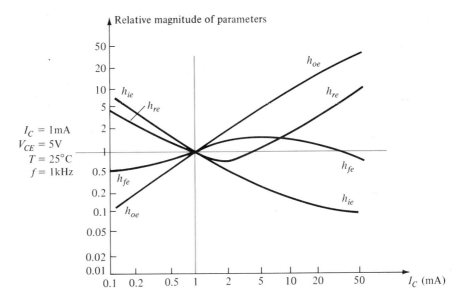

Figure 5.31 Hybrid parameter variations with collector current.

found is always indicated. For this particular situation the quiescent point is at the intersection of $V_{CE} = 5.0$ V and $I_C = 1.0$ mA. Since the frequency and temperature of operation will also affect the *h*-parameters, these quantities are also indicated on the curves. At 0.1 mA, h_{fe} is about 0.5 or 50% of its value at 1.0 mA, while at 3 mA, it is 1.5 or 150% of that value. In other words, h_{fe} has changed from a value of $0.5(50) = 25$ to $1.5(50) = 75$ with a change of I_C from 0.1 mA to 3 mA. In Section 5.13 we shall find that for the majority of applications it is a fairly good approximation to neglect the effects of h_{re} and h_{oe} in the equivalent circuit. Consider, however, the point of operation at $I_C = 50$ mA. The magnitude of h_{re} is now approximately 11 times that at the defined *Q*-point, a magnitude that may not permit eliminating this parameter from the equivalent circuit. The parameter h_{oe} is approximately 35 times the normalized value. This increase in h_{oe} will decrease the magnitude of the output resistance of the transistor to a point where it may approach the magnitude of the load resistor. There would then be no justification in eliminating h_{oe} from the equivalent circuit on an approximate basis.

In Fig. 5.32 the variation in magnitude of the h-parameters on a normalized basis has been indicated with change in collector voltage. This set of curves was normalized at the same operating point of the transistor discussed in Fig. 5.31 so that a comparison between the two sets of curves can be made. Note that h_{ie} and h_{fe} are relatively steady in magnitude, while h_{oe} and h_{re} are much larger to the left and right of the chosen operating point. In other words, h_{oe} and h_{re} are much more sensitive to changes in collector voltage than are h_{ie} and h_{fe}.

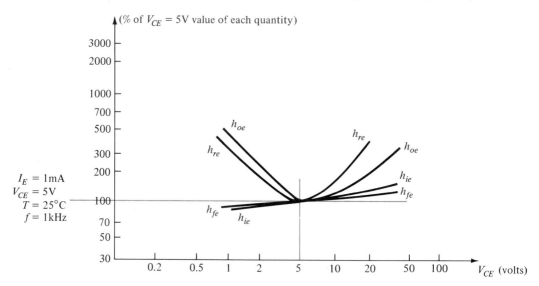

Figure 5.32 Hybrid parameter variations with collector-emitter potential.

Figure 5.33 Hybrid parameter variations with temperature.

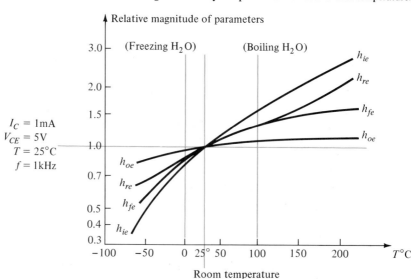

In Fig. 5.33 the variation in *h*-parameters has been plotted for changes in junction temperature. The normalization value is taken to be room temperature: $T = 25°C$. The horizontal scale is a linear scale rather than a logarithmic scale as was employed for Figs. 5.31 and 5.32. In general, all the parameters increase in magnitude with temperature. The parameter least affected, however, is h_{oe}, while the input impedance h_{ie} changes at the greatest rate. The fact that h_{fe} will change from 50% of its normalized value at $-50°C$ to 150% of its normalized value at $+150°C$ indicates clearly that the operating temperature must be carefully considered in the design of transistor circuits.

5.13 SMALL-SIGNAL ANALYSIS OF THE BASIC TRANSISTOR AMPLIFIER USING THE HYBRID EQUIVALENT CIRCUIT

The use of the hybrid equivalent circuit in the analysis of transistor networks will be introduced through the use of the network of Fig. 5.34. Note in the figure that the specification sheet for the transistor has provided the values of the hybrid parameters for the typical operating region of this device. If we remove the dc supply V_{CC} and replace it by a short circuit to ground and replace both capacitors by a short circuit, we obtain the network of Fig. 5.35. The connection of R_B

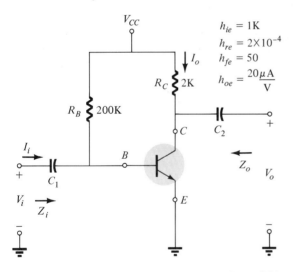

Figure 5.34

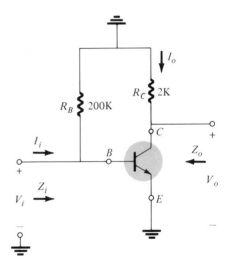

Figure 5.35 Network of Fig. 5.34 following the removal of the dc supply and replacing the coupling capacitors by a short-circuit impedance equivalent.

and R_C can certainly be separated, as shown in Fig. 5.36, and the hybrid equivalent circuit inserted between the proper terminals. The network can then be redrawn as shown in Fig. 5.37.

The analysis of the network as it now appears would be a long and cumbersome process. Fortunately, however, there are certain approximations we can make

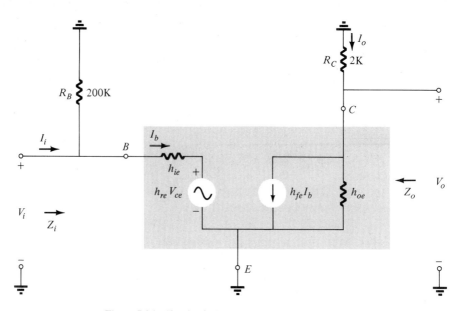

Figure 5.36 Circuit of Fig. 5.35 following the substitution of the small-signal hybrid equivalent circuit for the transistor.

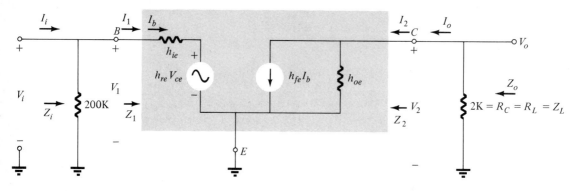

Figure 5.37 Redrawn circuit of Fig. 5.36.

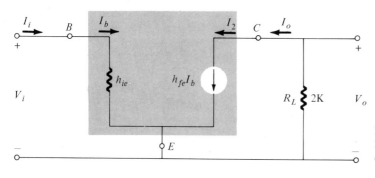

Figure 5.38 Reduced version of Fig. 5.37 following the application of some reasonable approximations.

that will make the analysis of this system considerably easier. First, and very important, the value of h_{re}, the feedback parameter, is usually sufficiently small to permit the removal of the feedback voltage source $h_{re}V_{ce}$ from the picture. This is shown in the reduced version of Fig. 5.38. In this case $1/h_{oe} = 1/20$ μmhos $= 50$ K appears in parallel with the 2-K load (R_C) resistor. Recall that for two resistors in parallel the parallel combination is always smaller than the smaller of the two. Further, if the ratio of the two is greater than 10:1, the larger can usually be ignored as a good approximation. In this case since the foregoing criterion is met (25:1), we can eliminate $1/h_{oe}$ from the diagram as appearing in Fig. 5.38. Further, since the removal of $h_{re}V_{ce}$ resulted in a parallel combination of the 200-K resistor and h_{ie}, we can remove the 200-K resistor as shown in Fig. 5.38. In Fig. 5.38 we see that $I_b = I_i$ and $I_o = h_{fe}I_b$. Solving for I_o, we find that

$$I_o = h_{fe}I_b = h_{fe}I_i$$

and the current gain

$$A_i = \frac{I_o}{I_i} \cong h_{fe} = 50$$

In other words, the ac current gain is approximately equal to the hybrid parameter h_{fe}. You may have some concern about the approximations that were made to reach this value. If the complete network of Fig. 5.37 were analyzed, the current gain obtained would be 48.1—a difference of little consequence when you consider that the resistor values and hybrid parameter values may be off by a greater percentage in the actual network. The point to be made is this—use approximations wherever possible and reasonable. The time and effort saved can be considerable and the results are in many ways as "accurate" as those obtained with the full model.

The input impedance (or resistance in this case) is simply h_{ie}. That is, $Z_i = h_{ie} = 1$ K versus 0.981 K if the complete network were investigated.

The output impedance is equal to $R_C = 2$ K versus 1.96 K with the complete model. It must be noted here that the output impedance is that impedance obtained between the output terminals with the source set to zero. This would result in $I_b = 0$ and $h_{fe}I_b = 0$, which would result in an open-circuit equivalent for the current source.

The voltage gain is the ratio of V_o to V_i. Since

$$V_o = -I_o R_C = -h_{fe} I_b R_C$$

and

$$V_i = I_b h_{ie}$$

$$A_v = \frac{V_o}{V_i} = -\frac{h_{fe} I_b R_C}{I_b h_{ie}} = -\frac{h_{fe} R_C}{h_{ie}}$$

For this example:

$$A_v = -\frac{(50)(2 \text{ K})}{1 \text{ K}} = -100$$

versus the -98 obtained for the complete model. The minus sign only indicates that the polarity of the output is the inverse of the input at each instant of time—

a 180° phase shift. In other words, when the input reaches a positive peak, the output attains a negative peak.

EXAMPLE 5.8 Determine the following for the network of Fig. 5.39:
(a) $A_v = V_o/V_i$.
(b) $A_i = I_o/I_i$.
(c) Z_i.
(d) Z_o.

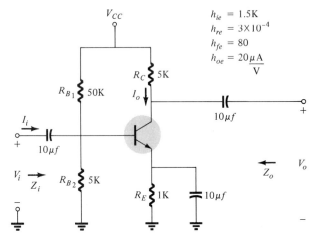

Figure 5.39 Circuit for Example 5.8.

Solution: Replacing the dc supplies and capacitors by short circuits will result in the circuit of Fig. 5.40. Figure 5.40 will appear as shown in Fig. 5.41 following the substitution of the approximate equivalent circuit.

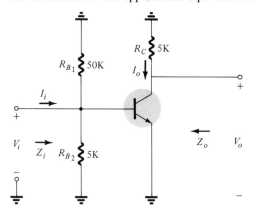

Figure 5.40 Circuit of Fig. 5.39 redrawn for small-signal ac analysis.

Figure 5.41 Approximate equivalent circuit for the network of Fig. 5.40.

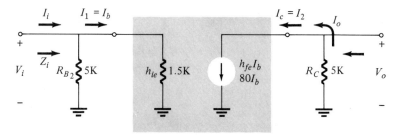

h_{oe} does not appear since

$$\frac{1}{h_{oe}} = \frac{1}{120 \, \mu A/V} = 50 \text{ K} \parallel 5 \text{ K} \cong 5 \text{ K} \; (10:1 \text{ ratio})$$

In addition, R_{B_1} does not appear since $R_{B_1} \parallel R_{B_2} = 50 \text{ K} \parallel 5 \text{ K} \cong 5 \text{ K} \; (10:1 \text{ ratio})$.

(a) A_v:

$$V_o = -I_o R_C = -h_{fe} I_b R_C$$

and

$$I_b = \frac{V_i}{h_{ie}}$$

so that

$$V_o = -h_{fe} \left(\frac{V_i}{h_{ie}}\right) R_C$$

and

$$A_v = \frac{V_o}{V_i} = -\frac{h_{fe}}{h_{ie}} R_C$$

as obtained for the fixed-bias network.

Substituting values, we obtain

$$A_v = -\frac{80}{1.5 \text{ K}} (5 \text{ K}) = \mathbf{-266.67}$$

as compared to 261.44 obtained using the complete model.

(b) A_i:

$$I_o = h_{fe} I_b$$

Current divider rule:

$$I_b = \frac{5 \text{ K} (I_i)}{5 \text{ K} + 1.5 \text{ K}} = 0.769 I_i$$

and

$$I_o = h_{fe}(0.769 I_i) = (80)(0.769) I_i = 61.52 I_i$$

with

$$A_i = \frac{I_o}{I_i} = \mathbf{61.52}$$

as compared to 55.71 with a complete model.

(c) Z_i:

$$Z_i \cong 5 \text{ K} \parallel 1.5 \text{ K} = \mathbf{1.15 \text{ K}}$$

as compared to 1.065 K, the solution with the complete model.

(d) Z_o:

The output impedance is defined by the condition $V_i = 0$. Therefore,

$$I_b = \frac{V_i}{h_{ie}} = 0 \quad \text{and} \quad h_{fe} I_b = 0 \quad \text{(an open-circuit equivalent)}$$

and

$$Z_o = \mathbf{5 \text{ K}}$$

as compared to 4.9 K obtained with a complete model. Take a moment to contemplate the degree of effort to obtain the results above as compared with what you might expect with a complete model.

EXAMPLE 5.9 For the common-base circuit of Fig. 5.42 with a load applied, find the following:
(a) $A_i = I_o/I_i$.
(b) $A_v = V_o/V_i$.
(c) Z_i.
(d) Z_o.

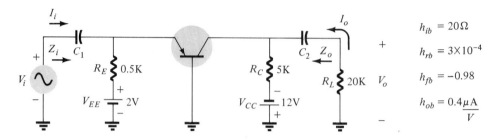

Figure 5.42 Circuit for Example 5.9.

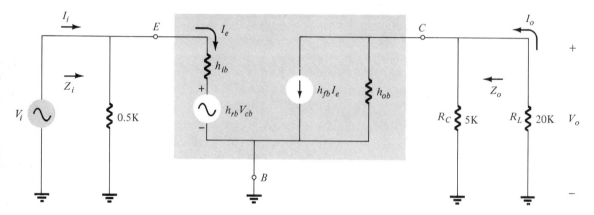

Figure 5.43 Circuit of Fig. 5.42 following the substitution of the hybrid equivalent circuit.

Solution: Replacing the dc supplies and coupling capacitors by short circuits will result in the configuration of Fig. 5.43.

Note, as mentioned above, that the common-base equivalent circuit has exactly the same format as the common-emitter configuration but with the appropriate common-base parameters. Consider that now $h_{re}V_{ce}$ is $h_{rb}V_{cb}$ and $h_{fe}I_b$ is now $h_{fb}I_e$. Since the configuration is the same and $1/h_{ob}$ is greater than $1/h_{oe}$ with h_{re} usually close in magnitude to h_{rb}, the approximations $(1/h_{ob} \cong \infty \ \Omega$ and $h_{rb} \cong 0)$ made for the common-emitter configuration are applicable here, also.

Applying $h_{rb} \cong 0$, we redraw the circuit (Fig. 5.44). Eliminating the 500-Ω and 500-K $(1/h_{ob})$ resistors due to the lower parallel resistor will result in the circuit of Fig. 5.45.
(a) $A_i = I_o/I_i$:

$$I_o = \frac{(5 \text{ K})(I_2)}{5 \text{ K} + 20 \text{ K}} = 0.2 I_2 \quad \text{(current divider rule)}$$

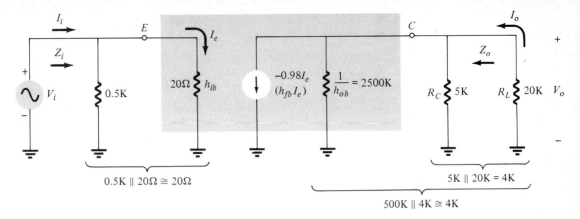

Figure 5.44 Redrawn circuit of Fig. 5.43 following the application of the approximation $h_{rb} \cong 0$.

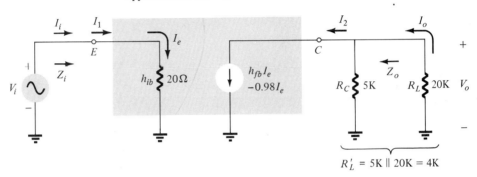

Figure 5.45 Circuit of Fig. 5.44 following the elimination (on an approximate basis) of certain parallel elements.

and $\quad I_2 = h_{fb}I_e = h_{fb}I_1 = h_{fb}I_i$ with A_i (transistor) $= \dfrac{I_2}{I_i} = h_{fb}$

Therefore, $\quad I_o = 0.2 I_2 = 0.2 h_{fb} I_i$

and $\quad A_i = \dfrac{I_o}{I_i} \cong 0.2(h_{fe}) = 0.2(-0.98) = \mathbf{-0.196}$

indicating as before that the current gain of the common-base configuration is always less than 1.

(b) $A_v = V_o/V_i$:

Using
$$R'_L = 5\text{ K} \| 20\text{ K} = R_C \| R_L$$
$$V_o = -I_2 R'_L = -h_{fb}I_e R'_L$$

and
$$I_e = I_i = \dfrac{V_i}{h_{ib}}$$

so that
$$V_o = -h_{fb}\left(\dfrac{V_i}{h_{ib}}\right)R'_L$$

with
$$A_v = \dfrac{V_o}{V_i} = -\dfrac{h_{fb}}{h_{ib}}R'_L$$

SEC. 5.13 SMALL-SIGNAL ANALYSIS OF THE BASIC TRANSISTOR AMPLIFIER

Note the similarities between this equation for voltage gain and that obtained for the common-emitter configurations. Consider also that the effect of the added load was only to change R_L to R_L', which is the parallel combination of R_C and the applied load R_L.

Substituting values, we get

$$A_v = -\frac{(-0.98)(4 \times 10^3)}{20} = 196$$

The voltage gain, therefore, can be significantly greater than 1 and results in an output that is *in phase* with the input (note the absence of the minus sign in the result).

(c) $Z_i \cong h_{ib} = 20\ \Omega.$
(d) $Z_o \mid_{v_i=0} \cong R_C = 5\ \text{K}.$

EXAMPLE 5.10 Find the following for the circuit of Fig. 5.46:
(a) $A_i = I_o/I_i.$
(b) $A_v = V_o/V_i.$
(c) $Z_i.$
(d) $Z_o.$

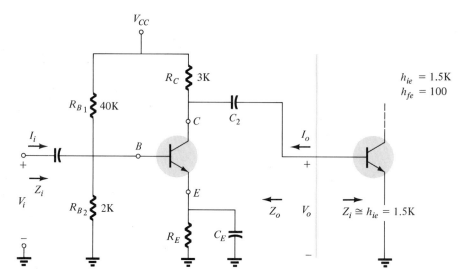

Figure 5.46 Circuit for Example 5.10.

Solution: Replacing the dc supplies and capacitors by short circuits and substituting the appropriate hybrid equivalent circuit will result in the configuration of Fig. 5.47. Note that the 2-K emitter resistor has been "shorted out" by the capacitor C_E.

Considering parallel elements will result in the configuration of Fig. 5.48.
(a) $A_i = I_o/I_i.$
From Fig. 5.48, $I_2 = 100 I_b$. That is, A_i (transistor) $= I_c/I_b = h_{fe}$. Applying the current divider rule to the input and output circuits, we get

$$I_b = \frac{(2\ \text{K})(I_i)}{2\ \text{K} + 1.5\ \text{K}} = 0.571 I_i$$

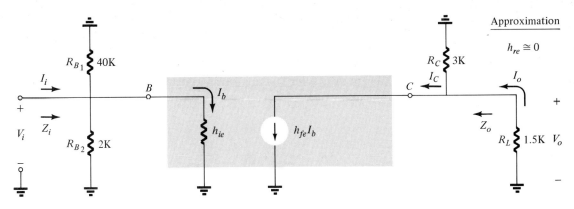

Figure 5.47 Circuit of Fig. 5.46 following the substitution of the approximate ($h_{re} \cong 0$) hybrid equivalent circuit.

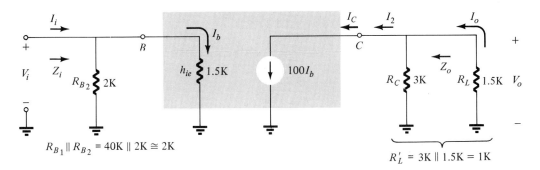

Figure 5.48 Circuit of Fig. 5.47 following the elimination (on an approximate basis) of certain parallel elements.

and
$$I_o = \frac{(3\text{ K})(I_2)}{3\text{ K} + 1.5\text{ K}} = 0.667 I_2$$

Substituting, we get
$$A_i = \frac{I_o}{I_i} = \left(\frac{I_o}{I_2}\right)\left(\frac{I_2}{I_i}\right) = \left(\frac{I_o}{I_2}\right)\left(\frac{I_2}{I_b}\right)\left(\frac{I_2}{I_i}\right)$$
$$= (0.667)(100)(0.571)$$
$$= 38.1$$

(b) $A_v = V_o/V_i$:

The configuration is similar to that obtained for both the CE and CB configurations and
$$A_v = \frac{V_o}{V_i} = -\frac{h_{fe} R'_L}{h_{ie}}$$
$$= -\frac{(100)(1 \times 10^3)}{1.5 \times 10^3}$$
$$= -66.7$$

(c) $Z_i \cong R_{B2} \| h_{ie} = 2\text{ K} \| 1.5\text{ K} = \mathbf{0.86\text{ K}}$.
(d) $Z_o|_{V_i=0} \cong R_C = \mathbf{3\text{ K}}$.

SEC. 5.13 SMALL-SIGNAL ANALYSIS OF THE BASIC TRANSISTOR AMPLIFIER

5.14 APPROXIMATE BASE, COLLECTOR, AND EMITTER EQUIVALENT CIRCUITS

In the following analysis it will prove very useful to know at a glance what the effect of loads and signal sources in another portion of the network will have on the base, collector, or emitter potential and current. In this section we find, on an approximate basis, the equivalent circuit "seen" looking into the base, collector, or emitter terminals of a transistor in the common-emitter configuration.

The approximations $h_{re} \cong 0$ and $1/h_{oe} \cong \infty \, \Omega$ (open circuit) will be used throughout this section. The full benefit of the circuits to be derived will be more apparent when the equivalent circuits have been obtained and they are applied in a few examples.

Consider the network of Fig. 5.49. It can be shown through a series of fairly logical steps that the network "seen" by the applied source V_1 is as shown in Fig. 5.50. That is,

$$Z_i = h_{ie} + (1 + h_{fe})R_E$$

and

$$I_b = I_i = \frac{V_1}{R_1 + h_{ie} + (1 + h_{fe})R_E}$$

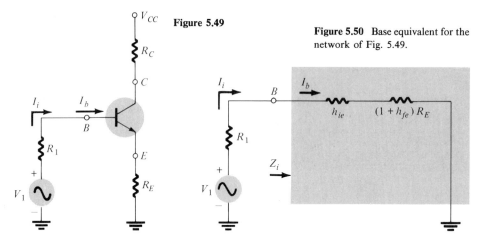

Figure 5.49

Figure 5.50 Base equivalent for the network of Fig. 5.49.

Typically, however, $h_{fe} \gg 1$ and $(1 + h_{fe})R_E \gg h_{ie}$, resulting in the following approximation:

$$\boxed{Z_i \cong h_{fe}R_E} \qquad (5.22)$$

In other words, a resistor in the emitter leg appears magnified by h_{fe} at the base of the system. This will result in a reduced I_b but an increased level of Z_i. Note that R_C is not transferred to the input network.

If we were interested in the network "seen" by the emitter terminal as we look back into the transistor configuration, we would obtain the equivalent network of Fig. 5.51. Note again that R_C does not appear and the input resistance elements R_1 and h_{ie} are divided by $(1 + h_{fe}) \cong h_{fe}$ and therefore substantially reduced in magnitude. The impendance "seen" is

$$Z_e \cong \frac{R_1 + h_{ie}}{h_{fe}} \qquad (5.23)$$

and the current

$$I_e \cong \frac{V_1}{\frac{R_1 + h_{ie}}{h_{fe}} + R_E} \qquad (5.24)$$

The collector network is simply as provided in Fig. 5.52, with $I_c = h_{fe}I_b$ and no reflection of R_E or V_1.

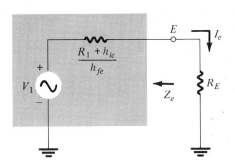

Figure 5.51 Emitter equivalent for the network of Fig. 5.49.

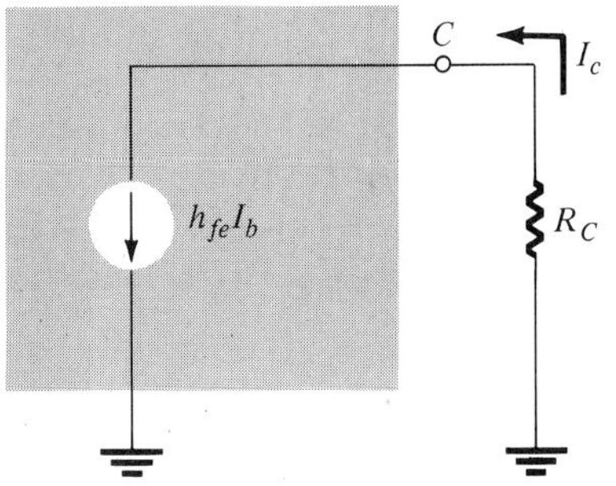

Figure 5.52 Collector equivalent for the network of Fig. 5.49.

The beneficial aspects of the equivalent circuits just derived will become obvious in the examples to follow and in later chapters. It would be wise to memorize these equivalent circuits for future use. They can be very powerful tools in analysis of transistor networks.

EXAMPLE 5.11 The transistor configuration of Fig. 5.53, called the *emitter follower*, is frequently used for impedance matching purposes; that is, it presents a high

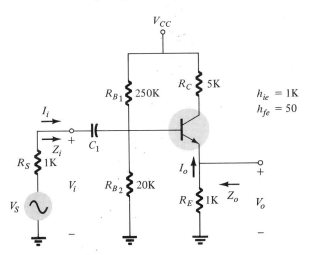

Figure 5.53 Circuit for Example 5.11.

impedance at the input terminals (Z_i) and a low output impedance (Z_o), rather than the reverse, which is typical of the basic transistor amplifier. The effect of the emitter follower circuit is much like that obtained using a transformer to match a load to the source impedance for maximum power transfer. The following analysis will reveal that the voltage gain of the emitter follower circuit is always less than 1.

For the circuit of Fig. 5.53, calculate the following:
(a) Z_i.
(b) Z_o.
(c) $A_{v_1} = V_o/V_s$ and $A_{v_2} = V_o/V_i$.
(d) $A_i = I_o/I_i$.

Solution: Eliminating the dc levels and replacing both capacitors by short circuits will result in the circuit of Fig. 5.54.

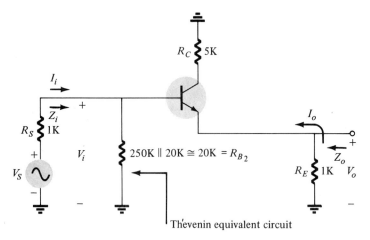

Figure 5.54 Circuit of Fig. 5.53 redrawn for small-signal ac analysis.

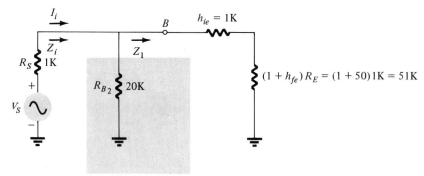

Figure 5.55 Substitution of the approximate base equivalent circuit into the circuit of Fig. 5.54.

(a) Using the base equivalent circuit (Fig. 5.55), we get

$$Z_1 = h_{ie} + (1 + h_{fe})R_2 = 52 \text{ K}$$

(certainly high compared to the typical input impedance $Z_i \cong h_{ie}$ for the basic transistor amplifier).

The resulting $Z_i =$ 20 K $\parallel$ 52 K = **14.5 K**

If we had used Eq. (5.22) and the conditions surrounding this approximation, the equivalent circuit would appear as shown in Fig. 5.56 and $Z_1 = h_{fe}R_E =$ 50 K and $Z_i =$ 20 K $\parallel$ 50 K $\cong$ **14.3 K.** This result will usually be so close to the result obtained using the complete approximate equivalent that we will use Eq. (5.22) throughout the following analysis unless otherwise noted.

(b) The Thévenin equivalent circuit of the portion of the network indicated in Fig. 5.54 will now be found so that the input circuit will have the basic configuration of Fig. 5.49. That is, R_1 and V_1 will be found to ensure that the proper values are substituted into the emitter equivalent circuit to be employed in the determination of Z_o.

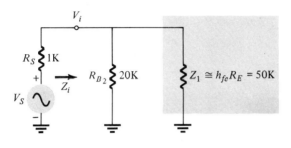

Figure 5.56 Approximation frequently used to determine the input impedance of a network with an unbypassed emitter resistor.

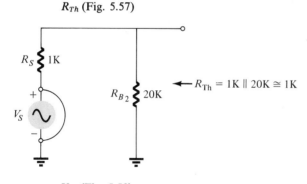

Figure 5.57 Determining R_{Th} for the portion of the circuit indicated in Fig. 5.54.

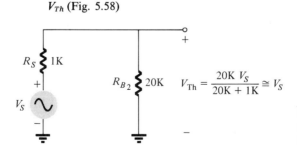

Figure 5.58 Determining V_{Th} for the portion of the circuit indicated in Fig. 5.54.

Substituting the Thévenin equivalent circuit into the circuit of Fig. 5.54 will result in the configuration of Fig. 5.59.

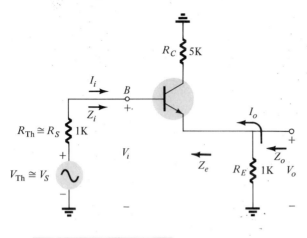

Figure 5.59 Circuit of Fig. 5.54 following the substitution of the Thévenin equivalent circuit.

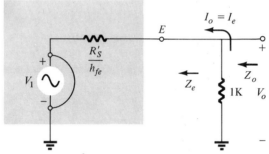

Figure 5.60 Substitution of the emitter equivalent circuit with V_1 set to zero.

Using the emitter equivalent circuit with $V_s = 0$ (as required by definition) to determine Z_o (Fig. 5.60), we get

$$Z_e = \frac{R_s'}{h_{fe}} = \frac{1\,K + 1\,K}{50} = 40\,\Omega$$

and
$$Z_o = Z_e \parallel R_L = 40 \parallel 1000 \cong 40\,\Omega$$

(c) A careful examination of Fig. 5.53 will reveal that the output voltage V_o is separated from the input voltage V_i by only the voltage drop V_{be} across the base-to-emitter junction. For ac operations and an unbypassed (no capacitor present) emitter resistor the approximation indicated by Eq. (5.25) is frequently employed.

$$\boxed{V_{be} \cong 0V} \quad \binom{\text{ac operations and}}{\text{unbypassed emitter resistor}} \quad (5.25)$$

Using this approximation, we get

$$V_o \cong V_i \quad \text{and} \quad A_{v_2} = \frac{V_o}{V_i} \cong 1$$

For $A_{v_1} = V_o/V_s$ we can turn to the network of Fig. 5.56, where

$$V_i = \frac{(50\,K \parallel 20\,K)(V_s)}{(50\,K \parallel 20\,K) + 1\,K} = 0.93\,V_s$$

and
$$A_{v_1} = \frac{V_o}{V_s} = \left[\frac{V_o}{V_i}\right]\left[\frac{V_i}{V_s}\right] = [\cong 1][0.98] \cong \mathbf{0.98}$$

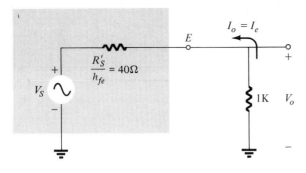

Figure 5.61 Circuit to be employed in determining A_{v_1}.

The validity of Eq. (5.25) can be demonstrated using the emitter equivalent circuit of Fig. 5.61, where

$$V_o = \frac{(1\text{ K})(V_s)}{1\text{ K} + 40} = 0.96 V_s$$

and
$$A_{v_1} = \frac{V_o}{V_s} = 0.96 \cong 0.93 \quad \text{(as determined above)}$$

(d) From the emitter equivalent circuit:

$$I_o = I_e = \frac{V_s}{1.040\text{ K}}$$

and from Fig. 5.56,

$$I_i = \frac{V_s}{R_s + Z_i} = \frac{V_s}{1\text{ K} + 14.3\text{ K}} = \frac{V_s}{15.3\text{ K}}$$

or $V_s = (I_i)(15.3\text{ K})$.

Substituting this result into the equation above, we get

$$I_o = -\frac{(I_i)(15.3\text{ K})}{1.040\text{ K}}$$

and
$$A_i = \frac{I_o}{I_i} = -\frac{15.3\text{ K}}{1.040\text{ K}} = -14.71$$

EXAMPLE 5.12 In an effort to clearly demonstrate the usefulness of the base, collector, and emitter circuits, we shall consider a simple *difference amplifier*. In its basic

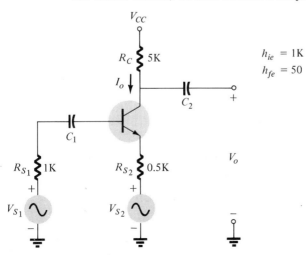

Figure 5.62 Circuit for Example 5.12—difference amplifier.

203

form, a difference amplifier is simply a network that will produce a signal that is the difference of the two applied signals.

Figure 5.62 is such a circuit. Note that a signal has been applied to both the base and emitter leg of the transistor.

Solution: Using the "collector equivalent circuit" after replacing the dc supply and capacitors by short circuits will result in the circuit of Fig. 5.63, where

$$V_o = (-I_o)(5 \text{ K}) = -(50I_b)(5 \text{ K})$$

Using the "base equivalent circuit" gives us Fig. 5.64. Note that V_{s_2} passes directly into the base without transformation. The base current

$$I_b \cong \frac{V_{s_1} - V_{s_2}}{R_{s_1} + h_{fe}R_{s_2}} = \frac{V_{s_1} - V_{s_2}}{1 \text{ K} + 25 \text{ K}} = \frac{V_{s_1} - V_{s_2}}{26 \text{ K}}$$

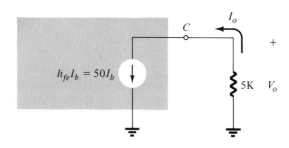

Figure 5.63 Application of the collector equivalent circuit to the circuit of Fig. 5.62.

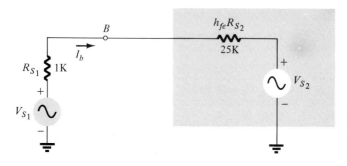

Figure 5.64 Application of the base equivalent circuit to the circuit of Fig. 5.62.

Substituting into the equation for V_o, we obtain

$$V_o = -(50I_b)(5 \text{ K}) = -50 \frac{V_{s_1} - V_{s_2}}{26 \text{ K}} 5 \text{ K} = \frac{-250(V_{s_1} - V_{s_2})}{26 \text{ K}}$$

and

$$V_o = 9.62(V_{s_1} - V_{s_2})$$

The collector potential V_o, therefore, is approximately 9.62 times the difference of the two applied signals.

5.15 AN ALTERNATIVE APPROACH

In recent years there has been an increasing interest in an approximate equivalent circuit for the transistor in which one of the parameters is determined by the dc operating conditions. You will possibly recall from the transistor specification sheet provided in Chapter 4 that the hybrid parameter h_{ie} was specified at a particular operating point. Figure 5.31 revealed a significant variation in h_{ie} with I_C ($\cong I_E$). The question then arises of what one would do with the provided value of h_{ie} if the conditions of operation (level of $I_C \cong I_E$) were different from those indicated on the specification sheet. The equivalent circuit derived below will permit the determination of an equivalent h_{ie} using the dc operating conditions of the network, thereby not limiting itself to the data on the device as provided by the manufacturer.

The derivation of the alternative equivalent circuit begins with a close examination of the input, and output characteristics, of the CB transistor configuration, as redrawn in Fig. 5.65, on an approximate basis. Note that straight-line segments are used to represent the collector characteristics and a single diode characteristic for the emitter circuit (neglecting the variation in the input characteristics with the change in V_{CB}), resulting in an equivalent circuit such as shown in Fig. 5.66b.

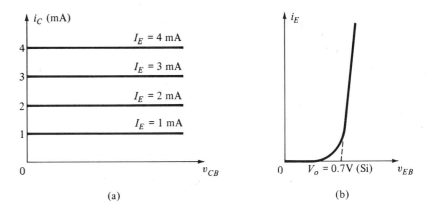

Figure 5.65 Approximate CB characteristics: (a) output; (b) input.

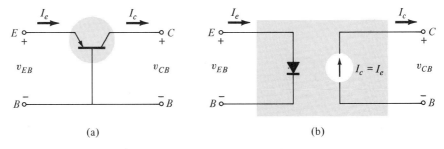

Figure 5.66 (a) CB configuration; (b) approximate CB equivalent circuit as defined by Fig. 5.65.

For ac conditions, therefore, the input impedance at the emitter of the CB transistor can be determined using Eq. (2.19) as introduced for the ac resistance of a diode. The factor r_B will be dropped to ensure that it does not affect the clarity of the introduction of this alternative technique. You will recall from Chapter 2 that in time it is conceivable that the factor r_B can be totally ignored with a negligible loss in accuracy if manufacturing techniques continue to improve. In time, when the alternative approach is developed, if you prefer to add a factor developed through experience for that transistor, then it can surely be introduced with little added confusion. For now we will define the input impedance for the CB configuration to be

$$r_e = \frac{26 \text{ mV}}{I_E \text{ (mA)}} \quad \text{(ohms)} \tag{5.26}$$

where I_E is the dc emitter current of the transistor. The emitter current is employed in Eq. (5.26) since it is the defined current of the diode in Fig. 5.65b. The result of the discussion above is the input equivalent circuit appearing in Fig. 5.67.

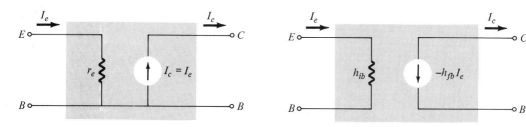

Figure 5.67 Approximate CB equivalent circuit.

Figure 5.68 Approximate CB hybrid equivalent circuit.

Figure 5.65a clearly indicates that the collector curves have been approximated to result in $I_C = I_E$ at any point on the characteristics. This would result in the output equivalent circuit appearing in Figs. 5.66 and 5.67. On an approximate basis, the alternative equivalent circuit is now defined. Note its similarities with the reduced hybrid equivalent circuit of Fig. 5.68. A comparison of the two clearly indicates that

$$\begin{aligned} h_{ib} &= r_e \\ h_{fb} &= 1 \end{aligned} \tag{5.27}$$

The following example will clarify the use of the alternative equivalent circuit.

EXAMPLE 5.13 For the network of Fig. 5.69, determine A_v, A_i, Z_i, and Z_o.

Solution
dc conditions:

$$I_E = \frac{V_{EE} - V_{BE}}{R_E} = \frac{10 - 0.7}{5 \text{ K}} = \frac{9.3}{5 \text{ K}} = 1.86 \text{ mA}$$

and

$$r_e = \frac{26 \text{ mV}}{I_E} = \frac{26 \text{ mV}}{1.86 \text{ mA}} \cong 14 \text{ }\Omega$$

ac conditions:
The network redrawn (Fig. 5.70):

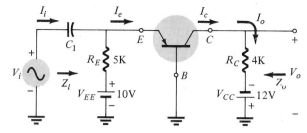

Figure 5.69 Network for Example 5.13.

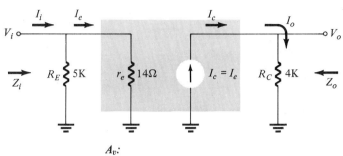

Figure 5.70 ac equivalent for the network of Fig. 5.69.

A_v:

$$V_o = I_c R_C = I_e R_C$$

and

$$I_e = \frac{V_i}{r_e}$$

so that

$$V_o = \frac{V_i}{r_e} R_C$$

and

$$\boxed{A_v = \frac{V_o}{V_i} = \frac{R_C}{r_e}} \qquad (5.28)$$

Compare this result to that obtained in Example 5.9, where

$$|A_v| = \frac{h_{fb}}{h_{ib}} R_L'$$

but in this case $R_L' = R_C$, $h_{fb} = 1$, $h_{ib} = r_e$ which through substitution will result in Eq. (5.28). Substituting values, we get

$$A_v = \frac{R_C}{r_e} = \frac{4\text{ K}}{14} = 285.71$$

A_i: Since $R_E \| r_e \cong r_e$,

$$I_o = I_c = I_e = I_i$$

and

$$\boxed{A_i \cong 1 \cong h_{fb}}$$

as obtained in Example 5.9.

Z_i:

$$\boxed{Z_i \cong r_e = h_{ib} = 14 \text{ }\Omega}$$

SEC. 5.15 AN ALTERNATIVE APPROACH

Z_o:

$$Z_o|_{V_i=0} = R_C = 4\text{ K}$$

For the common-emitter configuration appearing in Fig. 5.71a the input and output characteristics have been approximated by the set appearing in Fig. 5.71b and 5.71c, respectively. The base characteristics are again approximated to be those of a diode (the effect of V_{CE} on the characteristics is ignored) and

$$r_{ac} = \frac{26\text{ mV}}{I_B} \qquad (5.29)$$

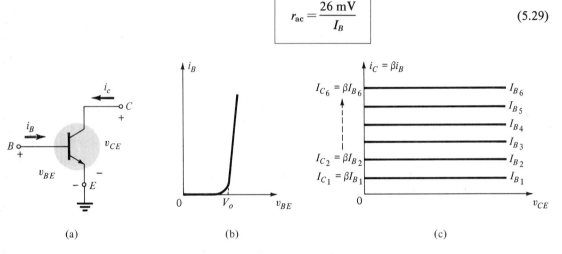

Figure 5.71 (a) CE configuration; (b) input characteristics; (c) output characteristics.

But $\qquad I_E \cong I_C = \beta I_B \quad \text{and} \quad I_B \cong \dfrac{I_E}{\beta}$

so that $\qquad r_{ac} = \dfrac{26\text{ mV}}{I_B} = \dfrac{26\text{ mV}}{I_E/\beta} = \beta\left(\dfrac{26\text{ mV}}{I_E}\right)$

or $\qquad \boxed{r_{ac} = \beta r_e} \qquad (5.30)$

Equation (5.30) has the same format as Eq. (5.22) ($Z_i \cong h_{fe} R_E$) used to reflect an emitter resistor to the base circuit. In this case Eq. (5.30) can be directly derived using the same equation [Eq. (5.22)] and Fig. 5.72a, where r_e appears as a resistor in the emitter leg.

For the situation of Fig. 5.72b,

$$\boxed{r_{ac} = \beta(r_e + R_E) \cong \beta R_E} \qquad (5.31)$$

The input circuit for the CE configuration is approximated, for the reasons discussed above, by the diode circuit appearing in Fig. 5.73, but the input impedance appears as βr_e in Fig. 5.74a since r_e is determined by I_E and not I_B. In Fig.

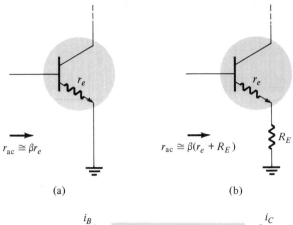

Figure 5.72 Determining r_{ac} for the CE configuration: (a) bypassed R_E; (b) unbypassed R_E.

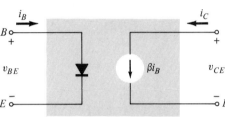

Figure 5.73 Approximate CE equivalent circuit.

5.71c the approximation was employed that β is the same throughout the device characteristics. This we know is absolutely untrue. However, its variation about the provided value for the typical application in the active region is assumed to be minimal and a fixed value a valid first approximation. For our analysis, we will consider it to be a constant at the provided value, resulting in the output equivalent circuits of Figs. 5.73 and 5.74a. From the equivalent circuits of Fig. 5.74 we can readily note that

$$\boxed{\begin{array}{c}\beta = h_{fe} \\ \beta r_e = h_{ie}\end{array}} \qquad (5.32)$$

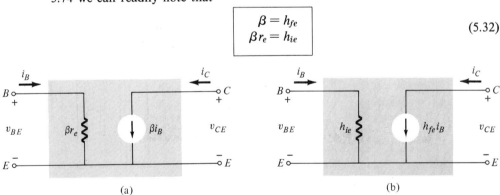

Figure 5.74 CE configuration: (a) alternate equivalent circuit; (b) approximate hybrid equivalent circuit.

For the various configurations examined in detail earlier the resulting equations can be quickly converted to a set having only β and r_e in place of the hybrid parameters using the relationships of Eq. (5.32). It is also important to keep in mind that Eq. (5.32) permits a direct determination of h_{ie} at bias points different from the provided condition.

SEC. 5.15 *AN ALTERNATIVE APPROACH*

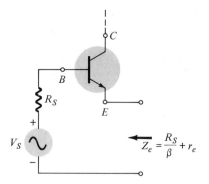

Figure 5.75 Emitter impedance Z_e.

For the case indicated in Fig. 5.75,

$$Z_e = \frac{R'_S}{h_{fe}} = \frac{R_S + h_{ie}}{h_{fe}} = \frac{R_S + \beta r_e}{\beta}$$

and

$$\boxed{Z_e = \frac{R_S}{\beta} + r_e} \qquad (5.33)$$

A few examples will clarify the use of the alternative CE equivalent circuit.

EXAMPLE 5.14 Determine A_v, A_i, Z_i, and Z_o for the network of Fig. 5.76.

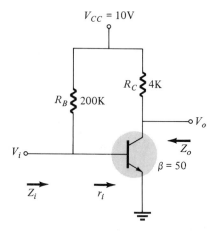

Figure 5.76 Network for Example 5.14.

Solution:

A_v: For a network such as shown in Fig. 5.76 I am sure we have reached the point where it should be unnecessary to redraw the network for each calculation. Therefore, for dc conditions:

$$I_B = \frac{V_{CC} - V_{BE}}{R_B} = \frac{10 - 0.7}{200\text{ K}} = \frac{9.3}{200\text{ K}} = 46.5\ \mu A$$

and

$$I_E \cong I_C = \beta I_B = 50(46.5 \times 10^{-6}) = 2.325\text{ mA}$$

so that

$$r_e = \frac{26\text{ mV}}{I_E} = \frac{26}{2.325} = 11.18\ \Omega$$

For ac conditions:
$$r_i = \beta r_e$$

and
$$I_b = \frac{V_i}{\beta r_e}$$

so that
$$V_o = -I_C R_C = \beta I_b R_C = \beta \left(\frac{V_i}{\beta r_e}\right) R_C = -\frac{R_C}{r_e} V_i$$

with
$$\boxed{A_v = \frac{V_o}{V_i} = -\frac{R_C}{r_e}} \qquad (5.34)$$

Substituting numbers; we get
$$A_v = -\frac{R_C}{r_e} = -\frac{4\text{ K}}{11.2} = \mathbf{357.14}$$

A_i:
$$R_B \| r_i = R_B \| \beta r_e \cong \beta r_e$$

therefore,
$$I_b \cong I_i$$

and
$$I_o = h_{fe} I_b = h_{fe} I_i$$

with
$$\boxed{A_i = \frac{I_o}{I_i} = h_{fe}} \qquad (5.35)$$

and
$$A_i = \mathbf{50}$$

Z_i:
$$\boxed{Z_i \cong \beta r_e} \qquad (5.36)$$
$$= (50)(11.2) = \mathbf{560\ \Omega}$$

Z_o:
$$\boxed{Z_o \cong R_C} \qquad (5.37)$$
$$= \mathbf{4\ K}$$

EXAMPLE 5.15 Determine A_v, A_i, Z_i, and Z_o for the network of Fig. 5.77.

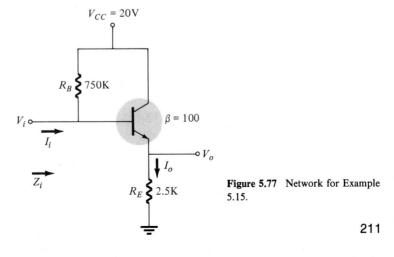

Figure 5.77 Network for Example 5.15.

Solution:
dc conditions:
From Section 5.7,

$$I_B = \frac{V_{CC} - V_{BE}}{R_B + \beta R_E} = \frac{20 - 0.7}{750\text{ K} + (100)(2.5\text{ K})} = \frac{19.3}{750\text{ K} + 250\text{ K}} = \frac{19.3}{1 \times 10^6}$$

$$= 19.3\ \mu A$$

$$I_E \cong I_C = \beta I_B = (100)(19.3 \times 10^{-6}) = 1.93\text{ mA}$$

and

$$r_e = \frac{26\text{ mV}}{I_E} = \frac{26}{1.93} = 13.47\ \Omega$$

ac conditions:
A reduced ac equivalent circuit appears in Fig. 5.78.

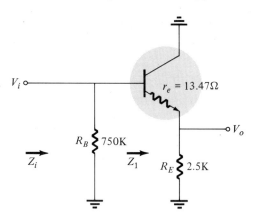

Figure 5.78 ac equivalent for the network of Fig. 5.77.

A_v: In a previous section we introduced the approximation $V_{be} \cong 0$ V when a network had an unbypassed emitter resistor. With this in mind $V_o = V_i$ and

$$\boxed{A_v \cong 1} \quad \text{(actually slightly less)} \quad (5.38)$$

A_i: The impedance Z_1:

$$Z_1 = \beta(r_e + R_E) = 100(13.47 + 2500)$$

Note here that r_e can realistically be ignored in comparison with R_E. Taking this approach, we get

$$Z_1 \cong \beta R_E = 100(2.5\text{ K}) = 250\text{ K}$$

and using Fig. 5.79, we get

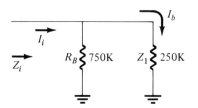

Figure 5.79 Determining the relationship between I_i and I_b.

$$I_b = \frac{(750\text{ K})(I_i)}{750\text{ K} + 250\text{ K}} = 0.75\, I_i$$

and
$$A_i = \frac{I_o}{I_i} = \left(\frac{I_b}{I_i}\right)\left(\frac{I_o}{I_b}\right) = (0.75)(\beta) = (0.75)(100) = \mathbf{75}$$

Z_i: From Fig. 5.79,

$$\boxed{Z_i \cong R_B \| \beta R_E} \qquad (5.39)$$

$$= 750\text{ K} \| 250\text{ K} = \mathbf{187.5\text{ K}}$$

Z_o: With V_i set to zero, R_B is effectively "shorted out" and in Eq. (5.33) $R_S = 0\,\Omega$. Therefore,

$$Z_e = \frac{R_S}{\beta} + r_e = 0 + 13.47 = 13.47\,\Omega$$

and
$$\boxed{Z_o = R_E \| r_e \cong r_e} = \mathbf{13.47\,\Omega} \qquad (5.40)$$

EXAMPLE: 5.16 Determine A_v, A_i, Z_i, and Z_o for the network of Fig. 5.80.

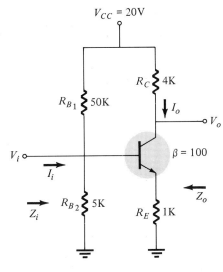

Figure 5.80 Network for Example 5.16.

Solution:
dc conditions:
From Section 5.6,

$$V_B \cong \frac{R_{B2}(V_{CC})}{R_{B2} + R_{B1}} = \frac{(5\text{ K})(20)}{5\text{ K} + 50\text{ K}} = \frac{5}{55}(20) = 1.818\text{ V}$$

and
$$I_E = \frac{V_B - V_{BE}}{R_E} = \frac{1.818 - 0.7}{1\text{ K}} = \frac{1.118}{1\text{ K}} = 1.118\text{ mA}$$

and
$$r_e = \frac{26\text{ mV}}{I_E} = \frac{26}{1.118} = 23.26\,\Omega$$

ac conditions:
The network is redrawn as shown in Fig. 5.81.

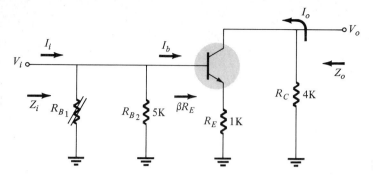

Figure 5.81 Network for Fig. 5.80.

A_v:

$$V_o = -I_o R_C = -I_c R_C = -I_e R_C$$

but

$$I_e = \beta I_b$$

with

$$I_b \cong \frac{V_i}{\beta R_E}$$

so that

$$I_e = \beta \left(\frac{V_i}{\beta R_E}\right) = \frac{V_i}{R_E}$$

and

$$V_o = -\left(\frac{V_i}{R_E}\right) R_C$$

with

$$\boxed{A_v = \frac{V_o}{V_i} = -\frac{R_C}{R_E}} \qquad (5.41)$$

The distinct advantage of this configuration is now obvious; it is β-independent. It is not concerned with the value of β ($= h_{fe}$), which will vary depending on the operating point and the particular transistor of a series used. As a consequence, however, there is a significant loss in gain with an unbypassed emitter resistor.

Substituting numbers, we get

$$A_v \cong -\frac{R_C}{R_E} = -\frac{4\text{ K}}{1\text{ K}} = -4$$

A_i: Since $R_{B_1} \| R_{B_2} \cong R_{B_2}$ as shown in Fig. 5.81,

$$I_b \cong \frac{R_{B_2} I_i}{R_{B_2} + \beta R_E}$$

and

$$I_o = I_c = \beta I_b = \beta \left(\frac{R_{B_2} I_i}{R_{B_2} + \beta R_E}\right)$$

with

$$A_i = \frac{I_o}{I_i} = \frac{\beta R_{B_2}}{R_{B_2} + \beta R_E} = \frac{R_{B_2}}{\frac{R_{B_2}}{\beta} + R_E}$$

$$\boxed{A_i \cong \frac{R_{B_2}}{R_E}} \qquad (5.42)$$

$$R_{B_1} \gg R_{B_2}$$

Substituting values, we get

$$A_i \cong \frac{5\text{ K}}{1\text{ K}} = 5$$

Z_i: From Fig. 5.81,

$$Z_i \cong R_{B_2} \| \beta R_E = 5\text{ K} \| 100\text{ K} \cong 5\text{ K}$$

and

$$\boxed{Z_i \cong R_{B_2}} \qquad (5.43)$$

Z_o: From Fig. 5.80,

$$Z_o = Z_{\text{trans.}} + Z_e \cong \infty\,\Omega + Z_e$$

and

$$\boxed{Z_o \cong \infty\,\Omega} \qquad (5.44)$$

For the following modifications (Fig. 5.82) of the network of Fig. 5.80 the results for A_v, A_i, Z_i, and Z_o are provided. The derivations will appear as exercises at the end of the chapter.

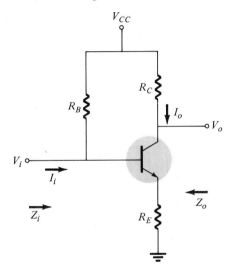

Figure 5.82

$$\boxed{A_v \cong -\frac{R_C}{R_E}} \qquad (5.45)$$

$$\boxed{A_i \cong \frac{\beta R_B}{R_B + \beta R_E}} \qquad (5.46)$$

$$\boxed{Z_i \cong R_B \| \beta R_E} \qquad (5.47)$$

$$\boxed{Z_o \cong \infty\,\Omega} \qquad (5.48)$$

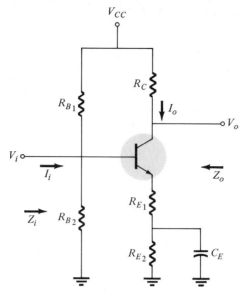

Figure 5.83

For the configuration of Fig. 5.83, the following results are obtained:

$$A_v = -\frac{R_C}{R_{E_1}} \quad (5.49)$$

$$A_i \cong \frac{R_{B_2}}{R_E} \bigg|_{R_{B_1} \gg R_{B_2}} \quad (5.50)$$

$$Z_i \cong R_{B_2} \| R_{E_1} \quad (5.51)$$

$$Z_o \cong \infty \; \Omega \quad (5.52)$$

5.16 SUMMARY TABLE

Table 5.2 (pp. 218–219) summarizes the various configurations examined in this chapter and a few that priorities will not permit covering here. The approximation h_{re}, $h_{oe} \cong 0$ has been applied in each case. The latter case of $1/h_{oe} \cong \infty \; \Omega$ is one that should be checked for each application. If $1/h_{oe}$ is close in magnitude to R_C, R_L, or their parallel combination $R_C \| R_L$, it cannot be dropped and its effects should be included with perhaps a hybrid equivalent circuit with just $h_{re} \cong 0$. If included, for the reasons above, it will cut both the voltage and current gain.

In addition, if a situation is encountered where R_E and r_e are comparable in value, then r_e cannot be dropped in equations such as $\beta(R_E + r_e)$. It must be included and the equations modified.

5.17 DARLINGTON COMPOUND CONFIGURATION

The Darlington circuit is a compound configuration that results in a set of improved amplifier characteristics. The configuration of Fig. 5.84 has a high input impedance with low output impedance and high current gain, all desirable characteristics for a current amplifier. We shall momentarily see, however, that the voltage gain will be less than 1 if the output is taken from the emitter terminal. A variation in the configuration can result in a trade-off between the output impedance and voltage gain.

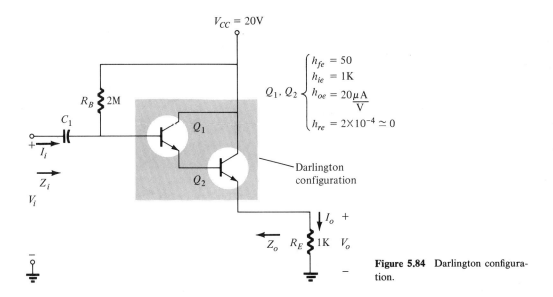

Figure 5.84 Darlington configuration.

The description of the biasing arrangement is similar to that of a single-stage emitter-follower configuration with current feedback. Note for the Darlington configuration that the emitter current of the first transistor is the base current for the second active device.

In its small-signal ac form the circuit will appear as shown in Fig. 5.85.

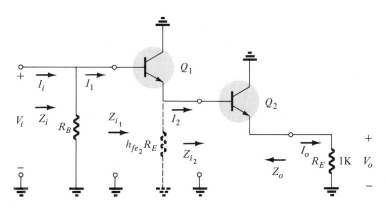

Figure 5.85 Darlington configuration of Fig. 5.84 redrawn to determine the small-signal ac response.

TABLE 5.2 Summary Table of Transistor Configurations (A_v, Z_i, Z_o)

(In each case $R'_L = R_L \| R_C$)	A_v	Z_i	Z_o
[Common-emitter, fixed bias: V_{CC}, R_B, R_C, R_L]	$-\dfrac{h_{fe}}{h_{ie}} R'_L$ $-\dfrac{R'_L}{r_e}$	$R_B \| h_{ie}$ $R_B \| \beta r_e$	R'_L
[Common-emitter, voltage-divider bias with bypassed R_E: V_{CC}, R_{B_1}, R_{B_2}, R_C, R_E, C_E, R_L]	$-\dfrac{h_{fe} R'_L}{h_{ie}}$ $-\dfrac{R'_L}{r_e}$	$R_{B_1} \| R_{B_2} \| h_{ie}$ $R_{B_1} \| R_{B_2} \| \beta r_e$	R'_L
[Emitter follower: V_{CC}, R_{B_1}, R_C, R_{B_2}, R_E, R_L, R_S]	$\cong 1$	$(R'_E = R_E \| R_L)$ $R_{B_1} \| R_{B_2} \| (h_{ie} + h_{fe} R'_E)$ $R_{B_1} \| R_{B_2} \| \beta(r_e + R'_E)$	$R'_S = R_S \| R_{B_1} \| R_{B_2}$ $R'_E \| \dfrac{R_S + h_{ie}}{h_{fe}}$ $R'_E \| \left(\dfrac{R'_S}{\beta} + r_e\right)$
[Common-base: V_{EE}, R_E, V_{CC}, R_C, R_L]	$-\dfrac{h_{fb}}{h_{ib}} R'_L$ $\cong -\dfrac{R'_L}{r_e}$	$R_E \| h_{ib}$ $R_E \| r_e$	R'_L

TABLE 5.2 (continued)

(In each case $R'_L = R_L \| R_C$)	A_v	Z_i	Z_o
Circuit 1: Voltage-divider bias with unbypassed R_E ($R_{B_1}, R_{B_2}, R_C, R_E, R_L$)	$-\dfrac{R'_L}{R_E}$	$R_{B_1} \| R_{B_2} \| (h_{ie} + h_{fe}R_E)$ $R_{B_1} \| R_{B_2} \| \beta(r_e + R_E)$	R'_L
Circuit 2: Fixed bias with R_{E_1} unbypassed, R_{E_2} bypassed by C_E	$-\dfrac{R'_L}{R_{E_1}}$	$R_B \| (h_{ie} + h_{fe}R_{E_1})$ $R_B \| \beta(r_e + R_{E_1})$	R'_L
Circuit 3: Collector-feedback bias with R_F	$-\dfrac{h_{fe}}{h_{ie}}R'_L$ $-\dfrac{R'_L}{r_e}$	$\dfrac{R_F}{|A_v|} \| h_{ie}$ $\dfrac{R_F}{|A_v|} \| \beta r_e$	$\cong R'_L$
Circuit 4: Collector-feedback with unbypassed R_E	$-\dfrac{R'_L}{R_E}$	$\dfrac{R_F}{|A_v|} \| h_{fe}R_E$ $\dfrac{R_F}{|A_v|} \| \beta R_E$	$\cong R'_L$

For the second stage:
$$Z_{i_2} \cong h_{fe_2} R_E$$

and
$$A_{i_2} = \frac{I_o}{I_2} = \frac{I_{e_2}}{I_{b_2}} \cong h_{fe_2}$$

On a *good* approximate basis, these equations cannot be applied to the first stage. The "fly in the ointment" is the closeness with which Z_{i_2} compares with $1/h_{oe_1}$. You will recall that $1/h_{oe_1}$ could be eliminated in the majority of situations because the load impedance $Z_L \ll 1/h_{oe_1}$. For the Darlington configuration the input impedance Z_{i_2} is close enough in magnitude to $1/h_{oe_1}$ to necessitate considering the effects of h_{oe_1}. It can be shown that for the single-stage grounded emitter transistor amplifier

$$A_i \cong \frac{h_{fe}}{1 + h_{oe} Z_L}$$

Applying the equation above to this situation, $Z_L = Z_{i_2} \cong h_{fe_2} R_E$ and

$$A_{i_1} = \frac{I_2}{I_1} = \frac{I_{c_1}}{I_{b_1}} \cong \frac{h_{fe_1}}{1 + h_{oe_1}(h_{fe_2} R_E)}$$

with
$$A_i = \frac{I_o}{I_1} = A_{i_1} A_{i_2} = \frac{h_{fe_1} h_{fe_2}}{1 + h_{oe_1}(h_{fe_2} R_E)} \quad (5.53)$$

For $h_{fe_1} = h_{fe_2}$ and $h_{oe_1} = h_{oe_2} = h_{oe}$,

$$A_i \cong \frac{h_{fe}^2}{1 + h_{oe} h_{fe} R_E} \quad (5.54)$$

For $h_{oe} h_{fe} R_E \leq 0.1$ a fairly good approximation (within 10%) is

$$A_i \cong h_{fe}^2 = \beta^2 \quad (5.55)$$

The current gain $A_{i_T} = I_o/I_i$, as defined by Fig. 5.85 can be determined through the use of the current divider rule:

$$I_1 = \frac{R_B I_i}{R_B + Z_{i_1}}$$

Since $Z_{i_2} \cong h_{fe_2} R_E$ is the "emitter resistor" of the first stage (note Fig. 5.85), the input impedance to the first stage is $Z_{i_1} \cong h_{fe_1}(Z_{i_2} \| 1/h_{oe_1})$ since $Z_{i_2} = h_{fe_2} R_{E_1}$, and $1/h_{oe_1}$ will appear in parallel in the small-signal equivalent circuit. The result is

$$Z_{i_1} \cong h_{fe_1} \left(h_{fe_2} R_E \bigg\| \frac{1}{h_{oe_1}} \right) = \frac{h_{fe_1} h_{fe_2} R_E (1/h_{oe_1})}{h_{fe_2} R_E + 1/h_{oe_1}}$$

and
$$Z_{i_1} = \frac{h_{fe_1} h_{fe_2} R_E}{h_{oe_1} h_{fe_2} R_E + 1} \quad (5.56)$$

which for $h_{fe_1} = h_{fe_2} = h_{fe}$ and $h_{oe_1} = h_{oe_2} = h_{oe}$ yields

$$\boxed{Z_{i_1} \cong \frac{h_{fe}^2 R_E}{1 + h_{oe} h_{fe} R_E}} \qquad (5.57)$$

For $h_{oe} h_{fe} R_E \le 0.1$

$$\boxed{Z_{i_1} \cong h_{fe}^2 R_E = \beta^2 R_E} \qquad (5.58)$$

Substituting the parameter values

$$h_{fe_1} = h_{fe_2} = h_{fe} = 50$$

with

$$h_{oe_1} = h_{oe_2} = h_{oe} = 20 \ \mu A/V$$

$$A_i = \frac{I_o}{I_1} \cong \frac{(h_{fe})^2}{1 + h_{oe} h_{fe} R_E} = \frac{(50)^2}{1 + (20 \times 10^{-6})(50)(1 \text{ K})}$$

$$= \frac{2500}{1+1} = \mathbf{1250}$$

and

$$Z_{i_1} \cong \frac{h_{fe}^2 R_E}{1 + h_{oe} h_{fe} R_E} = \frac{(50)^2 (1 \text{ K})}{2} = 1250 \text{ K} = \mathbf{1.25 \text{ M}}$$

so that for $R_B = 2 \text{ M}$,

$$\frac{I_1}{I_i} = \frac{R_B}{R_B + Z_{i_1}} = \frac{2 \text{ M}}{2 \text{ M} + 1.25 \text{ M}} = \frac{2}{3.25} = 0.615$$

and

$$A_{i_T} = \frac{I_o}{I_i} = \left[\frac{I_o}{I_1}\right]\left[\frac{I_1}{I_i}\right] = A_i \times \frac{I_1}{I_i}$$

$$= (1250)(0.615) = \mathbf{770}$$

with

$$Z_i = 2 \text{ M} \| Z_{i_1} = 2 \text{ M} \| 1.25 \text{ M} = \mathbf{770 \text{ K}}$$

Too frequently, the current gain of a Darlington circuit is assumed to be simply $A_i \cong h_{fe}^2$ without any regard to the output impedance $1/h_{oe}$. In this case $A_i \cong (h_{fe})^2 = 2500$. Certainly, 2500 versus 1250 is *not* a good approximation. The effect of h_{oe_1} must therefore be considered when the current gain of the first stage is determined.

The output impedance Z_o can be determined directly from the emitter-equivalent circuits, as follows.

For the first stage

$$\boxed{Z_{o_1} \cong \frac{R_{s_1} + h_{ie_1}}{h_{fe_1}}} \qquad (5.59)$$

$$= \frac{0 + 1 \text{ K}}{50} \cong \mathbf{20.0 \ \Omega}$$

and
$$\boxed{Z_{o_2} \cong \frac{(Z_{o_1} \| 1/h_{oe_1}) + h_{ie_2}}{h_{fe_2}}} \tag{5.60}$$

$$= \frac{(20.0 \| 50\text{ K}) + 1\text{ K}}{50} \cong \frac{20.0 + 1\text{ K}}{50} = \frac{1020}{50}$$

$$\cong 20.4\ \Omega$$

Note, as indicated in the introductory discussion, that the input impedance is high, output impedance low, and current gain high. We shall now examine the voltage gain of the system. Applying Kirchhoff's voltage law to the circuit of Fig. 5.84:

$$V_o = V_i - V_{be_1} - V_{be_2}$$

That the output potential is the input *less* the base-to-emitter potential of each transistor clearly indicates that $V_o < V_i$. It is closer in magnitude to 1 than to zero. On an approximate basis it is given by

$$\boxed{A_v \cong \frac{1}{1 + \dfrac{h_{ie_2}}{h_{fe_2} R_E}}} \tag{5.61}$$

as derived from the emitter equivalent circuit.

Substituting the numerical values of this general example, we get

$$A_v \cong \frac{1}{1 + \dfrac{1\text{ K}}{50\text{ K}}} = \frac{1}{1 + 0.2} = 0.98$$

The ratings and characteristics for a 10-A RCA *npn* Darlington power transistor are provided in Fig. 5.86. Some of the characteristics appear in Figs. 5.87 through 5.92. The data provided are for the complete device—the individual β values are not provided. Note in the characteristics that the level of V_{BE} is increased since it includes the drop across two transistors. Consider also that the minimum value of h_{fe} is 1000 at 1 kHz but drops to only 20 at 1 MHz. Frequency will obviously have a pronounced effect on its performance. Note in Fig. 5.87 that the collector current is in amperes and that a pulsed operation results in an increased level of current—the longer the pulse, the lower the permitted current. In Fig. 5.88 we note a drop in power rating starting with room temperature, and in Fig. 5.89 we find that the dc β is very sensitive to collector current. The effect of frequency on the ac gain is more carefully defined in Fig. 5.90. It begins a severe drop at about 0.1 MHz or 100 KHz. Note in Fig. 5.91 the increased level of V_o due to the two transistors and the base current in mA in Fig. 5.92.

As demonstrated by the dual Darlington array in Fig. 5.93, this compound configuration is also available in the IC package. Note the availability of two in the same package and the fact that the total h_{fe} for either pair is not simply the h_{fe} of either one squared. That is,

$$h_{fe}(\text{pair}) = 1300 \neq [h_{fe}(\text{single})]^2 = (82)^2 = 6724$$

Power Transistors

2N6383 2N6384 2N6385

JEDEC TO-3

10-Ampere, N-P-N Darlington Power Transistors

40-60-80 Volts, 100 Watts
Gain of 1000 at 5 A

TERMINAL CONNECTIONS
- Pin 1 - Base
- Pin 2 - Emitter
- Case - Collector
- Mounting Flange - Collector

Features:
- Operates from IC without predriver
- Low leakage at high temperature
- High reverse second-breakdown capability

Applications:
- Power switching
- Audio amplifiers
- Hammer drivers
- Series and shunt regulators

The 2N6383, 2N6384, and 2N6385• are monolithic n-p-n silicon Darlington transistors designed for low- and medium-frequency power applications. The double epitaxial construction of these devices provides good forward and reverse second-breakdown capability. Their high gain makes it possible for them to be driven directly from integrated circuits.

•Formerly RCA Dev. Nos. TA8349, TA8486, and TA8348.

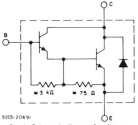

Fig. 1—Schematic diagram for all types.

MAXIMUM RATINGS, *Absolute-Maximum Values:*

		2N6385	2N6384	2N6383	
*	COLLECTOR-TO-BASE VOLTAGE	V_{CBO}	80	60	40	V
	COLLECTOR-TO-EMITTER VOLTAGE:					
	With external base-to-emitter resistance (R_{BE}) = 100Ω, sustaining	V_{CER}(sus)	80	60	40	V
	With base open, sustaining	V_{CEO}(sus)	80	60	40	V
*	With base reverse-biased V_{BE} = −1.5 V, R_{BB} = 100Ω	V_{CEX}	80	60	40	V
*	EMITTER-TO-BASE VOLTAGE	V_{EBO}	5	5	5	V
	COLLECTOR CURRENT:	I_C				
*	Continuous		10	10	10	A
	Peak		15	15	15	A
*	CONTINUOUS BASE CURRENT	I_B	0.25	0.25	0.25	A
*	TRANSISTOR DISSIPATION	P_T				
	At case temperatures up to 25°C		100	100	100	W
	At case temperatures above 25°C		←——— See Fig. 5.88 ———→			
*	TEMPERATURE RANGE:					
	Storage and Operating (Junction)		←——— −65 to +200 ———→			°C
*	PIN TEMPERATURE (During Soldering):					
	At distances ≥ 1/32 in. 0.8 mm from seating plane for 10 s max.		←——— 235 ———→			°C

*In accordance with JEDEC registration data format JS-6 RDF-2.

Figure 5.86 RCA NPN Darlington power transistors. (Courtesy RCA Solid State Division.)

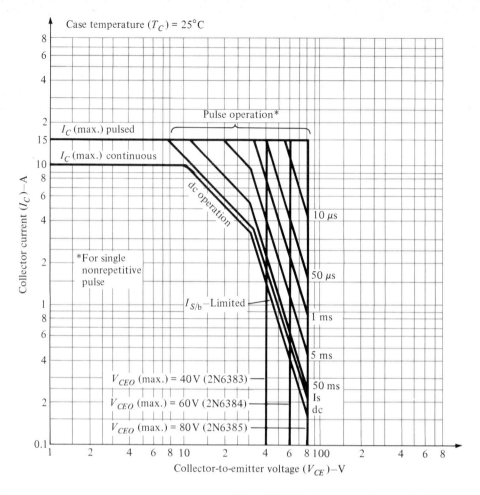

Figure 5.87 Maximum operating area for all types.

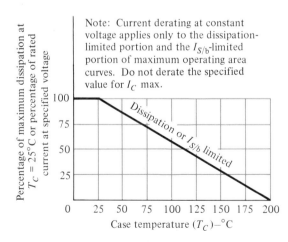

Figure 5.88 Derating curve for all types.

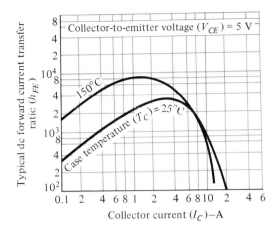

Figure 5.89 Typical dc-beta characteristics for all types.

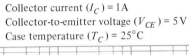

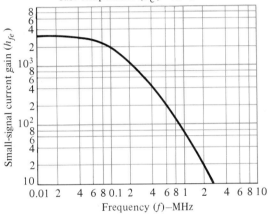

Figure 5.90 Typical small-signal gain for all types.

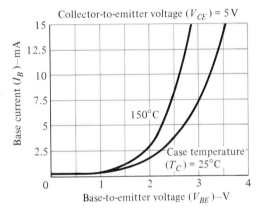

Figure 5.91 Typical input characteristics for all types.

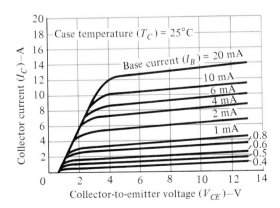

Figure 5.92 Typical output characteristics for all types.

CA3036
DUAL DARLINGTON ARRAY

- Two independent low-noise wide-band amplifier channels
- Particularly useful for preamplifier and low-level amplifier applications in single-channel and stereo systems
- Wide application in low-noise industrial instrumentation amplifiers
- Hermetically sealed, all-welded 10-lead TO-5-style metal package

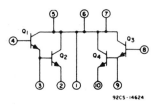

Fig. 1 — Schematic Diagram for CA3036.

ELECTRICAL CHARACTERISTICS, at $T_A = 25°C$

	CHARACTERISTICS	SYMBOLS	TEST CONDITIONS	LIMITS TYPE CA3036 Min.	Typ.	Max.	UNITS
For Each Transistor (Q_1, Q_2, Q_3, Q_4)	Collector-Cutoff Current	I_{CBO}	$V_{CB}=5V$, $I_E=0$	–	–	0.5	μA
	Collector-Cutoff Current	I_{CEO}	$V_{CE}=10V$, $I_B=0$	–	–	5	μA
	Collector-to-Emitter Breakdown Voltage	$V_{(BR)CEO}$	$I_C=1$ mA, $I_B=0$	15	20	–	V
	Collector-to-Base Breakdown Voltage	$V_{(BR)CBO}$	$I_C=10 \mu A$, $I_E=0$	30	44	–	V
	Emitter-to-Base Breakdown Voltage	$V_{(BR)EBO}$	$I_E=10 \mu A$, $I_C=0$	5	6	–	V
For Either Input Transistor (Q_1 or Q_3)	Static Forward Current-Transfer Ratio	h_{FE}	I_{C1} or $I_{C3}=1$ mA	30	82	–	–
For Either Darlington Pair (Q_1, Q_2 or Q_3, Q_4)	Emitter-to-Base Breakdown Voltage	$V_{(BR)EBO(D)}$	I_{E2} or $I_{E4}=10 \mu A$	10	12.6	–	V
	Static Forward Current-Transfer Ratio	$h_{FE(D)}$	$I_{C1}+I_{C2}$ or $I_{C3}+I_{C4}$ = 1 mA	1000	4540	–	–
For Each Input Transistor (Q_1 or Q_3)	Short-Circuit Forward Current-Transfer Ratio	h_{fe}	$f=1$ kHz, I_{C1} or $I_{C3}=1$ mA	–	82	–	–
	Short-Circuit Input Impedance	h_{ie}		–	2.6K	–	Ω
	Open-Circuit Output Admittance	h_{oe}		–	7	–	μmho
	Open-Circuit Reverse Voltage-Transfer Ratio	h_{re}		–	9.8×10^{-5}	–	–
For Either Darlington Pair (Q_1, Q_2 or Q_3, Q_4)	Short-Circuit Forward Current-Transfer Ratio	$h_{fe(D)}$	$f=1$ kHz, $I_{C1}+I_{C2}$ or $I_{C3}+I_{C4}$ = 1 mA	–	1300	–	–
	Short-Circuit Input Impedance	$h_{ie(D)}$		–	82K	–	Ω
	Open-Circuit Output Admittance	$h_{oe(D)}$		–	108	–	μmho
	Open-Circuit Reverse Voltage-Transfer Ratio	$h_{re(D)}$		–	2.7×10^{-3}	–	–
	Voltage Gain	$A_{(D)}$		–	26	–	dB
	Power Gain	$G_{p(D)}$		–	47	–	dB
	Noise Voltage See Fig. 3 for Test Circuit	E_N	$f=100$ Hz	–	0.2	3	μV(rms) / $\sqrt{f(Hz)}$
			$f=1$ kHz	–	0.05	0.3	
			$f=10$ kHz	–	0.012	0.1	
For Either Input Transistor (Q_1 or Q_3)	Forward Transfer Admittance	y_{fe}	$f=50$ MHz, I_{C1} or $I_{C3}=2$ mA	–	$0.68+j7.9$	–	mmho
	Input Admittance (Output Short-Circuited)	y_{ie}		–	$4.14+j5.95$	–	mmho
	Output Admittance (Input Short-Circuited)	y_{oe}		–	$1.94+j2.64$	–	mmho
	Reverse Transfer Admittance (Input Short-Circuited)	y_{re}		–	Negligible	–	mmho
For either Darlington Pair (Q_1, Q_2 or Q_3, Q_4)	Input Admittance (Output Short-Circuited)	$y_{ie(D)}$	$f=50$ MHz, $I_{C1}+I_{C2}$ or $I_{C3}+I_{C4}$ = 2 mA	–	$1.71+j2.8$	–	mmho
	Output Admittance (Input Short-Circuited)	$y_{oe(D)}$		–	$3.96+j2.6$	–	mmho
	Gain-Bandwidth Product	f_T		150	200	–	MHz

HIGHLIGHTS
- Matched transistors with emitter-follower outputs
- Low-noise performance
- 200-MHz gain-bandwidth product
- Operation from $-55°C$ to $+125°C$

APPLICATIONS
- Stereo phonograph preamplifiers
- Low-level stereo and single channel amplifier stages
- Low-noise, emitter-follower differential amplifiers
- Operational amplifier drivers

MAXIMUM RATINGS, Absolute-Maximum Values:
POWER DISSIPATION, P:
  Any one transistor . . . . 300 max. mW
  Total for array . . . . 600 max. mW
TEMPERATURE RANGE:
  Operating . . . . -55 to $+125$ °C
  Storage . . . . -65 to $+150$ °C
LEAD TEMPERATURE (During Soldering):
  At distance $1/16 \pm 1/32$ inch $(1.59 \pm 0.79$ mm$)$
  from case for 10 seconds max. . . . . . . +265 °C
The following ratings apply for each transistor in the array:
  Collector-to-Emitter Voltage, V_{CEO} . . . . 15 max. V
  Collector-to-Base Voltage, V_{CBO} . . . . . 30 max. V
  Emitter-to-Base Voltage, V_{EBO} . . . . . . 5 max. V
  Collector Current, I_C . . . . 50 max. mA

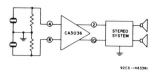

Fig. 2 - Block Diagram of Stereo System using CA3036 as Phono Preamplifier.

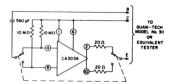

Fig. 3 - Noise Voltage Test Circuit for CA3036.

Figure 5.93 RCA dual Darlington array. (Courtesy RCA Solid State Division.)

In this case, the hybrid parameters and admittance parameters are provided. The admittance parameters, not presented in this book, simply represent the parameters of another equivalent network for the transistor. Note also the high value of input impedance for the pairs and the low value of the equivalent $1/h_{oe} = 9.3$ K. The dB gain in power is defined by Eq. (5.62):

$$A_{p(dB)} = 10 \log_{10} \frac{P_o}{P_i} \qquad (5.62)$$

and the voltage gain by

$$A_{v(dB)} = 20 \log_{10} \frac{V_o}{V_i} \qquad (5.63)$$

PROBLEMS

§ 5.3

1. (a) Calculate the dc bias voltages and currents for the *pnp* common-base bias circuit of Fig. 5.2. Assume transistor values of $\alpha = 0.985$, $V_{EB} = 0.7$ V. The circuit components are $R_E = 1$ K, $R_C = 3.9$ K, $V_{CC} = 12$ V, and $V_{EE} = 1.5$ V.
   (b) Sketch the load line for the characteristics of Fig. 5.2.
   (c) Find the Q-point and discuss whether you think it was a good choice.

2. Calculate the collector-base voltage for the *npn* common-base bias circuit of Fig. 5.4 if $R_E = 1.8$ K, $R_C = 2.7$ K, $V_{EE} = 9$ V, $V_{CC} = 22$ V, $\alpha = 0.995$, and $V_{BE} = 0.7$ V.

§ 5.4

3. For a fixed-bias common-emitter circuit, as in Fig. 5.5, calculate the bias currents and voltages for the following circuit values: $R_B = 150$ K, $R_C = 2.1$ K, $V_{CC} = 9$ V, $V_{BE} = +0.7$ V, and $\beta = 45$.

4. Using an *npn* fixed-bias circuit as in Fig. 5.7, calculate the collector-emitter bias voltage (V_{CE}) for the following circuit values: $R_B = 250$ K, $R_C = 1.8$ K, $V_{CC} = 12$ V, $V_{BE} = 0.7$ V, and $\beta = 70$.

§ 5.5

5. Calculate the dc bias voltages and currents for an emitter-stabilized bias circuit as in Fig. 5.8 for the following circuit values: $R_B = 47$ K, $R_E = 750$ Ω, $R_C = 0.5$ K, $V_{BE} = 0.7$ V, $\beta = 55$, and $V_{CC} = 18$ V.

6. Calculate the collector-emitter voltage (V_{CE}) for an *npn*, emitter-stabilized bias circuit as in Fig. 5.10 for the following circuit values: $R_B = 75$ K, $R_C = 0.5$ K, $R_E = 470$ Ω, $V_{BE} = 0.7$ V, $V_{CC} = 10$ V, $\beta = 80$, $C_E = 50$ μF, and $C_1 = C_2 = 10$ μF.

§ 5.6

7. Calculate the bias voltages and currents for a circuit as in Fig. 5.11 for the following

circuit values: $R_{B_1} = 56$ K, $R_{B_2} = 4.7$ K, $R_E = 750$ Ω, $R_C = 6.8$ K, $V_{CC} = 24$ V, $V_{BE} = 0.7$ V, and $\beta = 55$.

8. Calculate the collector voltage V_C for a bias circuit as in Fig. 5.11 for the following circuit values: $R_{B_1} = 12$ K, $R_{B_2} = 1.5$ K, $R_E = 1$ K, $R_C = 4.7$ K, $V_{CC} = 9$ V, $V_{BE} = +0.7$ V, and $\beta = 75$.

§ 5.7

9. For an emitter-follower circuit as in Fig. 5.14, calculate the dc bias currents and voltages for the following circuit values: $R_B = 240$ K, $R_E = 1.8$ K, $V_{CC} = 9$ V, $V_{BE} = 0.7$ V, and $\beta = 85$.

10. For an emitter-follower circuit as in Fig. 5.14, calculate the dc voltage across the emitter resistor for the following circuit values: $R_B = 91$ K, $R_E = 1.2$ K, $V_{CC} = 25$ V, $V_{BE} = 0.7$ V, and $\beta = 60$.

11. Calculate the emitter voltage (with respect to ground) for a bias circuit as in Fig. 5.16 for the following circuit values: $R_{B_1} = 270$ K, $R_{B_2} = 33$ K, $R_E = 4.7$ K, $V_{CC} = 15$ V, $V_{BE} = 0.7$ V, and $\beta = 50$.

§ 5.8

12. For a fixed-bias circuit such as that of Fig. 5.17 with circuit values of $R_B = 80$ K, $R_C = 2.5$ K, $V_{CC} = 20$ V, and $V_{BE} = 0.7$ V and transistor collector characteristic as shown in Fig. 5.94, do the following:

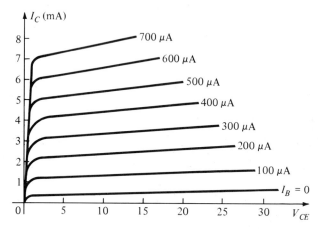

Figure 5.94

(a) Draw the dc load line.
(b) Obtain the quiescent operating point (Q-point).
(c) Find the operating point if R_C is changed to 10 K.
(d) Find the operating point if V_{CC} is changed to 15 V (R_C is 2.5 K).

13. Determine graphically the operating point for a fixed-bias circuit using a *pnp* transistor having a collector characteristic as in Fig. 5.95. The circuit values are as follows: $R_B = 150$ K, $R_C = 2$ K, $V_{CC} = -20$ V, and $V_{BE} = -0.3$ V.

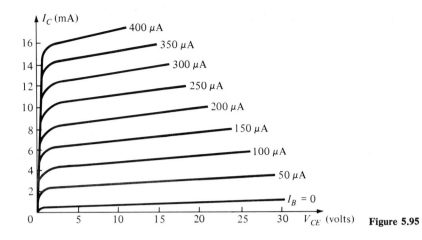

Figure 5.95

§ 5.9

14. (a) From memory, sketch a common-emitter *pnp* transistor configuration and insert the proper polarities for the required biasing potentials.
    (b) Repeat part (a) for an *npn* common-base configuration.

§ 5.11

15. (a) Determine the hybrid parameters for the network of Fig. 5.96.

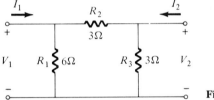

Figure 5.96

   (b) Sketch the hybrid equivalent circuit.

16. Sketch the complete hybrid equivalent circuit for the common-collector configuration and indicate the current directions as shown in Fig. 5.30a and b.

§ 5.12

17. For a change in I_C from 1 to 20 mA, which hybrid parameter exhibits the greatest change in Fig. 5.31? Which exhibits the least change?

18. For the range of V_{CE} from 1 to 50 V, which parameter in Fig. 5.32 exhibits the greatest change in value? Which exhibits the least change?

19. In Fig. 5.33, which parameter exhibits the least sensitivity to change in temperature? Which exhibits the most sensitivity?

§ 5.13

20. For the network of Fig. 5.97, determine the following:

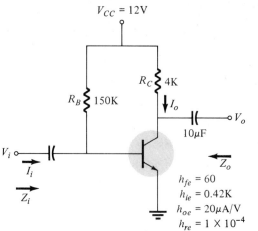

Figure 5.97

(a) Current gain $A_i = I_o/I_i$.
(b) Voltage gain $A_v = V_o/V_i$.
(c) Input impedance Z_i.
(d) Output impedance Z_o.
(e) Power gain A_p.
Apply any approximations that you feel are appropriate.

21. Repeat Problem 20 for the network of Fig. 5.98.

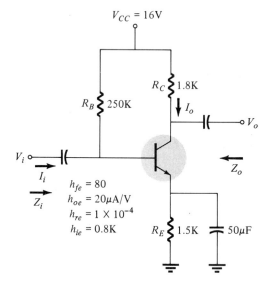

Figure 5.98

22. Repeat Problem 20 for the network of Fig. 5.99.

23. Repeat Problem 20 for the network of Fig. 5.100.

24. (a) Using an approximate equivalent circuit, determine the current gain $A_i = I_o/I_i$ and voltage gain $A_v = V_o/V_i$ for the network of Fig. 5.101.
    (b) Determine Z_i and Z_o.
    (c) Determine $A_v = V_o/V_s$.

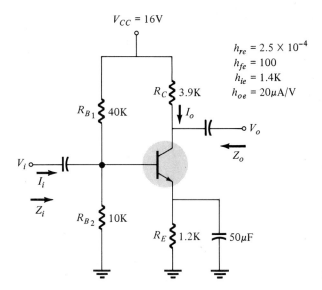

Figure 5.99

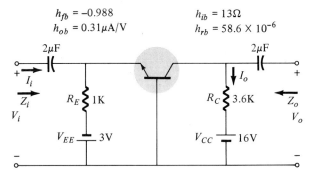

Figure 5.100

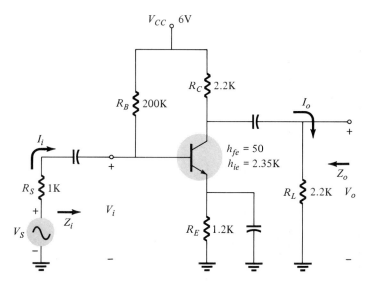

Figure 5.101

§ 5.14

25. Determine the input impedance Z_i, the output impedance Z_o, and the voltage gain $A_v = V_o/V_i$ for the network of Fig. 5.102.

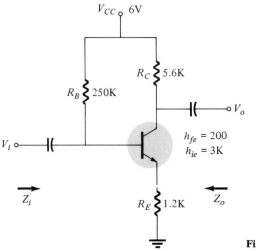

Figure 5.102

26. Determine Z_i, Z_o, A_v, and A_i for the network of Fig. 5.103.

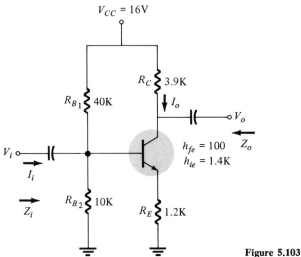

Figure 5.103

27. Determine Z_i, Z_o, A_v, and A_i for the network of Fig. 5.104.

28. Determine V_o (in terms of V_1 and V_2) for the network of Fig. 5.105.

§ 5.15

29. Using the approach introduced in Section 5.15 (use $r_B = 1.25 \, \Omega$), repeat Problem 20.

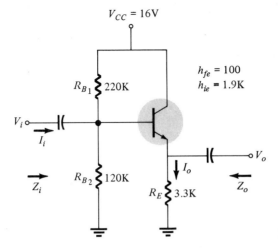

Figure 5.104

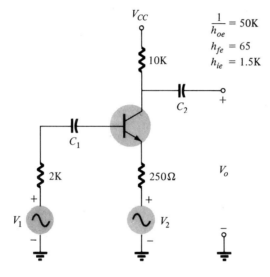

Figure 5.105

30. Using the approach introduced in Section 5.15 (use $r_B = 1.7\ \Omega$), repeat Problem 23.
31. Using the approach introduced in Section 5.15 (use $r_B = 1.1\ \Omega$), repeat Problem 25.
32. Using the approach introduced in Section 5.15 (use $r_B = 0.5\ \Omega$), repeat Problem 26.
33. Using the approach introduced in Section 5.15 (use $r_B = 1.7\ \Omega$), repeat Problem 27.
34. Determine A_v, A_i, Z_i, and Z_o for the network of Fig. 5.106.
35. Determine A_v, A_i, Z_i, and Z_o for the network of Fig. 5.107.
36. Derive the expressions for the network of Fig. 5.82.
37. Derive the expressions for the network of Fig. 5.83.

§ **5.17**

38. Determine A_i, Z_i, Z_o, and A_v for the Darlington configuration of Fig. 5.108.

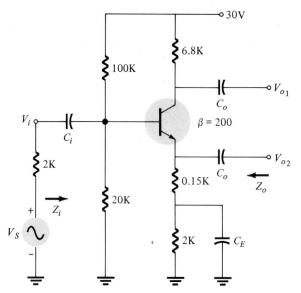

Figure 5.106

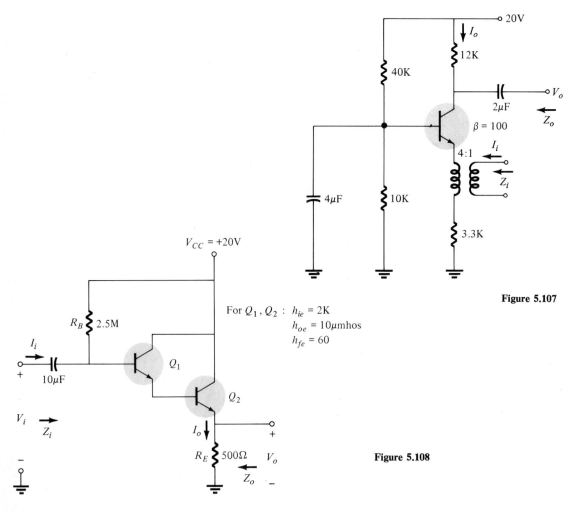

Figure 5.107

For Q_1, Q_2: $h_{ie} = 2K$
$h_{oe} = 10\mu mhos$
$h_{fe} = 60$

Figure 5.108

39. Repeat Problem 38 if a collector resistor of 2.2 K is added between the collector of Q_1 and V_{CC} and the output is taken off the collector of the Darlington configuration. I_o is the current through the added 2.2-K resistor.

40. Repeat Problem 38 if R_E is changed to 150 Ω.

41. Repeat Problem 39 if R_E is changed to 150 Ω.

42. (a) Referring to Fig. 5.87, determine the maximum collector current for I_C (continuous), if V_{CE} is 20 V.
    (b) Repeat part (a) for a 1-ms pulse duration with $V_{CE} = 30$ V.

43. (a) What is the maximum power dissipation from Fig. 5.88 at 125°C?
    (b) What is the derating factor?

44. (a) What is the maximum value of h_{FE} from Fig. 5.89 at a case temperature of 25°C?
    (b) What is the minimum value in the range 0.1 A to 10 A?
    (c) Is the difference significant between the results of parts (a) and (b)? Is the temperature stability of a system an important consideration?

45. (a) Referring to Fig. 5.90, what frequency range would appear to have a fairly level value of h_{fe}?
    (b) At what frequency will h_{fe} drop to 50% of its value at 10 kHz?

46. Referring to Fig. 5.91, what is a reasonable value for V_o at a case temperature of 25°C? Why is it different from the typical single silicon transistor unit?

47. If the maximum power dissipation is 70 W at a case temperature of 25°C, sketch the permissible region of operation on Fig. 5.92.

GLOSSARY

Transistor Biasing   Establishing the desired quiescent point through the proper choice of dc supplies and biasing resistors.

Operating Region   Defined by the maximum current and voltage ratings, maximum power curve, cutoff, and saturation region.

Small-Signal Analysis   The analysis of electronic systems that deals with very small levels of signals—that is, very small peak-to-peak variations, when compared to the scale used in the device terminal characteristics.

Two-Port Theory   The analysis of systems having two ports (pairs of terminals) through which the behavior and characteristics of the system can be defined.

Hybrid Parameters   A set of mixed parameters that can be used to model a two-port system.

Short-Circuit Input Impedance Parameter (h_i*)*   Defined by the ratio of the input voltage to the input current with the output voltage set to zero—measured in ohms.

Open-Circuit Reverse-Transfer Voltage Ratio Parameter (h_r*)*   Defined by the ratio of the input voltage to the output voltage with the input current set to zero—unitless.

Short-Circuit Forward-Transfer Current Ratio Parameter (h_f) Defined by the ratio of the output current to the input current with the output voltage set to zero—unitless.

Open-Circuit Output Conductance Parameter (h_o) Defined by the ratio of the output current to the output voltage with the input current equal to zero—measured in mhos.

Hybrid Equivalent Circuit A small-signal model for the transistor using a set of hybrid parameters defined by the operating conditions.

Darlington Configuration A compound configuration employing two transistors that will result in an improved level of output impedance, input impedance, and current gain.

CHAPTER 6

Field-Effect Transistors

6.1 INTRODUCTION

Field-effect transistors used in both discrete devices and integrated circuits are built either as junction FETs (JFETs) or as metal-oxide semiconductor FETs (MOSFETs). Either type can be built as *n*-channel (conduction of electrons) or *p*-channel (conduction of holes). FETs are voltage-controlled devices, allowing current to flow through the device as controlled by an applied voltage.

A FET is constructed quite simply, one reason for its great popularity. A length of either *n*- or *p*-type doped material forms a channel through which electrons move. The movement is that of free electrons in *n*-channel and holes in *p*-channel devices. Voltage applied to a gate terminal controls the passage of this electric charge through the channel, thereby providing control of the FET device operation.

The gate element at which the control voltage is applied is isolated from the channel through which the current occurs, providing quite high impedance at the gate. This high impedance is one of the important features of the FET. Many instruments specifically use the FET device at the input end (or front end) because of its high input impedance.

6.2 JFET CONSTRUCTION

The physical structure and basic operation of a junction-field-effect transistor (JFET) can be described using the representation of Fig. 6.1. As shown, the source and drain are terminals of the device connected by a path or channel of *n*-type

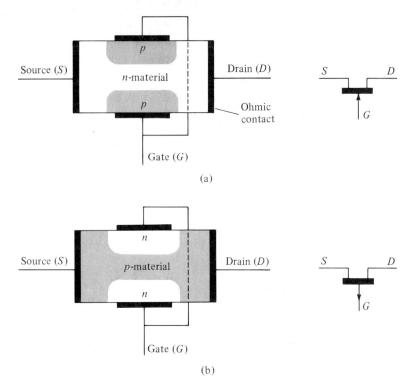

Figure 6.1 Physical structure and device symbol of a JFET: (a) n-channel; (b) p-channel.

material in an n-channel device or by p-type material in a p-channel device. In an n-channel device the gate is a region of p-type material (doping during fabrication provides an excess of holes) surrounding the path through the device channel. Ohmic contacts provide connection to either p-type or n-type material without rectification or semiconductor action occurring. The electrical symbol for the JFET device is shown in Fig. 6.1 alongside each device type. Although only a symbol, the relation to the device physical structure should be apparent. In the present structure source and drain appear to be the same and could be considered interchangeable. In general, construction does not provide for equal source and drain regions. The arrow of the electrical symbol is shown here on the gate lead and its direction indicates the channel type of the device. Considering the arrow and channel bar to represent a diode symbol, the symbol of Fig. 6.1a is p at the gate, n at the bar, therefore an n-channel device. Figure 6.1b shows the formation and symbol of a p-channel JFET.

Construction of a JFET is more often made using the structure of Fig. 6.2. Starting with a section of n-doped silicon material, a p-type region is diffused into the n-material, the surface of this structure then sealed off or *passivated* by a layer of silicon dioxide (SiO_2). Further construction opens *windows* in the silicon dioxide through which deposition of metal *(metallization)* provides ohmic contacts for the device's three terminals. In actual manufacture the n material is doped

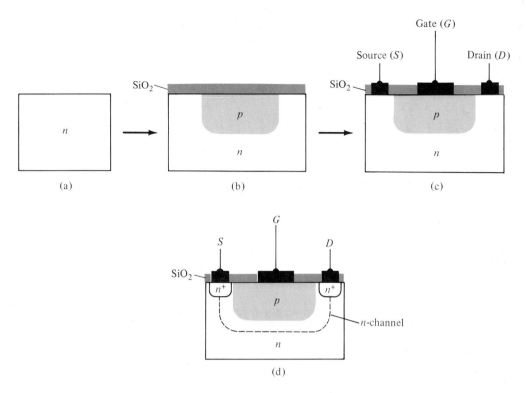

Figure 6.2 Construction of n-channel JFET.

more heavily (more charge concentration) in the n regions just below the ohmic contacts for source and drain as shown in Fig. 6.2d, the n^+ indicating the more heavily doped n region. A dashed line is shown to provide indication of how the electron charge moves between source and drain, the region from the p-type material to the bottom of the device forming the n-channel of the JFET.

6.3 JFET CHARACTERISTICS

To understand the electrical characteristics of the JFET we must first consider the basic operation of the device. Using the simple device structure of Fig. 6.1, the voltage supplies and resulting voltages and currents are shown in Fig. 6.3.

A battery or supply voltage applied in the drain-source path is shown connected from drain terminal to ground and labeled V_{DD} to indicate a battery in the drain path. A simple description of the device operation is that the voltage supply V_{DD} causes movement of the electrons in the n-channel, these electrons passing from source to drain to battery positive terminal (and back to the source through the battery). A positive current can be measured from the battery, this current I_D going from battery through drain, n-channel, source, and back to the battery. The same current passing through both source and drain is labeled as I_D. The device terminal nomenclature suggests that the current-carrying charge, electrons

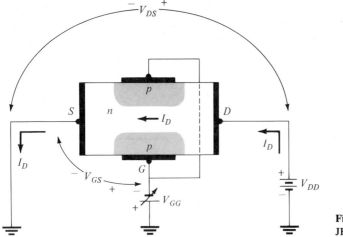

Figure 6.3 Basic operation of JFET.

in this case, come from the source and pass on to the drain. The current, of course, results from the free electrons in the *n*-doped channel moving as directed by the polarity of the battery voltage.

Before considering the controlling action resulting when the gate terminal is used, we can examine a basic feature of the device operation for the condition of gate-source voltage at 0 V. This condition, depicted in Fig. 6.4a, shows a channel current, I_D, due to the applied battery voltage, V_{DD}. As the value of V_{DD} is increased from 0 V, the amount of current through the channel increased, the maximum current being limited by the resistance of the *n* channel. A result of the voltage drop along the channel, increasing from 0 V at the source to V_{DD} at the drain, creates a reverse-bias region between the *n*-type channel and *p*-type gate material. As depicted in Fig. 6.4a, a back-biased region results, called a *depletion* region, since the reverse bias removes charge carriers from this location. The result of a depletion region is that the channel current, I_D, must pass through a narrow cross section of material. The narrower the effective cross section through which I_D passes, the larger the channel resistance.

As the battery voltage is increased, the incremental current increase becomes less due to the increased channel resistance, finally resulting in the channel becoming completely *pinched off,* as shown in Fig. 6.4b. When the drain current reaches this limiting or saturation level, no further increase in current can occur. Any further increase in V_{DD} will result in further increasing the voltage across the depletion region only, no further increase in channel current being possible. Figure 6.4c shows the FET operation in terms of the drain-source voltage, V_{DS}, and resulting drain current, I_D. As shown, increasing V_{DD} results in the voltage drop, V_{DS}, across drain to source, causing a current I_D. For small values of V_{DS}, I_D increases almost linearly, indicating a constant channel resistance. As the voltage V_{DS} increases, the channel pinches in and the channel resistance increases until the channel current, I_D, reaches a limiting or constant-current value, I_{DSS}. The nomenclature of I_{DSS} is that of a current from drain *(D)* to source *(S)* at the saturation *(S)* condition in the channel. Thus, I_{DSS} is the limiting or saturation current value occurring when the gate-source is at 0 V ($V_{GS} = 0$ V). The value of I_{DSS} is set at a particular level by the construction of the device, the doping level in the channel, and the width and length of the channel all being factors. Typical values of I_{DSS} are generally specified by the manufacturer. Otherwise, specific measurement of the desired parameter can be made.

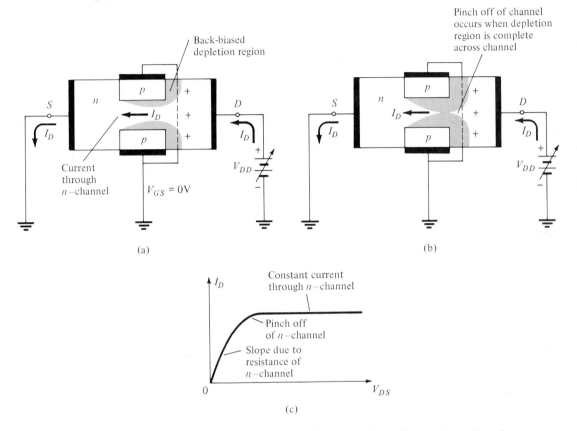

Figure 6.4 Pinch-off action due to channel current.

The gate acts to provide a voltage that additionally offsets the depletion region. For the *n*-channel JFET discussed here, negative voltage polarity applied from gate to source increases the depletion region, causing the current to close off the channel at lower values of current.

If V_{GS} is increased (more negative for an *n*-channel device), the channel will develop a depletion region and the current will saturate at lower levels as depicted in Fig. 6.5a. Sample curves are shown for $V_{GS} = 0$ V, -1 V, -2 V. At more negative gate-source voltages, a depletion region is formed in the *n* channel due to V_{GS} and increased by the channel current. For a *p*-channel JFET the channel saturation current is reduced from I_{DSS} (at $V_{GS} = 0$ V) as V_{GS} is made more positive (Fig. 6.5b).

While increasing channel current eventually develops a reverse bias across the gate-source, resulting in a depletion region which limits the channel saturation current, application of separate gate-source reverse-bias voltage will also add to increasing the depletion region. In fact, a large enough gate-source reverse-bias voltage could fully deplete the channel so that no current can pass from drain to source. As this reverse-bias voltage is increased, a voltage is reached at which no current can pass through the channel, called the gate-source pinch-off voltage, V_P. The values of pinch-off voltage, V_P, and channel saturation current (at $V_{GS} = 0$ V), I_{DSS}, are two of the important device parameters.

A plot of drain current, I_D, versus drain-source voltage, V_{DS}, as shown in Fig. 6.5 is called a drain characteristic. The drain characteristic of Fig. 6.5 shows that for an *n*-channel JFET, V_P is a negative voltage and that for a *p*-channel JFET V_P is positive.

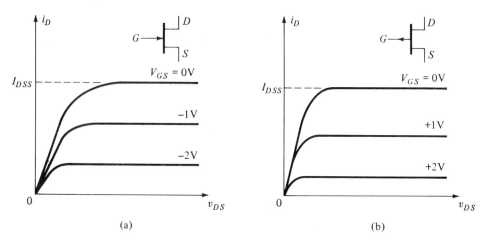

Figure 6.5 JFET drain characteristic.

Another form of device characteristic is the transfer characteristic, which is a plot of drain current, i_D, as a function of gate-source voltage, v_{GS}, for a constant value of drain-source voltage, v_{DS}. The transfer characteristic clearly shows the two important device parameters, V_P and I_{DSS}, as shown in Fig. 6.6. The transfer characteristic curve can be expressed mathematically by the equation

$$i_D = I_{DSS}\left(1 - \frac{v_{GS}}{V_P}\right)^2 \qquad (6.1)$$

Equation (6.1) or the JFET transfer characteristic curve depend on the two device parameters, V_P and I_{DSS} (V_P being negative for *n*-channel JFETs and positive for *p*-channel JFETs). Although Eq. (6.1) or a transfer curve plotted from it is an ideal representation of a JFET operation, they both describe actual device performance quite well.

Comparing Eq. (6.1) with Fig. 6.6, note that when $v_{GS} = 0$ V, the equation provides that $i_D = I_{DSS}$ and the curve crosses the vertical (i_D) axis line at the value of I_{DSS}. For $i_D = 0$ the equation provides that $v_{GS} = V_P$ and the transfer curve reaches the horizontal (v_{GS}) axis. The values of I_{DSS} and V_P listed by the manufacturer are typical values. Exact values for a particular JFET can be measured, if necessary.

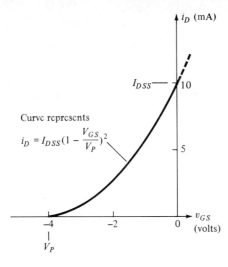

Figure 6.6 *n*-channel JFET transfer characteristic.

Construction of JFET Transfer Characteristic

The values of I_{DSS} and V_P are usually specified in the manufacturer's data sheet. The transfer curve plot can then be made using Eq. (6.1) as shown in Example 6.1.

EXAMPLE 6.1 Plot the transfer characteristic of an *n*-channel JFET having listed values of $I_{DSS} = 8$ mA and $V_P = -4$ V.

Solution: Obtain a few points for the plot as in the table below by selecting values of v_{GS} and calculating i_D using Eq. (6.1).

	v_{GS} (V)	i_D (mA)	
(0)	0	8	(I_{DSS})
(0.3 V_P)	−1.2	4	($\frac{1}{2}I_{DSS}$)
(0.5 V_P)	−2	2	($\frac{1}{4}I_{DSS}$)
(V_P)	−4	0	(0)

These points are plotted in Fig. 6.7.

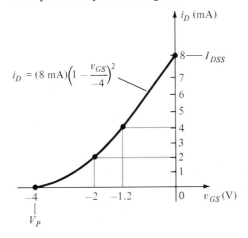

Figure 6.7 Transfer curve of JFET having $I_{DSS} = 8$ mA and $V_P = -4$ V.

243

Although the example is for an *n*-channel JFET it could be similarly carried out for a *p*-channel JFET with the same resulting voltage magnitudes, with opposite polarity, and same resulting current values, with opposite direction.

Drain Characteristic

A full description of the JFET operation can also be shown graphically by a drain characteristic as in Fig. 6.8. The value of I_{DSS} is read from the curves by noting the current level at which $V_{GS} = 0$ V provides a horizontal characteristic line. The value of V_P is not as clearly indicated, although it is obviously somewhat larger than the value of the lowest V_{GS} curve. A dashed curve is shown passing through the points at which current saturation occurs, the region to the left being that in which the device is below pinch-off and that to the right being that in which the device is above the pinch-off condition. The dashed line represents the condition of $V_{DS} = V_P - V_{GS}$. Although the dashed line is not generally part of a drain characteristic, it does show the value of V_P at the point where the curve touches the horizontal (v_{DS}) axis.

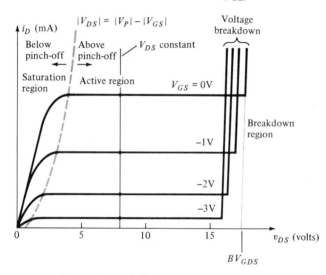

Figure 6.8 JFET drain and transfer characteristics.

The region of the curve to the right shows that as v_{DS} is increased, a point is reached after which the JFET goes into breakdown, the current increase no longer limited by the device. If the external circuit in which the JFET is connected provides a limit on the drain current, the device may not be damaged. Otherwise, the breakdown may result in permanent damage to the device.

The breakdown voltage, BV_{GDS}, is larger for lower values of gate-source bias voltage. Manufacturer's specifications provide a value of BV_{GDS} at a single value of V_{GS}, usually at $V_{GS} = 0$ V.

The region of the drain characteristic between the dashed curve and the breakdown curves is the device active region. The device is biased by the external circuitry so that it remains operating in the active region when used for amplifying signals. The regions of saturation and device cutoff (pinch-off) are used when operating the JFET in switching or digital circuits.

6.4 JFET OPERATION

Operating a JFET usually requires biasing the device to operate around some condition of drain current and drain-source voltage. For operation as an amplifier, the device is biased to operate within the active region. While the total operation of a circuit is the desired action, analysis of the circuit can be divided into consideration in the study of the dc bias and then the ac operation around this bias point.

DC Bias of JFET

Although the JFET can be biased using a number of circuit configurations, only a few of the more basic circuits will be considered here. Figure 6.9 shows a circuit to provide dc bias using two power supplies or batteries. Although practical circuits can achieve dc bias using a single supply voltage, the circuit of Fig. 6.9 is a simple circuit to be covered first.

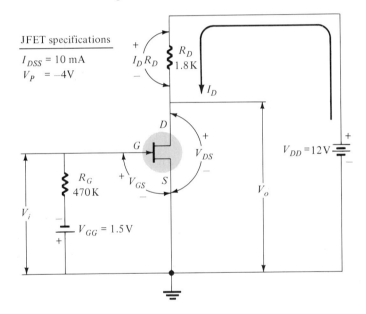

Figure 6.9 JFET amplifier circuit with fixed bias.

A signal to be amplified is applied as V_i across the gate-source with amplified output, V_o, taken across the drain-source. Resistor R_G provides a path for the ac signal but develops no dc voltage because no dc current passes through the reverse biased gate-source region. The dc voltage across the gate-source is that of the supply voltage, $V_{GS} = V_{GG} = -1.5$ V. Using this value in Eq. (6.1) determines the drain current

$$I_D = 10 \text{ mA}\left(1 - \frac{-1.5 \text{ V}}{-4 \text{ V}}\right)^2 = 3.9 \text{ mA}$$

Notice that the gate-source voltage, V_{GS}, sets the value of I_D, this current passing from the 12-V supply, V_{DD}, through the resistor, R_D, the drain-source channel, to ground (the minus terminal of the supply). As long as the JFET operates in its active region, the current I_D depends on V_{GS} and not on the value of resistor R_D.

A voltage drop develops across R_D due to the drain current I_D, as indicated in Fig. 6.9. The dc bias voltage from drain to source is that of the supply voltage *minus* the voltage drop across resistor R_D:

$$V_{DS} = V_{DD} - I_D R_D$$
$$= 12\text{V} - (3.9\text{ mA})(1.8\text{ K})$$
$$= 12\text{ V} - 7.02\text{ V} = 4.98\text{ V}$$

The JFET of Fig. 6.9 is therefore biased by the circuit to operate at

$$I_D = 3.9\text{ mA} \quad \text{and} \quad V_{DS} = 4.98\text{ V}$$

The significance of the bias point is that it sets a current through the device which can then be increased or decreased by an applied ac signal, the output voltage developed then varying around the bias voltage. In its basic operation a varying input voltage causes the gate-source voltage to vary around its bias value, the resulting varying channel current then producing a varying voltage drop across resistor R_D, providing an output voltage that varies at the frequency of the input signal. Depending on the amount of current change caused by the changing gate-source voltage for a selected value of R_D, the output voltage developed is typically many times larger than the input voltage—thus amplification occurs. The JFET parameter that provides information about this feature of the device is its transconductance, g_m, where g_m is defined as the change in channel current, Δi_D, caused by a change in gate-source voltage, Δv_{gs}, or mathematically,

$$g_m = \frac{\Delta i_d}{\Delta v_{gs}}$$

and has units of milliamperes per volt or millimhos.

The value of g_m can be obtained from the device values of I_{DSS} and V_P and the bias point set by the circuit. Mathematically, differentiating Eq. (6.1) results in[1]

$$\boxed{g_m = \frac{2I_{DSS}}{|V_P|}\left(1 - \frac{v_{gs}}{V_P}\right) = g_{mo}\left(1 - \frac{v_{gs}}{V_P}\right)} \tag{6.2}$$

where

$$\boxed{g_{mo} = \frac{2I_{DSS}}{|V_P|}} \tag{6.3}$$

Referring to the circuit of Fig. 6.9, we obtain

$$g_{mo} = \frac{2I_{DSS}}{|V_P|} = \frac{2(10\text{ mA})}{|-4\text{ V}|} = 5\text{ mmhos} = 5 \times 10^{-3}\text{ mho}$$

---

[1] $\dfrac{\partial i_D}{\partial v_{gs}} = g_m = -\dfrac{2I_{DSS}}{V_P}\left(1 - \dfrac{v_{gs}}{V_P}\right)$

and at the bias point ($V_{GS} = -1.5$ V)

$$g_m = g_{mo}\left(1 - \frac{V_{GS}}{V_P}\right) = 5 \text{ mmhos}\left(1 - \frac{-1.5 \text{ V}}{-4 \text{ V}}\right) = 3.125 \text{ mmhos}$$

The magnitude of the output voltage, V_o, in Fig. 6.9 can be calculated as

$$V_o = (g_m V_i)R_D$$

where $g_m V_i$ is the amount the channel current varies due to input voltage, V_i, and the device transconductance, g_m. An input of 10 mV would then produce

$$V_o = (g_m V_i)R_D = (3.125 \text{ mmhos})(10 \text{ mV})(1.8 \text{ K}) = 56.25 \text{ mV}$$

The circuit gain is then

$$A_v = \frac{V_o}{V_i} = \frac{56.25 \text{ mV}}{10 \text{ mV}} = 5.625$$

As the output is generally taken from the drain terminal with respect to ground, it will then be 180° opposite in phase from the input signal. The overall circuit gain can be directly calculated from

$$A_v = \frac{V_o}{V_i} = -g_m R_D \tag{6.4}$$

where the minus sign indicates only the phase inversion between input at gate-source and output from drain-source terminals. The magnitude of the gain is $g_m R_D$ (5.625, in the present example).

Since the gain depends on the device value of g_m and the resistor R_D, changing JFETs may cause the gain to be different, unless similar values of g_m are maintained.

One can obtain larger circuit gain by selecting a JFET with a larger value of g_{mo} and then biasing for largest g_m. The larger value of g_m is obtained when the device is biased at $V_{GS} = 0$ V [refer to Eq. (6.2)]. However, this would bias the device so that the input signal would vary v_{gs} above and below the bias point, 0 V. For an n-channel JFET the gate-source acts to deplete the channel, which remains reverse-biased (open-circuit) as long as the gate voltage is less positive than the source voltage. Once the gate voltage goes more positive, the p-n junction acts like a forward-biased diode and no control of channel current is achieved; therefore, no linear gain between input and output is possible.

Figure 6.10 shows some of the practical limitations to biasing the JFET. To obtain larger gain by biasing near 0 V limits the input voltage swing to values that do not drive v_{gs} positive. For very small input signals, bias near 0 V is possible, while for larger input signal operation, the bias point must be moved away from 0 V with resulting smaller values of g_m. Larger gain is also achieved by using larger values of R_D. Too large a value of R_D may result in a bias point out of the active region with the range of voltage swing limited by device saturation.

One main conclusion to take from the discussion above is that any gain desired cannot be achieved by merely picking a large enough value of R_D. The value of g_m also must be limited by the value of g_{mo} of the JFET selected and by the bias point at which it is operated. For JFETs, values of g_{mo} are usually no larger

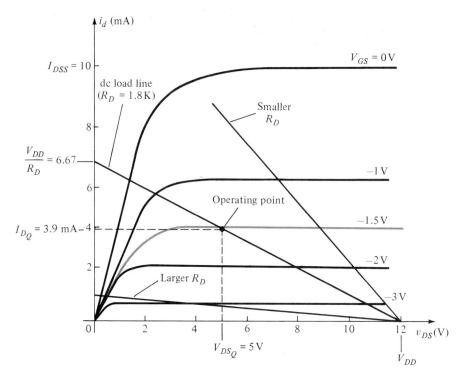

Figure 6.10 Operation and bias of JFET device.

than 10 mmhos, while R_D is typically no larger than 10 K, so that voltage gains of $g_m R_D = $ (10 mmhos) (10 K) = 100 should be considered large for a JFET amplifier circuit such as that of Fig. 6.9.

JFET Amplifier with Self-Bias

A more practical version of a JFET amplifier uses only a single voltage supply and uses a self-bias resistor, R_S, to obtain the gate-source bias voltage, as shown in Fig. 6.11a. The presence of resistor R_S results in a positive voltage, V_S, due to the voltage drop $I_D R_S$. Since the gate voltage, V_G, is 0 V (no dc current flows through gate or resistor R_G), the net voltage measured from gate (0 V) to source ($+V_S$) is a negative voltage, which is the gate-source bias voltage, V_{GS}. The gate-source bias relation

$$V_{GS} = 0 - I_D R_S = -I_D R_S \tag{6.5}$$

can be plotted on the transfer characteristic as shown in Fig. 6.11b. To plot the straight-line equation (6.5), select two values of I_D and calculate the corresponding value of V_{GS}. Two values are listed in the table shown in Fig. 6.11b. Connecting a straight line between these points intersects the device transfer characteristic at $I_D = 1.6$ mA and $V_{GS} = -2.4$ V, the resulting bias point of the circuit.

We should note that increasing R_S would lower the R_S bias line, resulting in a lower value of I_D at the bias point and a correspondingly larger value of V_{GS}. Reducing R_S would result in a steeper R_S line with higher I_D and lower V_{GS} values at the bias point.

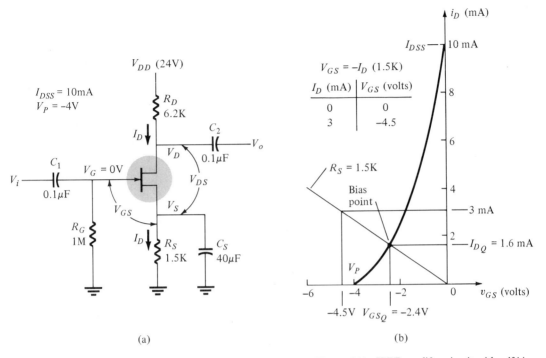

Figure 6.11 JFET amplifier circuit with self-bias.

Example 6.2 Determine the bias point of the circuit shown in Fig. 6.11a for $R_S = 1$ K.

Solution: The self-bias line is drawn on the characteristic of Fig. 6.11b from the 0 axis ($V_{GS} = 0$, $I_D = 0$) passing through a point of selected value I_D, say $I_D = 4$ mA, and corresponding voltage $V_{GS} = -I_D R_S = -(4 \text{ mA})(1 \text{ K}) = -4$ V. The intersection of this line with the device transfer characteristic is the bias point.

$$I_{D_Q} = 2.2 \text{ mA} \qquad V_{GS_Q} = -2.2 \text{ V}$$

We can then calculate

$$V_{DS_Q} = V_{DD} - I_{DQ}R_D - I_{DQ}R_S = 24 - 2.2 \text{ mA } (6.2 \text{ K} + 1 \text{ K})$$
$$= 24 - 15.8 = 8.2 \text{ V}$$

Example 6.3 Determine the operating point of the n-channel JFET amplifier shown in Fig. 6.12a.

Solution: The transfer characteristic required to solve for the bias point need not be provided as long as V_P and I_{DSS} are specified. It is fairly easy to sketch

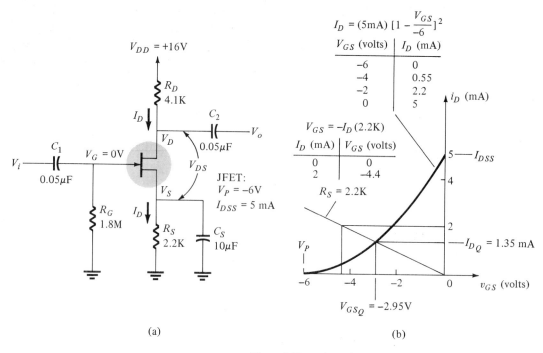

(a) (b)

Figure 6.12 *n*-channel JFET amplifier circuit for Example 6.3.

the transfer characteristic using the values of V_P and I_{DSS} provided and the transfer curve formula [Eq. (6.1)].

$$I_D = I_{DSS}\left(1 - \frac{V_{GS}}{V_P}\right)^2$$

A sketch on a sheet of graph paper can be made for a few points of selected V_{GS} and corresponding calculated values of I_D. For example, the present JFET would result in a transfer curve sketched from V_P (= −6 V) along the *x*-axis through points

$$V_{GS} = -4 \text{ V}: \quad I_D = 5 \text{ mA}\left(1 - \frac{-4}{-6}\right)^2 = 0.55\text{mA}$$

$$V_{GS} = -2 \text{ V}: \quad I_D = 5 \text{ mA}\left(1 - \frac{-2}{-6}\right)^2 = 2.2 \text{ mA}$$

to the point $I_D = I_{DSS} = 5$ mA along the *y*-axis.

Using this transfer characteristic (Fig. 6.12b), we can then plot the self-bias line for $R_S = 2.2$ K. Choosing a current, say $I_D = 2$ mA, we obtain

$$V_{GS} = -I_D R_S = -(2 \text{ mA})(2.2 \text{ K}) = -4.4 \text{ V}$$

The self-bias line then intersects the transfer curve at about

$$V_{GS_Q} = -2.95 \text{ V} \qquad I_{D_Q} = 1.35 \text{ mA}$$

from which we calculate

$$V_{DS_Q} = V_{DD} - I_{D_Q}(R_S + R_D) = 16 - 1.35 \text{ mA}(2.2 \text{ K} + 4.1 \text{ K}) = 7.5 \text{ V}$$

Another form of dc bias circuit is that shown in Fig. 6.13a. Except for the gate voltage being set other than 0 V, the determination of bias voltage and current proceeds as discussed previously. The bias circuit of Fig. 6.13 provides a greater dc bias stability than the bias circuit of Fig. 6.11. The value of V_G obtained from the voltage divider network is

$$V_G = V_{GG} = \frac{R_2}{R_1 + R_2} \cdot V_{DD} \tag{6.6}$$

and the bias voltage V_{GS_Q} is then

$$V_{GS_Q} = V_{GG} - I_D R_S \tag{6.7}$$

Example 6.4 Determine the dc bias of the JFET in the circuit of Fig. 6.13a.

Solution: The gate voltage is

$$V_G = V_{GG} = \frac{R_2}{R_1 + R_2} \cdot V_{DD} = \frac{280 \text{ K}}{2 \text{ M} + 280 \text{ K}} \cdot 16 \text{ V} = +2 \text{ V}$$

The result of a voltage drop $I_D R_S$ is a gate-source voltage

$$V_{GS} = V_G - V_S = +2 - I_D R_S$$

An R_S-bias line (see Fig. 6.13b) can be drawn corresponding to the circuit equation given above. For the JFET with $V_P = -4$ V and $I_{DSS} = 8$ mA we can also plot

Figure 6.13 (a) JFET bias circuit with voltage divider gate bias; (b) transfer characteristic and self-bias line for Example 6.4.

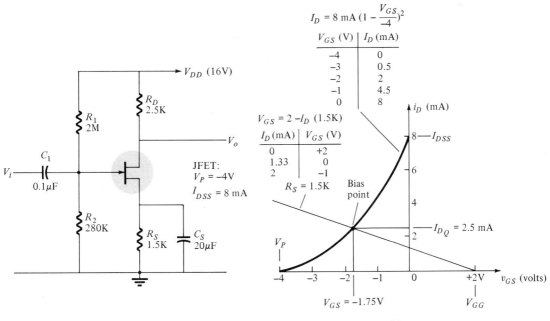

(a)

(b)

the transfer characteristic as in Fig. 6.13b, the intersection of self-bias line and transfer characteristic providing a dc bias at

$$I_{DQ} = 2.5 \text{ mA} \qquad V_{GSQ} = -1.75 \text{ V}$$

We can then calculate

$$V_{DQ} = V_{DD} - I_{DQ}R_D = 16 - 2.5 \text{ mA}(2.5 \text{ K}) = \mathbf{9.75 \text{ V}}$$

$$V_{SQ} = I_{DQ}R_D = 2.5 \text{ mA}(1.5 \text{ K}) = \mathbf{3.75 \text{ V}}$$

and $\qquad V_{DSQ} = V_{DQ} - V_{SQ} = 9.75 - 3.75 = \mathbf{6 \text{ V}}$

(Note that $V_{GSQ} = V_{GQ} - V_{SQ} = 2 - 3.75 = -1.75$ V, as expected.)

6.5 MOSFET CONSTRUCTION AND CHARACTERISTICS

A field-effect transistor can be constructed with the gate terminal insulated from the channel. The popular metal-oxide-semiconductor FET (MOSFET)[2] is constructed as either a *depletion* MOSFET (Fig. 6.14a) or an *enhancement* MOSFET (Fig. 6.14b). In the depletion-mode construction a channel is physically constructed and current between drain and source will result from a voltage connected across the drain-source. The enhancement MOSFET structure has *no* channel formed when the device is constructed. Voltage must be applied at the gate to develop a channel of charge carriers so that a current results when a voltage is applied across the drain-source terminals.

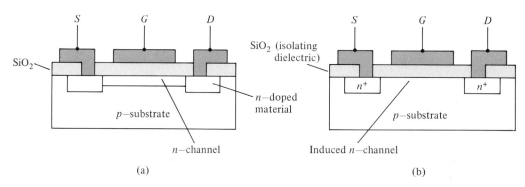

Figure 6.14 MOSFET construction: (a) depletion; (b) enhancement.

Depletion MOSFET

The n-channel depletion MOSFET device of Fig. 6.14a is formed on a p-substrate (p-doped silicon material used as the starting material onto which the FET structure is formed). The source and drain are connected by metal (aluminum) to n-doped source and drain regions which are connected internally by an n-doped channel region. A metal layer is deposited above the n-channel on a layer of silicon dioxide

---

[2] Also called insulated gate FET or IGFET.

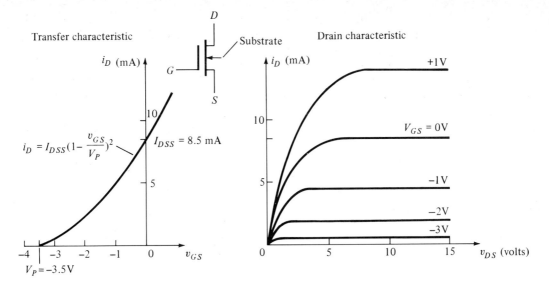

Figure 6.15 n-channel depletion MOSFET characteristics.

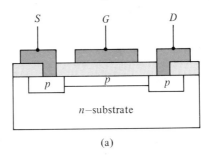

(a)

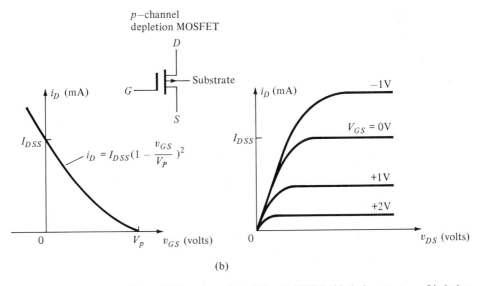

(b)

Figure 6.16 p-channel depletion MOSFET: (a) device structure; (b) device characteristic.

253

(SiO$_2$) which is an insulating layer. This combination of a *metal* gate on an *oxide* layer over a *semi*conductor substrate forms the depletion MOSFET device. For the *n*-channel device of Fig. 6.14a negative gate-source voltages push electrons out of the channel region to deplete the channel and a large enough negative gate-source voltage will pinch off the channel. Positive gate-source voltage, on the other hand, will result in an increase in the channel size (pushing away *p*-type carriers), allowing more charge carriers and therefore greater channel current to result.

An *n*-channel depletion MOSFET device characteristic is shown in Fig. 6.15. The device is shown to operate with either positive or negative gate-source voltage, negative values of V_{GS} reducing the drain current until the pinch-off voltage, after which no drain current occurs. The transfer characteristic is the same as that for a JFET for negative gate-source voltages and continues for positive values of V_{GS}. Since the gate is isolated from the channel for both negative and positive values of V_{GS}, the device can be operated with either polarity of V_{GS}—no gate current resulting in either case. The device schematic symbol in Fig. 6.15 shows the addition of a substrate terminal (in addition to gate, source, and drain leads) on which the device type is indicated, the arrow here indicating a *p*-substrate and thus *n*-channel device. A *p*-channel depletion MOSFET characteristic is shown in Fig. 6.16.

Enhancement MOSFET

The enhancement MOSFET of Fig. 6.17 has no channel between drain and source as part of the basic device construction. Application of a positive gate-source voltage will repel holes in the substrate region under the gate leaving a

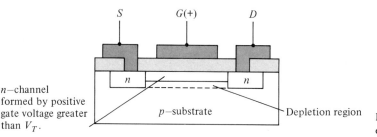

Figure 6.17 *n*-channel formed in enhancement MOSFET.

depletion region. When the gate voltage is sufficiently positive, electrons are attracted into this depletion region, making it then act as an *n*-channel between drain and source. The resulting *n*-channel enhancement MOSFET characteristic is shown in Fig. 6.18. There is no drain current until the gate-source voltage exceeds the threshold value, V_T. Positive voltages above this threshold value result in increased drain current, the transfer characteristic being described by[3]

$$i_D = K(v_{GS} - V_T)^2 \qquad (6.8)$$

where K, typically 0.3 mA/V^2, is a property of the device construction. Note

---

[3] Equation (6.8) is only valid for $|v_{GS}| > |V_T|$.

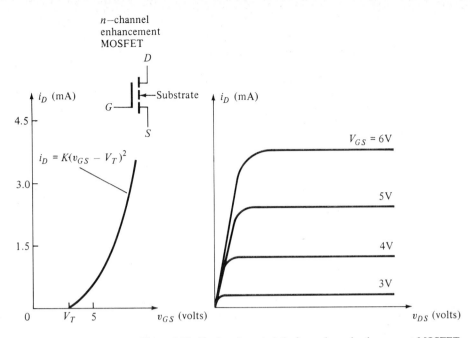

Figure 6.18 Device characteristic for *n*-channel enhancement MOSFET.

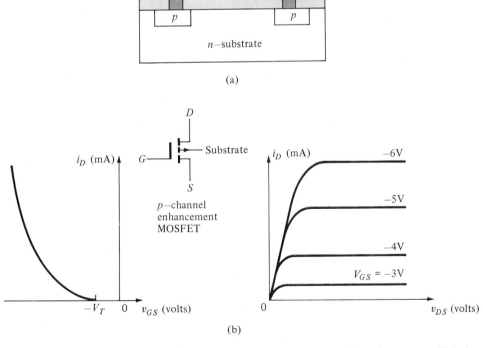

Figure 6.19 *p*-channel enhancement MOSFET: (a) device structure; (b) device characteristic.

that no value I_{DSS} can be associated with an enhancement MOSFET because no drain current occurs with $V_{GS} = 0$ V. Although the enhancement MOSFET is more restricted in operating range than is the depletion device, the enhancement device is very useful in large-scale integrated circuits in which the simpler construction and smaller size make it a suitable device. The enhancement schematic symbol shows a broken line between drain and source indicating that there is no initial channel for the enhancement device. The substrate terminal arrow shows a p-substrate and an n-channel. P-channel enhancement MOSFETs can also be constructed, the device and characteristic being shown in Fig. 6.19.

6.6 MOSFET OPERATION

Depletion MOSFET Circuit

The n-channel depletion MOSFET amplifier circuit shown in Fig. 6.20a is the same as that of a JFET device (except for the larger value of R_G possible with MOSFET devices). Since the gate-source voltage can go positive in this circuit, it is possible to bias the device at only slightly negative gate-source voltages.

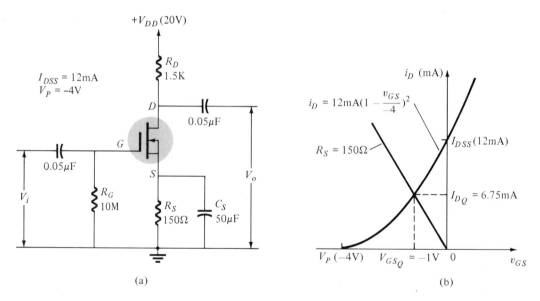

Figure 6.20 n-channel depletion MOSFET amplifier circuit and device characteristic.

For the circuit of Fig. 6.20 we obtain the dc bias values as follows:

Self-bias load line for $R_S = 150\ \Omega$, drawn on Fig. 6.20, provides bias point at $V_{GS_Q} = -1$ V and $I_{D_Q} = 6.75$ mA.
The drain voltage is then

$$V_{D_Q} = V_{DD} - I_{D_Q} R_D = 20 - (6.75\ \text{mA})(1.5\ \text{K}) = 9.88\ \text{V}$$

and
$$V_{DS_Q} = V_{D_Q} - V_{S_Q} = 9.88 - 1 = 8.88 \text{ V}$$

At this bias point the value of g_m is [Eqs. (6.2) and (6.3)]

$$g_m = \frac{2I_{DSS}}{|V_P|}\left(1 - \frac{V_{GS}}{V_P}\right) = \frac{2(12 \times 10^{-3})}{|-4|}\left(1 - \frac{-1}{-4}\right)$$
$$= 4.5 \times 10^{-3} = 4.5 \text{ mmhos}$$

Using Eq. (6.4), the ac gain is then

$$A_v = -g_m R_D = -(4.5 \times 10^{-3})(1.5 \times 10^3) = \mathbf{-6.75}$$

A few additional examples will show the calculations of dc bias values and ac gain for a depletion MOSFET circuit.

EXAMPLE 6.5 Determine the voltage gain of the amplifier shown in Fig. 6.21, and the output voltage, V_o.

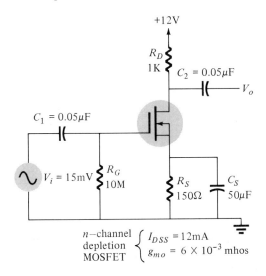

Figure 6.21 Depletion MOSFET amplifier circuit.

Solution: The dc bias can be obtained as previously considered, resulting in

$$V_{GS_Q} = -1 \text{ V} \qquad I_{D_Q} = 6.75 \text{ mA}$$

Using Eqs. (6.2) and (6.3), we calculate g_m to be

$$g_m = 6 \times 10^{-3}\left(1 - \frac{-1}{-4}\right) = 4.5 \text{ mmhos}$$

The circuit voltage gain is then [using Eq. (6.4)]

$$A_v = -g_m R_D = -(4.5 \times 10^{-3})(1 \times 10^3) = \mathbf{-4.5}$$

The output voltage is then

$$V_o = A_v V_i = -4.5(15 \text{ mV}) = \mathbf{-60 \text{ mV}}$$

EXAMPLE 6.6 Calculate the ac voltage gain of the depletion MOSFET amplifier of Fig. 6.22.

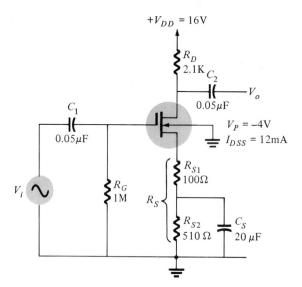

Figure 6.22 Depletion MOSFET amplifier for Example 6.6.

Solution: Dc bias analysis using the MOSFET device transfer characteristic and a self-bias line of $R_S = 610\ \Omega$ results in

$$V_{GS_Q} = -1.95\ \text{V} \quad \text{and} \quad I_{D_Q} = 3.2\ \text{mA}$$

At this bias point g_m is

$$g_m = g_{m0}\left(1 - \frac{V_{GS_Q}}{V_P}\right) = \frac{2(12 \times 10^{-3})}{|-4|}\left(1 - \frac{-1.95}{-4}\right) = 3.1\ \text{mmhos}$$

The voltage gain is then calculated using a more general form of Eq. (6.4):[4]

$$A_v = \frac{-g_m R_D}{1 + g_m R_{S1}} = \frac{-(3.1 \times 10^{-3})(2.1 \times 10^3)}{1 + (3.1 \times 10^{-3})(100)} = -5$$

Enhancement MOSFET Bias Circuit

A popular dc bias arrangement for an enhancement device is shown in Fig. 6.23a. Using the drain-source voltage as gate-source bias voltage is achieved by connecting $R_G = 10\ \text{M}$ from drain to gate. Since there is no gate current, there is no voltage drop across R_G and $V_{DS} = V_{GS}$. The resulting bias point for $R_D = 2\ \text{K}$ (and $V_{DD} = 20\ \text{V}$) can be obtained using the device transfer characteristic (see Fig. 6.23b). The device transfer characteristic can be drawn on graph paper using

$$I_D = K(V_{GS} - V_T)^2 = 0.3(V_{GS} - 3)^2$$

---

[4]
$$V_{gs} = V_i - (g_m V_{gs})R_{S1}$$
$$(1 + g_m R_{S1})V_{gs} = V_i$$
$$V_o = -g_m V_{gs} R_D$$
$$A_v = \frac{V_o}{V_i} = \frac{-g_m R_D}{1 + g_m R_{S1}} = \frac{-g_m R_D}{1 + g_m R_{S1}}$$

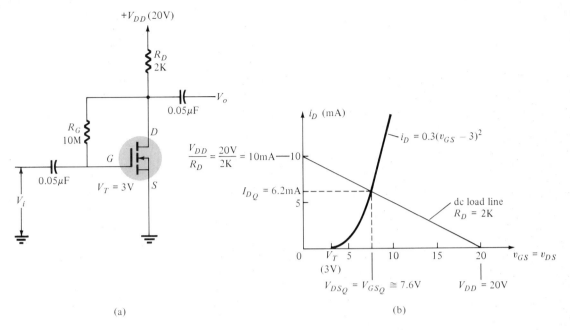

Figure 6.23 n-channel enhancement MOSFET dc bias circuit and characteristic.

The circuit dc load line can also be drawn on the same graph, since

$$V_{GS} = V_{DS} = V_{DD} - I_D R_D = 20 - (2\text{ K})I_D$$

The intersection of load line and device characteristic provides the operating point shown in Fig. 6.23b to be

$$V_{GS_Q} = V_{DS_Q} = 7.6 \text{ V} \qquad I_{D_Q} = 6.2 \text{ mA}$$

EXAMPLE 6.7 Calculate the voltage gain of the enhancement MOSFET amplifier circuit of Fig. 6.24.

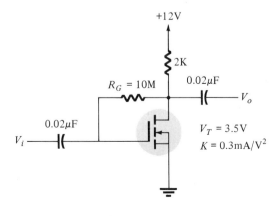

Figure 6.24 Enhancement MOSFET amplifier circuit.

SEC. 6.6 MOSFET OPERATION

Solution: The dc bias point can be determined to be $V_{GS_Q} = V_{DS_Q} = 6.7$ V at $I_{D_Q} = 3.1$ mA. The value of g_m at the bias point can be obtained from[5]

$$g_m = 2K(V_{DS_Q} - V_T) \tag{6.9}$$

$$g_m = 2(0.3 \times 10^{-3})(6.7 - 3.5) = 1.9 \text{ mmhos}$$

Using Eq. (6.4), we obtain

$$A_v = -g_m R_D = -1.9 \times 10^{-3}(2 \times 10^3) = -3.8$$

PROBLEMS

§ 6.3

1. Sketch the transfer characteristic of a JFET having $I_{DSS} = 12$ mA and $V_P = -4$ V.

2. Sketch the transfer characteristic of a p-channel JFET having $I_{DSS} = 10$ mA and $V_P = +2.5$ V.

§ 6.4

3. Calculate the operating point of a JFET circuit such as the one shown in Fig. 6.9 using the JFET device of Problem 1.

4. Calculate the operating point of a JFET such as that shown in the circuit of Fig. 6.11 using the JFET device of Problem 1.

5. Calculate the operating point, V_{DS_Q}, and I_{D_Q} for the JFET shown in Fig. 6.25.

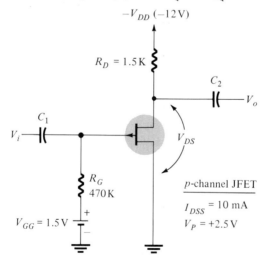

Figure 6.25 JFET circuit for Problem 6.5.

---

[5] Since $i_D = K(v_{GS} - V_T)^2$,

$$g_m = \left. \frac{\partial i_D}{\partial v_{GS}} \right|_{v_{DS}=\text{constant}} = 2K(v_{GS_Q} - V_T)$$

6. Determine the drain voltage, V_D, and drain-source voltage, V_{DS}, for the JFET of Problem 1 used in the circuit of Fig. 6.12a.

7. Determine the operating point of the circuit shown in Fig. 6.26.

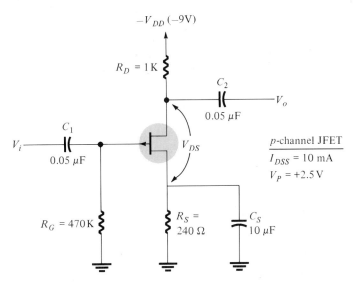

Figure 6.26 JFET circuit for Problem 6.7.

8. Calculate the ac voltage gain of the amplifier circuit of Fig. 6.11 biased at $V_{GS_Q} = -2.4$ V as shown, with $g_{mo} = 5000$ μmhos.

9. Calculate the ac voltage gain of the amplifier circuit of Fig. 6.11 with the JFET replaced by one having $I_{DSS} = 12$ mA and $V_P = -4$ V.

10. Calculate the ac voltage gain of the amplifier circuit of Fig. 6.26.

11. Repeat Problem 10 for R_S replaced by $R_S = 180$ Ω.

12. Calculate the output ac voltage from the circuit of Fig. 6.26 with $V_i = 50$ mV rms.

§ 6.6

13. Calculate the operating point of a depletion MOSFET amplifier circuit such as that shown in Fig. 6.20, replacing R_S with 100 Ω and R_D with 1.2 K.

14. Determine the operating point for a p-channel depletion MOSFET in a self-bias circuit such as that of Fig. 6.26 with $R_S = 120$ Ω and $R_D = 1.8$ K. Device parameters are $I_{DSS} = 12$ mA and $V_P = +4$ V, and supply voltage is $V_{DD} = -24$ V.

15. Determine the operating point of a circuit such as that shown in Fig. 6.21 using a MOSFET with $I_{DSS} = 8$ mA and $V_P = -4$ V.

16. Determine the operating point for a circuit as in Fig. 6.21 using a p-channel MOSFET ($I_{DSS} = 12$ mA, $V_P = +4$ V) and $V_{DD} = -12$ V.

17. Calculate the dc bias voltage, V_{GS}, for the amplifier circuit of Fig. 6.22 using a MOSFET with $I_{DSS} = 8$ mA and $V_P = -4$ V.

18. Determine the dc operating point of a p-channel depletion MOSFET ($I_{DSS} = 10$ mA, $V_P = +4$ V) using a circuit as in Fig. 6.22 with $V_{DD} = -16$ V.

19. Calculate the ac voltage gain of the amplifier circuit in Fig. 6.21 for $R_D = 1.5$ K and $R_S = 100$ Ω.

20. Calculate the ac voltage gain of the amplifier circuit of Fig. 6.21 for $R_D = 1.2$ K and $R_S = 120$ Ω.

21. Calculate the ac voltage gain of the amplifier circuit of Fig. 6.22 with $R_D = 1.8$ K.

22. Calculate the output ac voltage from the circuit of Fig. 6.22 for $R_D = 1.8$ K, with $V_i = 80$ mV, rms.

23. Calculate the dc bias of the amplifier circuit of Fig. 6.23 for $R_D = 3.9$ K.

24. Calculate the ac voltage gain of the circuit of Fig. 6.23.

25. Calculate the output ac voltage of the circuit of Fig. 6.23 for input of 50 mV and $R_D = 1.8$ K.

GLOSSARY

Breakdown Voltage The maximum voltage that can be applied across a pair of device terminals.

DC Bias Circuit voltage and current conditions to operate a device in a desired region of its linear operation.

Depletion MOSFET A metal-oxide-semiconductor field-effect transistor constructed with a drain-source channel that is depleted or reduces charge carriers by voltage applied across the gate-source terminals.

Depletion Region A device junction region wherein free charge carriers have been removed by applied voltage.

Enhancement MOSFET A metal-oxide-semiconductor field-effect transistor in which the channel is created (or enhanced) by voltage applied across the gate-source.

FET A field-effect-transistor semiconductor device constructed for control of a channel current by a voltage applied across the drain-source terminals.

I_{DSS} The drain current occurring with 0 V applied across the gate-source (maximum channel current of a JFET device).

JFET A junction-field-effect transistor constructed with a *p-n* gate-source boundary controlled by reverse-bias voltage.

n-Channel The drain-source region of an FET device doped with an excess of electrons.

p-Channel The drain-source region of an FET device doped with an excess of holes.

Pinch-off Reverse bias of the gate-source, resulting in no channel current.

Reverse Bias The voltage applied across a gate-source of polarity such as to reduce the current in the channel.

Saturation An operating condition at which no further increase in the channel current occurs as the drain-source voltage is increased.

Self-bias A circuit-bias connection permitting a single supply voltage to provide for the channel current and the gate-source bias voltage.

Transfer Characteristic A device characteristic plot of the drain current versus the gate-source voltage.

V_P The pinch-off voltage of an FET device.

CHAPTER 7

pnpn and Miscellaneous Devices

7.1 INTRODUCTION

In this chapter a number of important devices not discussed in detail in earlier chapters will be introduced. The two-layer semiconductor diode has led the way to three-, four-, and even five-layer devices. A family of four-layer *pnpn* devices will first be considered (SCR, SCS, GTO, LASCR, Shockley diode, diac, and triac), followed by an increasingly important device—the UJT (unijunction transistor). Those four-layer devices with a control mechanism are commonly referred to as *thyristors,* although the term is most frequently applied to the SCR (the semiconductor equivalent of the *thyratron* vacuum tube). The chapter will close with an introduction to the power FET, the phototransistor, opto-isolators, and the PUT (programmable unijunction transistor).

pnpn DEVICES

7.2 SILICON-CONTROLLED RECTIFIER (SCR)

Within the family of *pnpn* devices the silicon-controlled rectifier (SCR) is unquestionably of the greatest interest today. It was first introduced in 1956 by a group of Bell Telephone Laboratory engineers. A few of the more common areas of application for SCRs include relay controls, time delay circuits, regulated power suppliers, static switches, motor controls, choppers, inverters, cyclo-converters, battery chargers, protective circuits, heater controls, and phase controls.

In recent years, SCRs have been designed to *control* powers as high as 10 MW with individual ratings as high as 2000 A at 1800 V. Its frequency range of application has also been extended to about 50 kHz, permitting some high-frequency applications such as induction heating and ultrasonic cleaning.

7.3 BASIC SILICON-CONTROLLED RECTIFIER (SCR) OPERATION

As the terminology indicates the SCR is a rectifier constructed of silicon material which has a third terminal for control purposes. Silicon was chosen because of its high temperature and power capabilities. The basic operation of the SCR is different from the fundamental two-layer semiconductor diode in that a third terminal, called a *gate,* determines when the rectifier switches from the open-circuit to the short-circuit state. It is not enough simply to forward-bias the anode-to-cathode region of the device. In the conduction region the dynamic resistance of the SCR is typically 0.01 to 0.1 Ω. The reverse resistance is typically 100 K or more.

The graphic symbol for the SCR is shown in Fig. 7.1 with the corresponding connections to the four-layer semiconductor structure. As indicated in Fig. 7.1a, if forward conduction is to be established, the anode must be positive with respect to the cathode. This is not, however, a sufficient criterion for turning the device on. A pulse of sufficient magnitude must also be applied to the gate to establish a turn-on gate current, represented symbolically by I_{GT}.

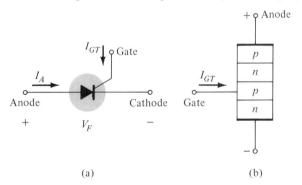

Figure 7.1 (a) SCR symbol; (b) basic construction.

A more detailed examination of the basic operation of an SCR is best effected by splitting the four-layer *pnpn* structure of Fig. 7.1b into two three-layer transistor structures as shown in Fig. 7.2a and then considering the resultant circuit of Fig. 7.2b.

Note that one transistor for Fig. 7.2 is an *npn* device while the other is a *pnp* transistor. For discussion purposes, the signal shown in Fig. 7.3a will be applied to the gate of the circuit of Fig. 7.2b. During the interval $0 \rightarrow t_1$, $V_{\text{gate}} = 0$ V, the circuit of Fig. 7.2b will appear as shown in Fig. 7.3b ($V_{\text{gate}} = 0$ V is equivalent to the gate terminal being grounded as shown in the figure). For $V_{BE_2} = V_{\text{gate}} = 0$ V, the base current $I_{B_2} = 0$ and I_{C_2} will be approximately I_{CO}. The base current of Q_1, $I_{B_1} = I_{C_2} = I_{CO}$, is too small to turn Q_1 on. Both transistors are therefore in the OFF state, resulting in a high impedance between

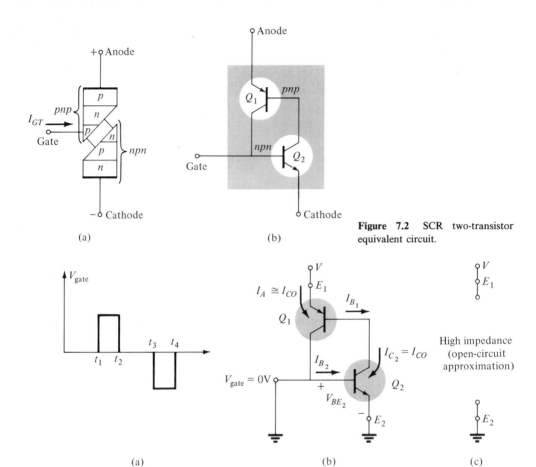

Figure 7.2 SCR two-transistor equivalent circuit.

Figure 7.3 OFF state of the SCR.

the collector and emitter of each transistor and the open-circuit representation for the controlled rectifier as shown in Fig. 7.3c.

At $t = t_1$ a pulse of V_G volts will appear at the SCR gate. The circuit conditions established with this input are shown in Fig. 7.4a. The potential V_G was chosen sufficiently large to turn Q_2 on $(V_{BE_2} = V_G)$. The collector current of Q_2 will then rise to a value sufficiently large to turn Q_1 on $(I_{B_1} = I_{C_2})$. As Q_1 turns on, I_{C_1} will increase, resulting in a corresponding increase in I_{B_2}. The increase in base current for Q_2 will result in a further increase in I_{C_2}. The net result is a regenerative increase in the collector current of each transistor. The resulting anode-to-cathode resistance $[R_{SCR} = V/(I_A - \text{large})]$ is then very small, resulting in the short-circuit representation for the SCR as indicated in Fig. 7.4b. The regenerative action described above results in SCRs having typical turn-on times of 0.1 to 1 μs. However, higher power devices in the 100- to 400-A range may have 10- to 25-μs turn-on times.

In addition to gate triggering, SCRs can also be turned on by significantly raising the temperature of the device or raising the anode-to-cathode voltage to the breakover value shown on the characteristics of Fig. 7.7.

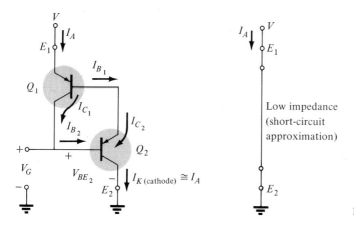

Figure 7.4 ON state of the SCR.

The next question of concern is: How long is the turn-off time and how is turn-off accomplished? An SCR *cannot* be turned off by simply removing the gate signal, and only a special few can be turned off by applying a negative pulse to the gate terminal as shown in Fig. 7.3a at $t = t_3$. The two general methods for turning off an SCR are categorized as the *anode current interruption* and the *forced commutation technique*. The two possibilities for current interruption are shown in Fig. 7.5.

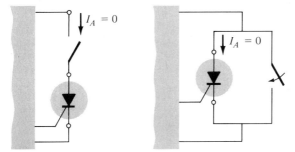

Figure 7.5 Anode current interruption.

In Fig. 7.5a, I_A is zero when the switch is opened (series interruption) while in Fig. 7.5b the same condition is established when the switch is closed (shunt interruption). Forced commutation is the "forcing" of current through the SCR in the direction opposite to forward conduction. There are a wide variety of circuits for performing this function, a number of which can be found in the manuals of major manufacturers in this area. One of the more basic types is shown in Fig. 7.6. As indicated in the figure, the turn-off circuit consists of an *npn* transistor, a dc battery V_B, and a pulse generator. During SCR conduction the transistor is in the "off state"; that is, $I_B = 0$ and the collector-to-emitter impedance is very high (for all practical purposes an open circuit). This high impedance will isolate the turn-off circuitry from affecting the operation of the SCR. For turn-off conditions, a positive pulse is applied to the base of the transistor, turning it heavily on, resulting in a very low impedance from collector to emitter (short-circuit representation). The battery potential will then appear directly across the SCR as shown in Fig. 7.6b, forcing current through it in the reverse direction for turn-off. Turn-off times of SCRs are typically 5 to 30 μs.

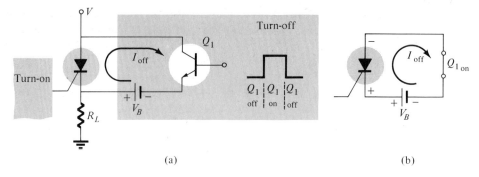

Figure 7.6 Forced commutation technique.

7.4 SCR CHARACTERISTICS AND RATINGS

The characteristics of an SCR are provided in Fig. 7.7 for various values of gate current. The currents and voltages of usual interest are indicated on the characteristic. A brief description of each follows.

1. *Forward breakover voltage* $V_{(BR)F*}$ is that voltage above which the SCR enters the conduction region. The asterisk (*) is a letter to be added that is dependent on the condition of the gate terminal as follows:

   O = open-circuit from G to K

   S = short-circuit from G to K

   R = resistor from G to K

   V = fixed bias (voltage) from G to K

2. *Holding current* (I_H) is that value of current below which the SCR switches from the conduction state to the forward blocking region under stated conditions.
3. *Forward and reverse blocking regions* are the regions corresponding to the open-circuit condition for the controlled rectifier which *block* the flow of charge (current) from anode to cathode.
4. *Reverse breakdown voltage* is equivalent to the Zener or avalanche region of the fundamental two-layer semiconductor diode.

It should be immediately obvious that the SCR characteristics of Fig. 7.7 are very similar to those of the basic two-layer semiconductor diode except for the horizontal offshoot before entering the conduction region. It is this horizontal jutting region that gives the gate control over the response of the SCR. For the characteristic having the solid line in Fig. 7.7 ($I_G = 0$) V_F must reach the largest required breakover voltage before the "collapsing" effect will result and the SCR can enter the conduction region corresponding to the *on* state. If the gate current is increased to I_{G_1}, as shown in the same figure, by applying a bias voltage to the gate terminal the value of V_F required for the conduction is considerably less. Note also that I_H drops with increase in I_G. If increased to I_{G_2} the SCR will fire at very low values of voltage and the characteristics begin to approach those of the basic *p-n* junction diode. Looking at the characteristics in a completely different sense, for a particular V_F voltage, say V_{F_1} (Fig. 7.7), if the gate current is increased from $I_G = 0$ to $I_G = I_{G_1}$, the SCR will fire.

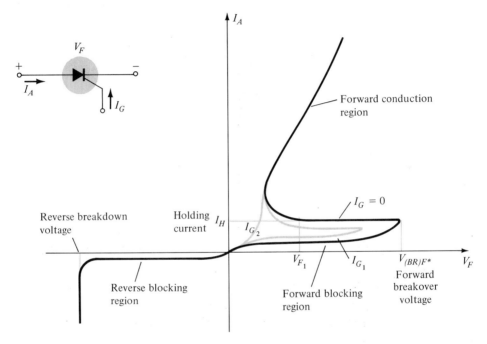

Figure 7.7 SCR characteristics.

The gate characteristics are provided in Fig. 7.8. The characteristics of Fig. 7.8b are an expanded version of the shaded region of Fig. 7.8a. In Fig. 7.8a the three gate ratings of greatest interest, P_{GFM}, I_{GFM}, and V_{GFM} are indicated. Each is included on the characteristics in the same manner employed for the transistor. Except for portions of the shaded region, any combination of gate current and voltage that falls within this region will fire any SCR in the series of components for which these characteristics are provided. Temperature will determine which sections of the shaded region must be avoided. At $-65°C$ the minimum current that will trigger the series of SCRs is 80 mA, while at $+150°C$ only 20 mA are required. The effect of temperature on the minimum gate voltage is usually not indicated on curves of this type since gate potentials of 3 V or more are usually obtained easily. As indicated on Fig. 7.8b, a minimum of 3 V is indicated for all units for the temperature range of interest.

Other parameters usually included on the specification sheet of an SCR are the turn-on time (t_{on}), turn-off time (t_{off}), junction temperature (T_j), and case temperature (T_C), all of which should by now be, to some extent, self-explanatory.

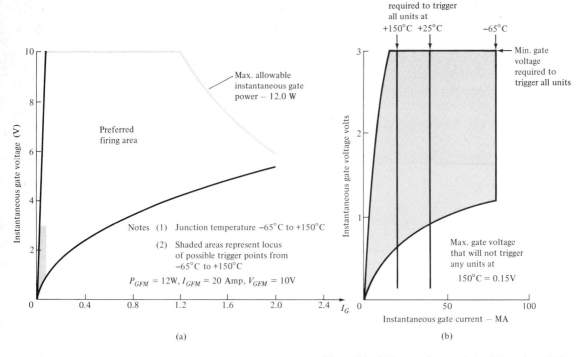

Figure 7.8 SCR gate characteristics (GE series—C38).

Figure 7.9 (a) Alloy-diffused SCR pellet; (b) thermal fatigue-free SCR construction. (Courtesy General Electric Company.)

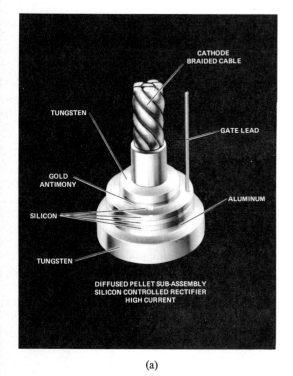

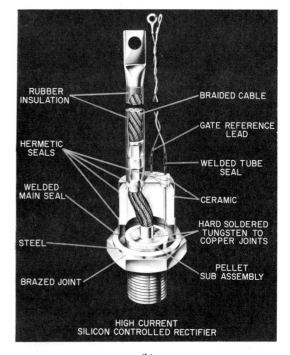

7.5 SCR CONSTRUCTION AND TERMINAL IDENTIFICATION

The basic construction of the four-layer pellet of an SCR is shown in Fig. 7.9a. The complete construction of a thermal-fatigue-free, high-current SCR is shown in Fig. 7.9b. Note the position of the gate, cathode, and anode terminals. The pedestal acts as a heat sink by transferring the heat developed to the chassis on which the SCR is mounted. The case construction and terminal identification of SCRs will vary with the application. Other case-construction techniques and the terminal identification of each are indicated in Fig. 7.10.

Figure 7.10 SCR case construction and terminal identification. (*Top* and *middle,* courtesy General Electric Company; *bottom,* courtesy International Rectifier Corporation, Inc.)

7.6 SCR APPLICATIONS

A few of the possible applications for the SCR are listed in the introduction to the SCR (Section 7.2). In this section we consider five: a static switch, phase control system, battery charger, temperature controller, and single-source emergency lighting system.

A half-wave *series static switch* is shown in Fig. 7.11a. If the switch is closed as shown in Fig. 7.11b, a gate current will flow during the positive portion of the input signal, turning the SCR on. Resistor R_1 limits the magnitude of the gate current. When the SCR turns on, the anode-to-cathode voltage (V_F) will drop to the conduction value, resulting in a greatly reduced gate current and very little loss in the gate circuitry. For the negative region of the input signal the SCR will turn off, since the anode is negative with respect to the cathode. The diode D_1 is included to prevent a reversal in gate current.

The waveforms for the resulting load current and voltage are shown in Fig. 7.11b. The result is a half-wave rectified signal through the load. If less than 180° conduction is desired, the switch can be closed at any phase displacement during the positive portion of the input signal. The switch can be electronic, electromagnetic, or mechanical, depending on the application.

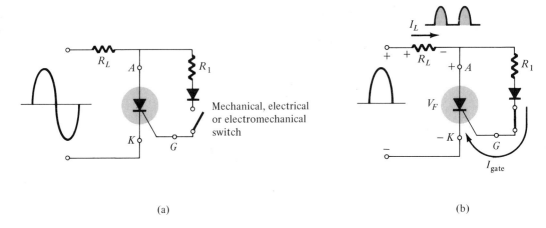

(a)   (b)

Figure 7.11 Half-wave series static switch.

A circuit capable of establishing a conduction angle between 90° and 180° is shown in Fig. 7.12a. The circuit is similar to that of Fig. 7.11a except for the addition of a variable resistor and the elimination of the switch. The combination of the resistors R and R_1 will limit the gate current during the positive portion of the input signal. If R_1 is set to its maximum value, the gate current may never reach turn-on magnitude. As R_1 is decreased from the maximum the gate current will increase for the same input voltage. In this way, the required turn-on gate current can be established in any point between 0° and 90° as shown in Fig. 7.12b. If R_1 is low, the SCR will fire almost immediately, resulting in the same action as that obtained from the circuit of Fig. 7.11a (180° conduction). However, as indicated above, if R_1 is increased, a larger input voltage (positive)

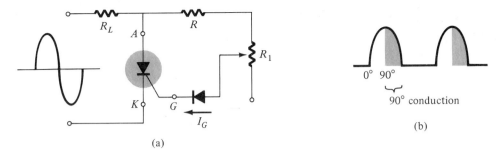

Figure 7.12 Half-wave variable-resistance phase control.

will be required to fire the SCR. As shown in Fig. 7.12b, the control cannot be extended past a 90° phase displacement since the input is its maximum at this point. If it fails to fire at this and lesser values of input voltage on the positive slope of the input, the same response must be expected from the negatively sloped portion of the signal waveform. The operation here is normally referred to in technical terms as *half-wave variable resistance phase control*. It is an effective method of controlling the rms current and therefore power to the load.

A third popular application of the SCR is in a *battery-charging regulator*. The fundamental components of the circuit are shown in Fig. 7.13. You will note that the control circuit has been blocked off for discussion purposes.

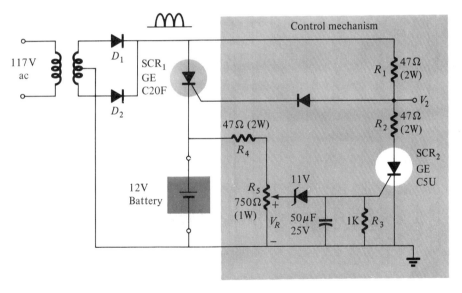

Figure 7.13 Battery-charging regulator.

As indicated in the figure, D_1 and D_2 establish a full-wave rectified signal across SCR_1 and the 12-V battery to be charged. At low battery voltages SCR_2 is in the off state for reasons to be explained shortly. With SCR_2 open, the SCR_1 controlling circuit is exactly the same as the series static switch control discussed earlier in this section. When the full-wave rectified input is sufficiently large to produce the required turn-on gate current (controlled by R_1), SCR_1 will turn on and charging of the battery will commence. At the start of charging, the low battery voltage will result in a low voltage V_R as determined by the simple voltage divider circuit. Voltage V_R is in turn too small to cause 11.0-V Zener conduction. In the off state, the Zener is effectively an open-circuit maintaining SCR_2 in the off state since the gate current is zero. The capacitor C_1 is included to prevent any voltage transients in the circuit from accidentally turning on SCR_2. Recall from your fundamental study of circuit analysis that the voltage cannot instantaneously change across a capacitor. In this way C_1 prevents transient effects from affecting the SCR.

As charging continues, the battery voltage rises to a point where V_R is sufficiently high to both, turn on the 11.0-V Zener and fire SCR_2. Once SCR_2 has fired, the short-circuit representation for SCR_2 will result in a voltage divider circuit determined by R_1 and R_2 that will maintain V_2 at a level too small to turn SCR_1 on. When this occurs, the battery is fully charged and the open-circuit state of SCR_1 will cut off the charging current. Thus, the regulator recharges the battery whenever the voltage drops and prevents overcharging when fully charged.

The schematic diagram of a 100-W heater control using an SCR appears in Fig. 7.14. It is designed such that the 100-W heater will turn on and off as determined by thermostats. Mercury-in-glass thermostats are very sensitive to temperature change. In fact, they can sense changes as small as 0.1°C. It is limited in application, however, in that it can only handle very low levels of current—below 1 mA. In this application, the SCR serves as a current amplifier in a load-switching element. It is not an amplifier in the sense that it magnifies the current level of the thermostat. Rather it is a device whose higher current level is controlled by the behavior of the thermostat.

It should be clear that the bridge network is connected to the ac supply through the 100-W heater. This will result in a full-wave rectified voltage across the SCR. When the thermostat is open, the voltage across the capacitor will charge to a gate firing potential through each pulse of the rectified signal. The charging time constant is determined by the RC product. This will trigger the SCR during each half-cycle of the input signal, permitting a flow of current to the heater.

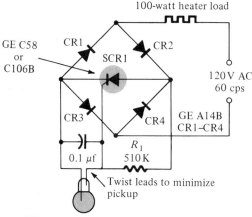

Figure 7.14 Temperature controller. (Courtesy General Electric Semiconductor Products Division.)

As the temperature rises, the conductive thermostat will short-circuit the capacitor, eliminating the possibility of the capacitor charging to the firing potential and triggering the SCR. The 510-K resistor will then contribute to maintaining a very low current (less than 250 μA) through the thermostat.

The last application for the SCR to be described is shown in Fig. 7.15. It is a single-source emergency lighting system that will maintain the charge on a 6-V battery to ensure its availability and also provide dc energy to a bulb if there is a power shortage.

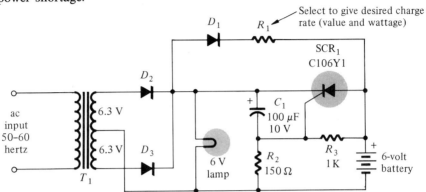

Figure 7.15 Single-source emergency lighting system. (Courtesy General Electric Semiconductor Products Division.)

A full-wave rectified signal will appear across the 6-V lamp due to diodes D_2 and D_1. The capacitor C_1 will charge to a voltage slightly less than a difference between the peak value of the full-wave rectified signal and the dc voltage across R_2 established by the 6-V battery. In any event, the cathode of SCR_1 is higher than the anode and the gate-to-cathode voltage is negative, ensuring that the SCR is nonconducting. The battery is being charged through R_1 and D_1 at a rate determined by R_1. Charging will only take place when the anode of D_1 is more positive than its cathode. The dc level of the full-wave rectified signal will ensure that the bulb is lit when the power is on. If the power should fail, the capacitor C_1 will discharge through D_1, R_1, and R_3 until the cathode of SCR_1 is less positive than the anode. At the same time the junction of R_2 and R_3 will become positive and establish sufficient gate-to-cathode voltage to trigger the SCR. Once fired, the 6-V battery would discharge through the SCR_1 and energize the lamp and maintain its illumination.

Once power is restored the capacitor C_1 will recharge and reestablish the nonconducting state of SCR_1 as described above.

7.7 SILICON-CONTROLLED SWITCH

The silicon-controlled switch (SCS), like the silicon-controlled rectifier, is a four-layer *pnpn* device. All four semiconductor layers of the SCS are available due to the addition of an anode gate, as shown in 7.16a. The graphic symbol and transistor equivalent circuit are shown in the same figure.

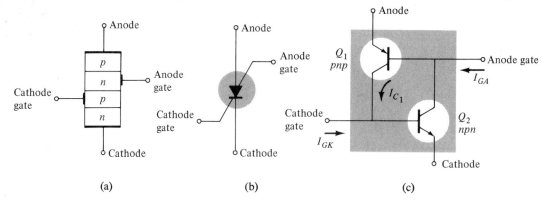

Figure 7.16 Silicon-controlled switch (SCS): (a) basic construction; (b) graphic symbol; (c) equivalent transistor circuit.

The characteristics of the device are essentially the same as those for the SCR. The effect of an anode gate current is very similar to that demonstrated by the gate current in Fig. 7.7. The higher the anode gate current, the lower the required anode-to-cathode voltage to turn the device on.

The anode gate connection can be used to turn the device either on or off. To turn on the device, a negative pulse must be applied to the anode gate terminal, while a positive pulse is required to turn off the device. The need for the type of pulse indicated above can be demonstrated using the circuit of Fig. 7.16c. A negative pulse at the anode gate will forward-bias the base-to-emitter junction of Q_1, turning it on. The resulting heavy collector current I_{C_1} will turn on Q_2, resulting in a regenerative action and the on state for the SCS device. A positive pulse at the anode gate will reverse-bias the base-to-emitter junction of Q_1, turning if off, resulting in the open-circuit off state of the device. In general, the triggering (turn-on) anode gate current is larger in magnitude than the required cathode gate current. For one representative SCS device, the triggering anode gate current is 1.5 mA while the required cathode gate current is 1 μA. The required turn-on gate current at either terminal is affected by many factors. A few include the operating temperature, anode-to-cathode voltage, load placement, and type of cathode, gate-to-cathode or anode gate-to-anode connection (short-circuit, open-circuit, bias, load, etc.). Tables, graphs, and curves are normally available for each device to provide the type of information indicated above.

Three of the more fundamental types of turn-off circuits for the SCS are shown in Fig. 7.17. When a pulse is applied to the circuit of Fig. 7.17a, the transistor conducts heavily, resulting in a low impedance ($\cong$ short-circuit) characteristic between collector and emitter. This low-impedance branch diverts anode current away from the SCS, dropping it below the holding value and consequently turning it off. Similarly, the positive pulse at the anode gate of Fig. 7.17b will turn the SCS off by the mechanism described earlier in this section. The circuit of Fig. 7.17c can be turned either off *or* on by a pulse of the proper magnitude at the cathode gate. The turn-off characteristic is possible only if the correct value of R_A is employed. It will control the amount of regenerative feedback, the magnitude of which is critical for this type of operation. Note the variety of positions in

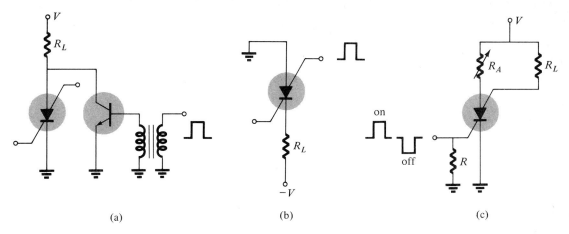

Figure 7.17 SCS turn-off techniques.

which the load resistor R_L can be placed. There are a number of other possibilities that can be found in any comprehensive semiconductor handbook or manual.

An advantage of the SCS over a corresponding SCR is the reduced turn-off time, typically within the range 1 to 10 μs for the SCS and 5 to 30 μs for the SCR.

Some of the remaining advantages of the SCS over an SCR include increased control and triggering sensitivity and a more predictable firing situation. At present, however, the SCS is limited to low power, current, and voltage ratings. Typical maximum anode currents range from 100 to 300 mA with dissipation (power) ratings of 100 to 500 mW.

A few of the more common areas of application include a wide variety of computer circuits (counters, registers, and timing circuits) pulse generators, voltage sensors, and oscillators. One simple application for an SCS as a voltage-sensing device is shown in Fig. 7.18. It is an alarm system with n inputs from various stations. Any single input will turn that particular SCS on, resulting in an energized alarm relay and light in the anode gate circuit to indicate the location of the input (disturbance).

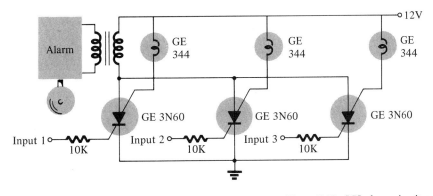

Figure 7.18 SCS alarm circuit.

SEC. 7.7 SILICON-CONTROLLED SWITCH

One additional application of the SCS is in the alarm circuit of Fig. 7.19. R_s represents a temperature, light, or radiation sensitive resistor, that is, an element whose resistance will decrease with the application of any of the three energy sources listed above. The cathode gate potential is determined by the divider relationship established by R_s and the variable resistor. Note that the gate potential is at approximately 0 volts if R_s equals the value set by the variable resistor, since both resistors will have 12 V across them. However, if R_s decreases, the potential of the junction will increase until the SCS is forward-biased, causing the SCS to turn on and energize the alarm relay.

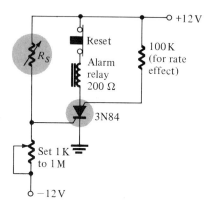

Figure 7.19 Alarm circuit. (Courtesy General Electric Semiconductor Products Division.)

The 100-K resistor is included to reduce the possibility of accidental triggering of the device through a phenomena known as *rate effect*. It is caused by the stray capacitance levels between gates. A high-frequency transent can establish sufficient base current to turn the SCS on accidentally. The device is reset by pressing the reset button, which in turn opens the conduction path of the SCS and reduces the anode current to zero.

Sensitivity to resistors R_s that increase in resistance due to the application of any of the three energy sources described above can be accommodated by simply interchanging the location of R_s and the variable resistor.

The terminal identification of an SCS is shown in Fig. 7.20 with a packaged SCS.

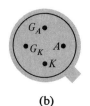

(a) (b)

Figure 7.20 Silicon-controlled switch (SCS): (a) device; (b) terminal identification. (Courtesy General Electric Company.)

7.8 GATE TURN-OFF SWITCH

The gate turn-off switch (GTO) is the third *pnpn* device to be introduced in this chapter. Like the SCR, however, it has only three external terminals, as indicated in Fig. 7.21a. Its graphic symbol is also shown in Fig. 7.21b. Although the graphic symbol is different from either the SCR or the SCS, the transistor equivalent is exactly the same and the characteristics are similar.

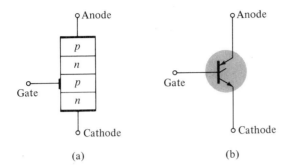

Figure 7.21 Gate turn-off switch (GTO): (a) basic construction; (b) symbol.

The most obvious advantage of the GTO over the SCR or SCS is the fact that it can be turned on *or* off by applying the proper pulse to the cathode gate (without the anode gate and associated circuitry required for the SCS). A consequence of this turn-off capability is an increase in the magnitude of the required gate current for triggering. For an SCR and GTO of similar maximum rms current ratings, the gate-triggering current of a particular SCR is 30 μA, while the triggering current of the GTO is 20 mA. The turn-off current of a GTO is slightly larger than the required triggering current. The maximum rms current and dissipation ratings of GTOs manufactured today is limited to about 3 A and 20 W, respectively.

A second very important characteristic of the GTO is improved switching characteristics. The turn-on time is similar to the SCR (typically 1 μs), but the turn-off time of about the *same* duration (1 μs) is much smaller than the typical turn-off time of an SCR (5 to 30 μs). The fact that the turn-off time is similar to the turn-on time rather than considerably larger permits the use of this device in high-speed applications.

A typical GTO and its terminal identification are shown in Fig. 7.22. The GTO gate input characteristics and turn-off circuits can be found in a comprehensive

Figure 7.22 Typical GTO and its terminal identification. (Courtesy General Electric Company.)

279

manual or specification sheet. The majority of the SCR turn-off circuits can also be used for GTOs.

Some of the areas of application for the GTO include counters, pulse generators, multivibrators, and voltage regulators. Figure 7.23 is an illustration of a simple sawtooth generator employing a GTO and a Zener diode.

When the supply is energized, the GTO will turn on, resulting in the short-circuit equivalent from anode to cathode. The capacitor C_1 will then begin to charge toward the supply voltage as shown in Fig. 7.23. As the voltage across the capacitor C_1 charges above the Zener potential, a reversal in gate-to-cathode voltage will result, establishing a reversal in gate current. Eventually, the negative gate current will be large enough to turn the GTO off. Once the GTO turns off, resulting in the open-circuit representation, the capacitor C_1 will discharge through the resistor R_3. The discharge time will be determined by the circuit time constant $\tau = R_3 C_1$. The proper choice of R_3 and C_1 will result in the sawtooth waveform of Fig. 7.23. Once the output potential V_o drops below V_Z, the GTO will turn on and the process will repeat.

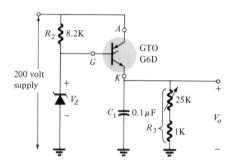
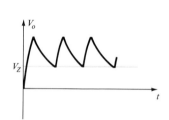

Figure 7.23 GTO sawtooth generator.

7.9 LIGHT-ACTIVATED SCR

The next in the series of *pnpn* devices is the light-activated SCR (LASCR). As indicated by the terminology, it is an SCR whose state is controlled by the light falling upon a silicon semiconductor layer of the device. The basic construction of an LASCR is shown in Fig. 7.24a.

As indicated in Fig. 7.24a, a gate lead is also provided to permit triggering the device using typical SCR methods. Note also in the figure that the mounting surface for the silicon pellet is the anode connection for the device.

The graphic symbols most commonly employed for the LASCR are provided in Fig. 7.24b. The terminal identification and typical LASCRs are shown in Fig. 7.25a.

Some of the areas of application for the LASCR include optical light controls, relays, phase control, motor control, and a variety of computer applications. The maximum current (rms) and power (gate) ratings for LASCRs commercially available today are about 3 A and 0.1 W. The characteristics (light triggering) of a representative LASCR are provided in Fig. 7.25b. Note in this figure that

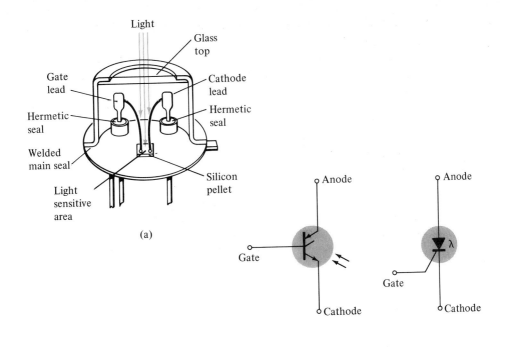

Figure 7.24 Light-activated SCR (LASCR): (a) basic construction; (b) symbols.

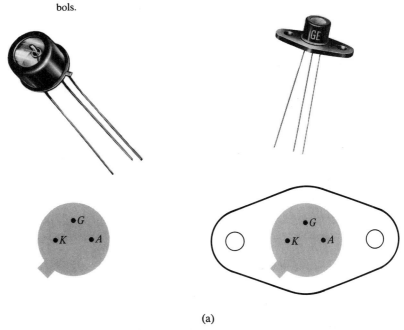

(a)

Figure 7.25 LASCR: (a) appearance and terminal identification. (Courtesy General Electric Company.)

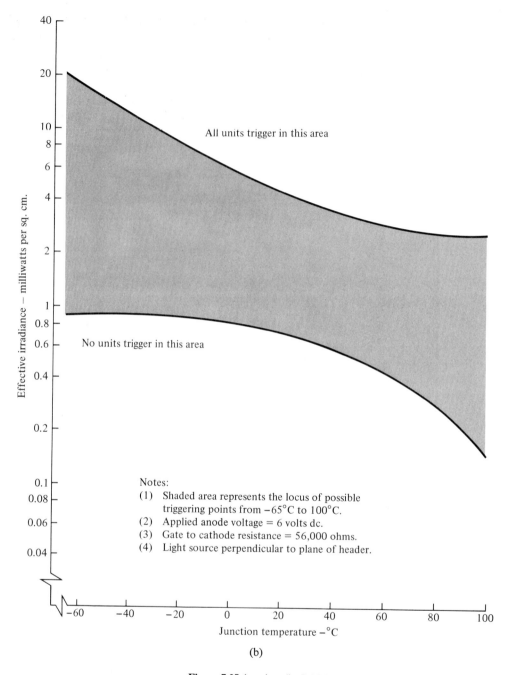

Figure 7.25 (continued) LASCR: (b) light-triggering characteristics.

an increase in junction temperature results in a reduction in light energy required to activate the device.

One interesting application of an LASCR is in the AND and OR circuits of Fig. 7.26. Only when light falls on LASCR$_1$ *and* LASCR$_2$ will the short-circuit representation for each be applicable and the supply voltage appear across the load. For the OR circuit, light energy applied to LASCR$_1$ *or* LASCR$_2$ will result in the supply voltage appearing across the load.

The LASCR is most sensitive to light when the gate terminal is open. Its sensitivity can be reduced and controlled somewhat by the insertion of a gate resistor, as shown in Fig. 7.26.

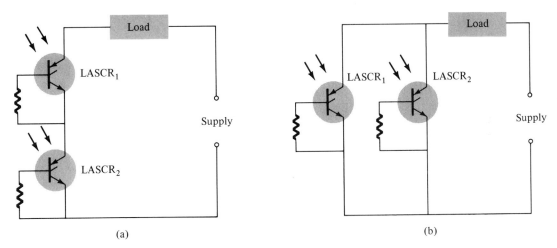

Figure 7.26 LASCR optoelectronic logic circuitry: (a) AND gate—input to LASCR$_1$ *and* LASCR$_2$ required for energization of the load; (b) OR gate—input to either LASCR$_1$ *or* LASCR$_2$ will energize the load.

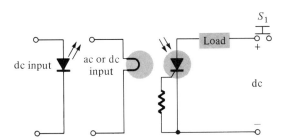

Figure 7.27 Latching relay. (Courtesy General Electric Semiconductor Products Division.)

A second application of the LASCR appears in Fig. 7.27. It is the semiconductor analog of an electromechanical relay. Note that it offers complete isolation between the input and switching element. The energizing current can be passed through a light-emitting diode or a lamp, as shown in the figure. The incident light will cause the LASCR to turn on and permit a flow of current through the load as established by the dc supply. The LASCR can be turned off using the reset switch S_1. This system offers the additional advantages over an electromechanical switch of long life, microsecond response, small size, and the elimination of contact bounce.

7.10 SHOCKLEY DIODE

The Shockley diode is a four-layer *pnpn* diode with only two external terminals, as shown in Fig. 7.28a with its graphic symbol. The characteristics (Fig. 7.28b) of the device are exactly the same as those encountered for the SCR with $I_G = 0$. As indicated by the characteristics, the device is in the off state (open-circuit representation) until the breakover voltage is reached, at which time avalanche conditions develop and the device turns on (short-circuit representation).

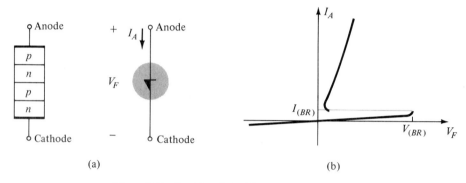

Figure 7.28 Shockley diode: (a) basic construction and symbol; (b) characteristics.

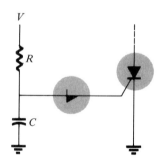

Figure 7.29 Shockley diode application—trigger switch for an SCR.

One common application of the Shockley diode is shown in Fig. 7.29, where it is employed as a trigger switch for an SCR. When the circuit is energized, the voltage across the capacitor will begin to change toward the supply voltage. Eventually, the voltage across the capacitor will be sufficiently high to first turn on the Shockley diode and then the SCR.

7.11 DIAC

The diac is basically a two-terminal parallel-inverse combination of semiconductor layers that permits triggering in either direction. The characteristics of the device, presented in Fig. 7.30a, clearly demonstrate that there is a breakover voltage in either direction. This possibility of an on condition in either direction can be used to its fullest advantage in ac applications.

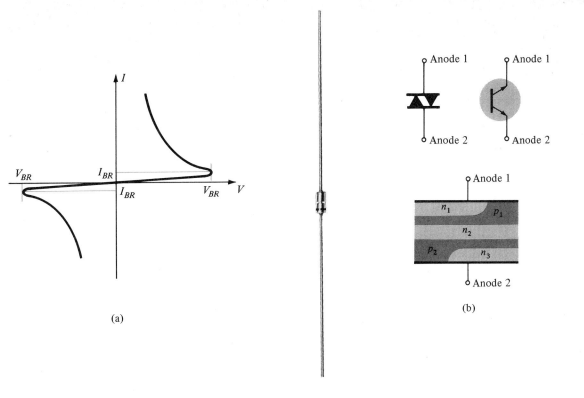

Figure 7.30 Diac: (a) characteristics; (b) symbols and basic construction. (Courtesy General Electric Company.)

The basic arrangement of the semiconductor layers of the diac is shown in Fig. 7.30b, along with its graphic symbol. Note that neither terminal is referred to as the cathode. Instead, there is an anode 1 (or electrode 1) and an anode 2 (or electrode 2). When anode 1 is positive with respect to anode 2, the semiconductor layers of particular interest are $p_1 n_2 p_2$ and n_3. For anode 2 positive with respect to anode 1 the applicable layers are $p_2 n_2 p_1$ and n_1.

For the unit appearing in Fig. 7.30, the breakdown voltages are very close in magnitude but may vary from a minimum of 28 V to a maximum of 42 V. They are related by the following equation provided in the specification sheet:

$$V_{BR_1} = V_{BR_2} \pm 10\% \ V_{BR_2} \tag{7.1}$$

The current levels (I_{BR_1} and I_{BR_2}) are also very close in magnitude for each device. For the unit of Fig. 7.30, both current levels are about 200 μA = 0.2 mA.

The use of the diac in a proximity detector appears in Fig. 7.31. Note the use of an SCR in series with the load and the programmable unijunction transistor (to be described shortly) connected directly to the sensing electrode.

As the human body approaches the sensing electrode, the capacitance between the electrode and ground will increase. The programmable UJT (PUT) is a device

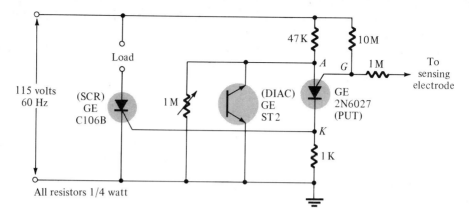

Figure 7.31 Proximity detector or touch switch. (Courtesy General Electric Semiconductor Products Division.)

that will fire (enter the short-circuit state) when the anode voltage (V_A) is at least 0.7 V (for silicon) greater than the gate voltage (V_G). Before the programmable device turns on, the system is essentially as shown in Fig. 7.32. As the input voltage rises, the diac voltage V_G will follow as shown in the figure until the firing potential is reached. It will then turn on and the diac voltage will drop substantially, as shown. Note that the diac is in essentially an open-circuit state until it fires. Before the capacitive element is introduced, the voltage V_G will be the same as the input. As indicated in the figure, since both V_A and V_G follow the input, V_A can never be greater than V_G by 0.7 V and turn on the device. However, as the capacitive element is introduced, the voltage V_G will begin to lag the input voltage by an increasing angle, as indicated in the figure. There is therefore a point established where V_A can exceed V_G by 0.7 V and cause the programmable device to fire. A heavy current will flow through the PUT at this point, raising the voltage V_K and turning on the SCR. A heavy SCR current

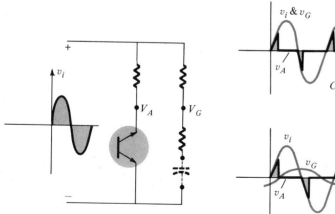

Figure 7.32 Effect of capacitive element on the behavior of the network of Fig. 7.31.

can then flow through the load, reacting to the presence of the approaching individual.

A second application of the diac appears in the next section (Fig. 7.34) as we consider an important power control device: the triac.

7.12 TRIAC

The triac is fundamentally a diac with a gate terminal for controlling the turn-on conditions of the bilateral device in either direction. In other words, for either direction the gate current can control the action of the device in a manner very similar to that demonstrated for an SCR. The characteristics, however, of the triac in the first and third quadrants are somewhat different from those of the diac, as shown in Fig. 7.33c. Note the holding current in each direction not present in the characteristics of the diac.

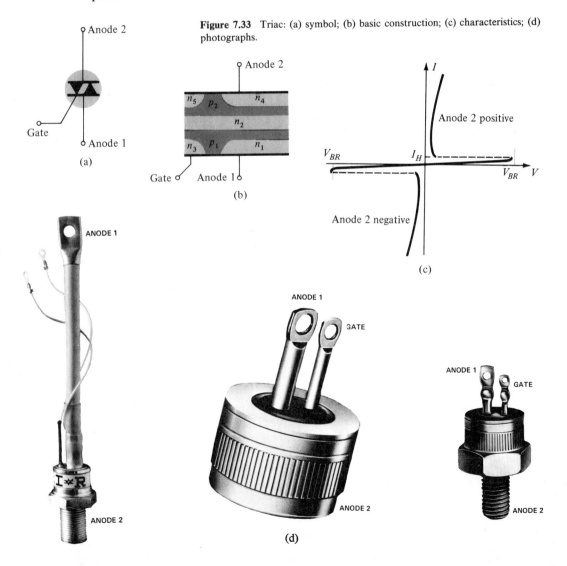

Figure 7.33 Triac: (a) symbol; (b) basic construction; (c) characteristics; (d) photographs.

The graphic symbol for the device and the distribution of the semiconductor layers are provided in Fig. 7.33 with photographs of the device. For each possible direction of conduction there is a combination of semiconductor layers whose state will be controlled by the signal applied to the gate terminal.

One fundamental application of the triac is presented in Fig. 7.34. In this capacity, it is controlling the ac power to the load by switching on and off during the positive and negative regions of input sinusoidal signal. The action of this circuit during the positive portion of the input signal is very similar to that encountered for the Shockley diode in Fig. 7.29. The advantage of this configuration is that during the negative portion of the input signal the same type of response will result, since both the diac and triac can fire in the reverse direction. The resulting waveform for the current through the load is provided in Fig. 7.34. By varying the resistor R the conduction angle can be controlled. There are units available today that can handle in excess of 10-kW loads.

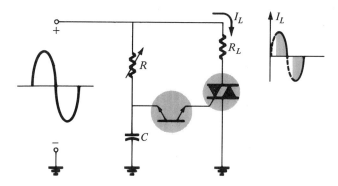

Figure 7.34 Triac application: phase (power) control.

The information appearing in a data sheet for the CSB20, 40, and 60 series triacs from Unitrode appears in Figs. 7.35 through 7.39. It has been found that the solid-state thyristor performs better than the electromechanical relay in a number of applications, such as in appliances and industrial controls, and is now approaching the same price. The units use a beryllia substrate to improve the thermal conductance and bring the thermal resistance down to 1.2°C/W (maximum). Note the high values of off-state voltage and the high one-cycle surge current. Note also that the holding current is now a higher level at 50 mA and the gate-trigger current is 50 to 80 mA.

Once the device is in the on state, note the similarities between the characteristics of Fig. 7.36 and those of a typical p-n semiconductor diode. In Fig. 7.37 we find that the maximum level of current flow decreases rapidly with the number of cycles that that flow is permitted to exist. Figure 7.38 depicts how the maximum power dissipation is related to the full-cycle on-state current, while Fig. 7.39 demonstrates how the maximum case temperature decreases with increase in on-state current.

(a) Absolute Maximum Ratings

	CSB20	CSB40	CSB60
Repetitive peak off-state voltage, V_{DRM}	200V	400 V	600 V
On-state current $I_{T(RMS)}$ (at $T_c = 65°C$ and conduction angle of 360°)		25 A	
Peak one cycle surge (non-rep.) on-state current, I_{TSM}		250 A	
Peak gate power, P_{GM}		16 W	
Average gate power, $P_{G(AV)}$		0.5 W	
Rate of on-state current, di/dt (at $V_{DM} = V_{DRM}$, $I_{GT} = 175$ mA, $t_r = 0.1$ μs)		125 A/μs	
Storage temperature range		−40°C to +150°C	
Operating temperature range		−40°C to +110°C	
Isolation voltage, flange to terminal		2500 V RmS	

(b) Electrical Specifications (at 25°C Unless Noted)

Test	Symbol	Min.	Typical	Max.	Units	Test Conditions
Off-state current	I_{DRM}	—	—	2.0	mA	V_{DRM} = rating, T_C = 100°C
Gate trigger current	I_{GT}	—	—	50	mA	$\{V_D = 12$ V Quadrants 1, 3 (+ +, − −)
		—	—	80		$V_D = 12$ V Quadrants 2, 4 (+ −, − +)
Gate trigger voltage	V_{GT}	—	—	2.5	V	$V_D = 12$ V
Peak on-state voltage	V_{TM}	—	—	1.9	V	I_{TM} = 28 A Peak
Holding current	I_H	—	—	50	mA	$V_D = 12$ V
Critical rate of rise— off-state voltage	dv/dt	20	50	—	V/μS	V_{DRM} = rating, T_C = 100°C
Critical rate of rise— commutated off-state voltage	$dv/dt_{(c)}$	3	10	—	V/μS	I_T = rating, V_{DRM} = rating, T_C = 65°C
Steady-state thermal resistance (junction-to-case)	$R\theta_{JC}$	—	—	1.1	°C/W	Steady state

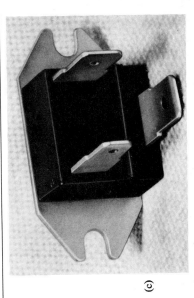

(c)

Figure 7.35 Unitrode CBS 20,40,60 series triacs. (Courtesy Unitrode.)

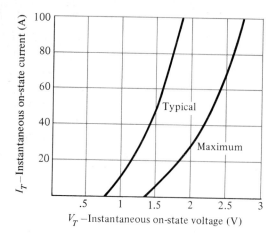

Figure 7.36 I-V on-state characteristics. (Courtesy Unitrode.)

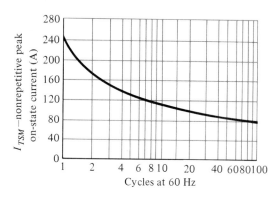

Figure 7.37 Maximum allowable nonrepetitive peak on-state current following rated load conditions.

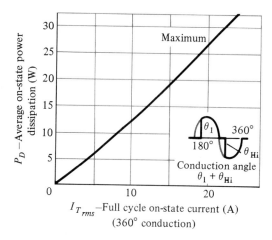

Figure 7.38 Maximum conduction power dissipation versus on-state current (50 or 60 Hz).

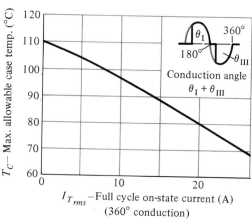

Figure 7.39 Maximum allowable case temperature versus on-state current (50 or 60 Hz).

OTHER DEVICES

7.13 UNIJUNCTION TRANSISTOR

Recent interest in the unijunction transistor (UJT) has, like that for the SCR, been increasing at an exponential rate. Although first introduced in 1948, the device did not become commercially available until 1952. The low cost per unit, combined with the excellent characteristics of the device, have warranted its use in a wide variety of applications. A few include oscillators, trigger circuits, sawtooth generators, phase control, timing circuits, bistable networks, and voltage- or current-regulated supplies. The fact that this device is, in general, a low-power absorbing device under normal operating conditions is a tremendous aid in the continual effort to design relatively efficient systems.

The UJT is a three-terminal device having the basic construction of Fig. 7.40. A slab of lightly doped (increased resistance characteristic) n-type silicon material has two base contacts attached to both ends of one surface and an aluminum rod alloyed to the opposite surface. The p-n junction of the device is formed at the boundary of the aluminum rod and the n-type silicon slab. The single p-n junction accounts for the terminology unijunction. It was originally called a duo (double) base diode due to the presence of two base contacts. Note in Fig. 7.40 that the aluminum rod is alloyed to the silicon slab at a point closer to the base 2 contact than the base 1 contact and that the base 2 terminal is made positive with respect to the base 1 terminal by V_{BB} volts. The effect of each will become evident in the paragraphs to follow.

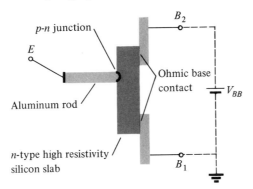

Figure 7.40 Unijunction transistor (UJT): basic construction.

The symbol for the unijunction transistor is provided in Fig. 7.41. Note that the emitter leg is drawn at an angle to the vertical line representing the slab of n-type material. The arrowhead is pointing in the direction of conventional current (hole) flow when the device is in the forward-biased, active, or conducting state.

The circuit equivalent of the UJT is shown in Fig. 7.42. Note the relative simplicity of this equivalent circuit: two resistors (one fixed, one variable) and a single diode. The resistance R_{B_1} is shown as a variable resistor since its magnitude will vary with the current I_E. In fact, for a representative unijunction transistor, R_{B_1} may vary from 5 K down to 50 Ω for a corresponding change of I_E from 0 to 50 μA. The interbase resistance R_{BB} is the resistance of the device between

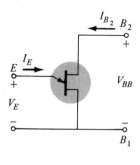

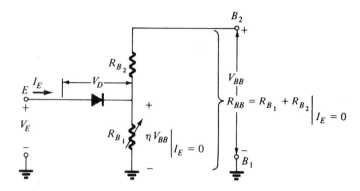

Figure 7.41 Symbol and basic biasing arrangement for the unijunction transistor.

Figure 7.42 UJT equivalent circuit.

terminals B_1 and B_2 when $I_E = 0$. In equation form,

$$R_{BB} = (R_{B_1} + R_{B_2})\big|_{I_E = 0} \tag{7.2}$$

(R_{BB} is typically within the range 4 to 10 K.) The position of the aluminum rod of Fig. 7.40 will determine the relative values of R_{B_1} and R_{B_2} with $I_E = 0$. The magnitude of $V_{R_{B1}}$ (with $I_E = 0$) is determined by the voltage divider rule in the following manner:

$$V_{R_{B_1}} = \frac{R_{B_1} V_{BB}}{R_{B_1} + R_{B_2}} = \eta V_{BB}\bigg|_{I_E = 0} \tag{7.3}$$

The Greek letter η (eta) is called the *intrinsic stand-off* ratio of the device and is defined by

$$\eta = \frac{R_{B_1}}{R_{B_1} + R_{B_2}}\bigg|_{I_E = 0} \tag{7.4}$$

For applied emitter potentials (V_E) greater than $V_{R_{B_1}} = \eta V_{BB}$ by the forward voltage drop of the diode, V_D (0.35 → 0.70 V) the diode will fire, assume that the short-circuit representation (on an ideal basis), and I_E will begin to flow through R_{B_1}. In equation form the emitter firing potential is given by

$$V_P = \eta V_{BB} + V_D \tag{7.5}$$

The characteristics of a representative unijunction transistor are shown for $V_{BB} = 10$ V in Fig. 7.43. Note that for emitter potentials to the left of the peak point, the magnitude of I_E is never greater than I_{EO} (measured in microamperes). The current I_{EO} corresponds very closely with the reverse leakage current I_{CO} of the conventional bipolar transistor. This region, as indicated in the figure, is called the cutoff region. Once conduction is established at $V_E = V_P$, the emitter potential

V_E will drop with increase in I_E. This corresponds exactly with the decreasing resistance R_{B_1} for increasing current I_E, as discussed earlier. This device, therefore, has a *negative resistance* region which is stable enough to be used with a great deal of reliability in the areas of application listed earlier. Eventually, the valley point will be reached, and any further increase in I_E will place the device in the saturation region. In this region the characteristics approach that of the semiconductor diode in the equivalent circuit of Fig. 7.42.

The decrease in resistance in the active region is due to the holes injected into the *n*-type slab from the aluminum *p*-type rod when conduction is established. The increased hole content in the *n*-type material will result in an increase in the number of free electrons in the slab, producing an increase in conductivity *(G)* and a corresponding drop in resistance $(R\downarrow = 1/G\uparrow)$. Three other important parameters for the unijunction transistor are I_P, V_V, and I_V. Each is indicated on Fig. 7.43. They are all self-explanatory.

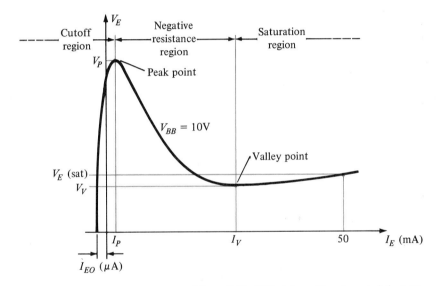

Figure 7.43 UJT static emitter characteristic curve.

The emitter characteristics as they normally appear are provided in Fig. 7.44. Note that I_{EO} (μA) is not in evidence since the horizontal scale is in milliamperes. The intersection of each curve with the vertical axis is the corresponding value of V_P. For fixed values of η and V_D, the magnitude of V_P will vary as V_{BB}, that is,

$$V_P\uparrow = \underbrace{\eta V_{BB}\uparrow + V_D}_{\text{fixed}}$$

A typical set of specifications for the UJT is provided in Fig. 7.45b. The discussion of the last few paragraphs should make each quantity readily recognizable. The terminal identification is provided in the same figure with a photograph of a representative UJT. Note that the base terminals are opposite each other

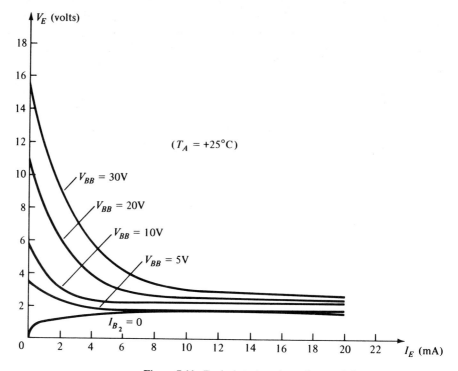

Figure 7.44 Typical static emitter characteristic curves for a UJT.

absolute maximum ratings: (25°C)

Power Dissipation	300 mw
RMS Emitter Current	50 ma
Peak Emitter Current	2 amperes
Emitter Reverse Voltage	30 volts
Interbase Voltage	35 volts
Operating Temperature Range	−65°C to +125°C
Storage Temperature Range	−65°C to +150°C

electrical characteristics: (25°C)

		Min.	Typ.	Max.
Intrinsic Standoff Ratio $(V_{BB} = 10V)$	η	0.56	0.65	0.75
Interbase Resistance $(V_{BB} = 3V, I_E = 0)$	R_{BB}	4.7	7	9.1
Emitter Saturation Voltage $(V_{BB} = 10V, I_E = 50\text{ ma})$	$V_{E(SAT)}$		2	
Emitter Reverse Current $(V_{BB} = 30V, I_{B1} = 0)$	I_{EO}		0.05	12
Peak Point Emitter Current $(V_{BB} = 25V)$	I_P		0.4	5
Valley Point Current $(V_{BB} = 20V, R_{B2} = 100\Omega)$	I_V	4	6	

(a)   (b)   (c)

Figure 7.45 UJT: (a) appearance; (b) specification sheet; (c) terminal identification. (Courtesy General Electric Company.)

while the emitter terminal is between the two. In addition, the base terminal to be tied to the higher potential is closer to the extension on the lip of the casing.

One rather common application of the UJT is in the triggering of other devices such as the SCR. The basic elements of such a triggering circuit are shown in Fig. 7.46. The resistor R_1 must be chosen to ensure that the load line determined by R_1 passes through the device characteristics to the right of the peak point but to the left of the valley point. If the load line fails to pass to the right of the peak point, the device cannot turn on. An equation for R_1 that will ensure a turn-on condition can be established if we consider the peak point at which $I_P = I_{R_1}$ and $V_E = V_P$. (The equality $I_P = I_{R_1}$ is valid since the charging current of the capacitor, at this instant, is zero; that is, the capacitor is at this particular instant changing from a charging to a discharging state.) Then $V - I_P R_1 = V_P$ or $(V - V_P)/I_P = R_1$.

To ensure firing,

$$\boxed{\frac{V - V_P}{I_P} > R_1} \qquad (7.6)$$

At the valley point $I_E = I_V$ and $V_E = V_V$, so that to ensure turning off,

$$\boxed{\frac{V - V_V}{I_V} < R_1} \qquad (7.7)$$

For the typical values of $V = 30$ V, $\eta = 0.5$, $V_V = 1$ V, $I_V = 10$ mA, $I_P = 10$ μA, and $R_{BB} = 5$ K.

$$\frac{V - V_P}{I_P} = \frac{30 - [0.5(30) + 0.5]}{10 \times 10^{-6}} = \frac{14.5}{10 \times 10^{-6}} = 1.45 \text{ M} > R_1$$

and

$$\frac{V - V_V}{I_V} = \frac{30 - 1}{10 \times 10^{-3}} = 2.9 \text{ K} < R_1$$

Therefore, 1.45 MΩ $> R_1 >$ 2.9 K.

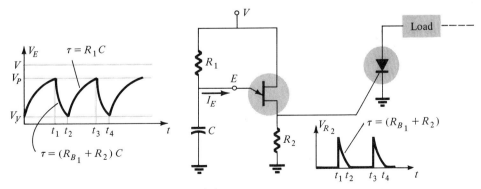

Figure 7.46 UJT triggering of an SCR.

The range for R_1 is therefore extensive. The resistor R_2 must be chosen small enough to ensure that the SCR is not turned on by the interbase current I_{BB} that will flow through R_2 when $I_E = 0$.

The capacitor C will determine, as we shall see, the time interval between triggering pulses and the time span of each pulse.

At the instant the dc supply voltage V is applied, the voltage V_E will charge toward V volts since the emitter circuit of the UJT is in the open-circuit state. The time constant of the charging circuit is $R_1 C$. When $V_E = V_P$, the UJT will enter the conduction state and the capacitor C will discharge through R_{B_1} and R_2 at a rate determined by the time constant $(R_{B_1} + R_2)C$. This time constant is much smaller than the former, resulting in the patterns of Fig. 7.46. Once V_E decays to V_V, the UJT will turn off and the charging phase will repeat itself. Since I_{R_2} and V_{R_2} are related by Ohm's law (linear relationship), the waveform for I_{R_2} appears the same as for V_{R_2}. The positive pulse of V_{R_2} is designed to be sufficiently large to turn the SCR on.

If we remove the SCR from Fig. 7.46, we have the basic construction of a *UJT relaxation oscillator*—that is, a source of a continually repeating waveform such as appears at V_E in Fig. 7.46. The frequency of the generated waveform is given approximately by

$$f \cong \frac{1}{R_1 C \log_e (1/1 - \eta)} \tag{7.8}$$

The conditions for oscillation are the same as described for the network of Fig. 7.46.

7.14 V-FET

In recent years there has been increasing interest in raising the power limits of the FET. One technique that appears to be taking hold in the commercial market is the V-FET construction appearing in Fig. 7.47a. The basic construction of the typical MOSFET appears in Fig. 7.47b for comparison purposes. Most noticeable are the four *diffused* layers in the V-FET as compared to the three regions of the MOSFET developed through *photolithographic* methods. The term V-FET is derived primarily from the fact that the drain-to-source current follows a "*v*ertical" path rather than the horizontal path of the conventional MOSFET. Obviously, the V-type construction of the gate could also suggest this terminology. Increased currents are possible with the V-FET, due to the significantly reduced channel length (1:3) (x versus y in Fig. 7.47), the availability of two current paths from the lower drain to the separated source, and the fact that many other Vs of construction can be introduced resulting in a saw-edge pattern in the top layer and a significant increase in the number of paths from drain to source as shown in Fig. 7.48. In the past FETs have been limited to the milliampere range with relatively low power ratings (milliwatts to few watts). Now the Siliconix Corporation of California has introduced to the commercial market an *n*-channel, enhancement-mode, 60-W, 2-A V-FET.

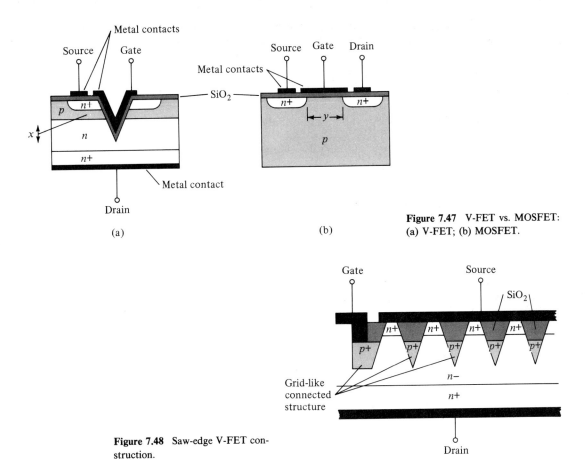

Figure 7.47 V-FET vs. MOSFET: (a) V-FET; (b) MOSFET.

Figure 7.48 Saw-edge V-FET construction.

In Fig. 7.47a, the added *n*-type region will result in improved drain-to-source breakdown voltages and lower parasitic capacitance levels. For the V-FET the short channel results in a straight-line (linear) relationship between I_D and V_G.

Other important characteristics of the V-FET as compared to the bipolar transistor include a negative temperature coefficient to remove the concern about thermal runaway, a very low leakage current (a few nanoampere), and higher switching speeds (the VMP-1 produced by Siliconix can switch 2 A in 5 ns).

Further information on this commercial unit (VMP-1) include gate threshold voltages from 0.8 to 1.8 V, a maximum gate voltage of 10 V, and a drain-to-source breakdown voltage of 60 V. Its minimum g_m is 200 mmhos.

7.15 PHOTOTRANSISTORS

The fundamental behavior of photoelectric devices was introduced earlier with the description of the photodiode. This discussion will now be extended to include the phototransistor, which has a photosensitive collector-base *p-n* junction. The current induced by photoelectric effects is the base current of the transistor.

If we assign the notation I_λ for the photoinduced base current, the resulting collector current, on an approximate basis, is

$$I_c \cong h_{fe} I_\lambda \tag{7.9}$$

A representative set of characteristics for a phototransistor is provided in Fig. 7.49 with the symbolic representation of the device. Note the similarities between these curves and those of a typical bipolar transistor. As expected, an increase in light intensity corresponds with an increase in collector current. To develop

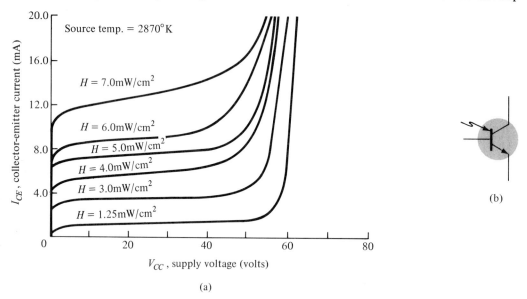

Figure 7.49 Phototransistor: (a) collector characteristics (MRD300); (b) symbol.

a greater degree of familiarity with the light-intensity unit of measurement, milliwatts per square centimeter, a curve of base current versus flux density appears in Fig. 7.50a. Note the exponential increase in base current with increasing flux density. In the same figure a sketch of the phototransistor is provided with the terminal identification and the angular alignment.

Some of the areas of application for the phototransistor include punch-card readers, computer logic circuitry, lighting control (highways, etc.), level indication, relays, and counting systems.

A high-isolation AND gate is shown in Fig. 7.51 using three phototransistors and three LEDs (light-emitting diodes). The LEDs are semiconductor devices that emit light at an intensity determined by the forward current through the device. With the aid of discussions in Chapter 2, the circuit behavior should be relatively easy to understand. The terminology "high isolation" simply refers to the lack of an electrical connection between the input and output circuits.

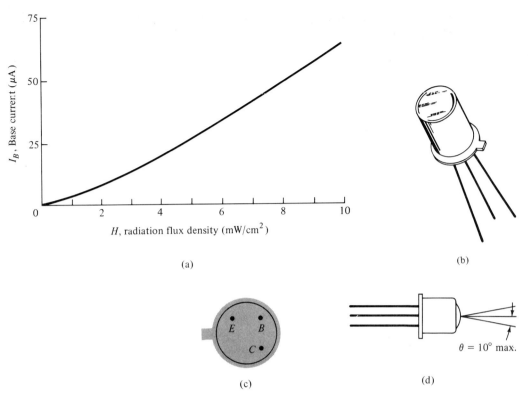

Figure 7.50 Phototransistor: (a) base current vs. flux density; (b) device; (c) terminal identification; (d) angular alignment. (*b*, courtesy Motorola, Inc.)

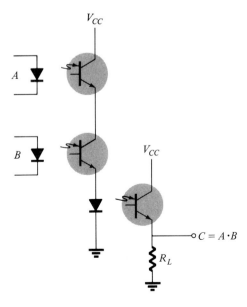

Figure 7.51 High-isolation AND gate employing phototransistors and light-emitting diodes (LEDs).

7.16 OPTO-ISOLATORS

The *opto-isolator* is a device that incorporates many of the characteristics described in the preceding section. It is simply a package that contains both an infrared LED and a photo-detector such as a silicon diode, transistor Darlington pair, or SCR. The wavelength response of each device is tailored to be as identical as possible to permit the highest measure of coupling possible. In Fig. 7.52, two possible chip configurations are provided, with a photograph of each. There is a transparent insulating cap between each set of elements embedded in the structure (not visible) to permit the passage of light. They are designed with response times so small that they can be used to transmit data in the megahertz range.

ISO-LIT 1   ISO-LIT Q1

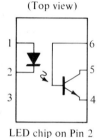

(Top view)

LED chip on Pin 2
PT chip on Pin 5

Pin No.	Function
1	anode
2	cathode
3	nc
4	emitter
5	collector
6	base

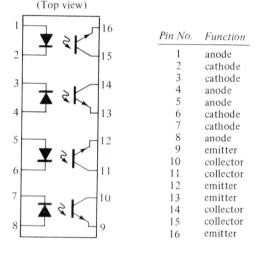

Pin No.	Function
1	anode
2	cathode
3	cathode
4	anode
5	anode
6	cathode
7	cathode
8	anode
9	emitter
10	collector
11	collector
12	emitter
13	emitter
14	collector
15	collector
16	emitter

Figure 7.52 Two Litronix opto-isolators. (Courtesy Litronix, Inc.)

The maximum ratings and electrical characteristics for the IL-1 model are provided in Fig. 7.53. Note that I_{CEO} is measured in nanoamperes and that the power dissipation of the LED and transistor are about the same.

The typical optoelectronic characteristic curves for each channel are provided in Figs. 7.54 to 7.58. Note the very pronounced effect of temperature on the output current at low temperatures but the fairly level response at or above room temperature (25°C). As mentioned earlier, the level of I_{CEO} is improving steadily with improved design and construction techniques (the lower the better). In Fig. 7.54 we do not reach 1 μA until the temperature rises above 75°C. The transfer characteristics of Fig. 7.55 compare the input LED current (which establishes the luminous flux) to the resulting collector current of the output transistor (whose base current is determined by the incident flux). In fact, Fig. 7.56 demonstrates

(a) Maximum Ratings

Gallium arsenide LED (each channel) IL-1
    Power dissipation @ 25°C   200 mW
    Derate linearly from 25°C   2.6 mW/°C
    Continuous forward current   150 mA

Detector silicon phototransistor (each channel) IL-1
    Power dissipation @ 25°C   200 mW
    Derate linearly from 25°C   2.6 mW/°C
    Collector-emitter breakdown voltage   30 V
    Emitter-collector breakdown voltage   7 V
    Collector-base breakdown voltage   70 V

Package IL-1
    Total package dissipation at 25°C ambient (LED plus detector)   250 mW
    Derate linearly from 25°C   3.3 mW/°C
    Storage temperature   −55°C to +150°C
    Operating temperature   −55°C to +100°C

(b) Electrical Characteristics per Channel (at 25°C Ambient)

Parameter	Min.	Typ.	Max.	Units	Test Conditions
Gallium arsenide LED					
Forward voltage		1.3	1.5	V	$I_F = 60$ mA
Reverse current		0.1	10	μA	$V_R = 3.0$ V
Capacitance		100		pF	$V_R = 0$
Phototransistor detector					
BV_{CEO}	30			V	$I_C = 1$ mA
I_{CEO}		5.0	50	nA	$V_{CE} = 10$ V, $I_F = 0$
Collector-emitter capacitance		2.0		pF	$V_{CE} = 0$
BV_{ECO}	7			V	$I_E = 100$ μA
Coupled characteristics					
dc current transfer ratio	0.2	0.35			$I_F = 10$ mA, $V_{CE} = 10$ V
Capacitance, input to output		0.5		pF	
Breakdown voltage	2500			V	DC
Resistance, input to output		100		GΩ	
V_{SAT}			0.5	V	$I_C = 1.6$ mA, $I_F = 16$ mA
Propagation delay					
$t_{D\ ON}$		6.0		μs	$R_L = 2.4$ K, $V_{CE} = 5$ V
$t_{D\ OFF}$		25		μs	$I_F = 16$ mA

Figure 7.53 The Litronix IL-1 opto-isolator.

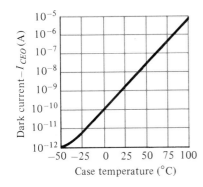
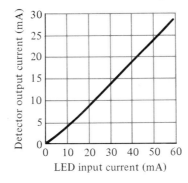

Figure 7.54 *(left)* Dark current (I_{CEO}) vs. temperature.

Figure 7.55 *(right)* Transfer characteristics.

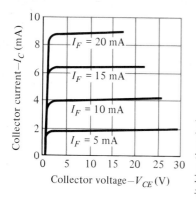

Figure 7.56 Detector output characteristics.

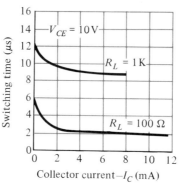

Figure 7.57 Switching time vs. collector current.

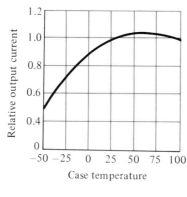

Figure 7.58 Relative output vs. temperature.

that the V_{CE} voltage affects the resulting collector current only very slightly. It is interesting to note in Fig. 7.57 that the switching time of an opto-isolator decreases with increased current, while for many devices it is exactly the reverse. Consider that it is only 2 μs for a collector current of 6 mA and a load R_L of 100 Ω. The relative output versus temperature appears in Fig. 7.58.

The schematic representation for a transistor coupler appears in Fig. 7.53. The schematic representations for a photo-diode, photo-Darlington, and photo-SCR opto-isolator appear in Fig. 7.59.

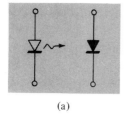

(a)

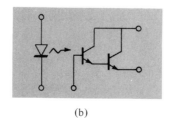

(b)

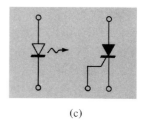

(c)

Figure 7.59 Opto-isolators: (a) photodiode; (b) photo-Darlington; (c) photo-SCR.

7.17 PROGRAMMABLE UNIJUNCTION TRANSISTOR

Although there is a similarity in name, the actual construction and mode of operation of the programmable unijunction transistor (PUT) is quite different from the unijunction transistor. The fact that the I–V characteristics and applications of each are similar prompted the choice of labels.

As indicated in Fig. 7.60, it is a four-layer *pnpn* device with a gate connected directly to the sandwiched *n*-type layer. The symbol for the device and the basic biasing arrangement appears in Fig. 7.61. As the symbol suggests, it is essentially

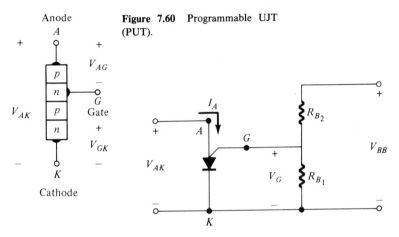

Figure 7.60 Programmable UJT (PUT).

Figure 7.61 Basic biasing arrangement for the PUT.

an SCR with a control mechanism that permits a duplication of the characteristics of the typical SCR. The term "programmable" is applied because R_{BB}, η, and V_P as defined for the UJT can be controlled through the resistors R_{B_1}, R_{B_2}, and the supply voltage V_{BB}. Note in Fig. 7.61 that through an application of the voltage divider rule when $I_G = 0$:

$$V_G = \frac{R_{B_1}}{R_{B_1} + R_{B_2}} V_{BB} = \eta V_{BB} \tag{7.10}$$

where

$$\eta = \frac{R_{B_1}}{R_{B_1} + R_{B_2}}$$

as defined for the UJT.

The characteristics of the device appear in Fig. 7.62. As noted on the diagram, the off state (I low, V between 0 and V_P) and the on state ($I \geq I_V$, $V \geq V_V$) are separated by the unstable region as occurred for the UJT. That is, the device cannot stay in the unstable state—it will simply shift to either the off or on stable states.

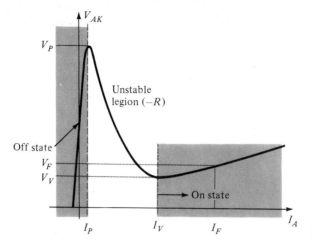

Figure 7.62 PUT characteristics.

The firing potential *(V_P)* or voltage necessary to "fire" the device is given by

$$V_P = \eta V_{BB} + V_D \tag{7.11}$$

as defined for the UJT. However, V_P represents the voltage drop V_{AG} in Fig. 7.60 (the forward voltage drop across the conducting diode). For silicon V_D is typically 0.7 V. Therefore,

$$V_P = \eta V_{BB} + V_D = \eta V_{BB} + V_{AG}$$

$$V_P = \eta V_{BB} + 0.7 \quad \text{silicon} \tag{7.12}$$

We noted above, however, that $V_G = \eta V_{BB}$ with the result that

$$V_P = V_G + 0.7 \quad \text{silicon} \tag{7.13}$$

Recall that for the UJT both R_{B_1} and R_{B_2} represent the bulk resistance and ohmic base contacts of the device—both inaccessible. In the development above, we note that R_{B_1} and R_{B_2} are external to the device permitting an adjustment of η and hence V_G above. In other words, a measure of control on the level of V_P required to turn on the device.

Although the characteristics of the PUT and UJT are similar, the peak and valley currents of the PUT are typically lower than those of a similarly rated UJT. In addition, the minimum operating voltage is also less for a PUT.

If we take a Thévenin equivalent of the network to the right of gate terminal in Fig. 7.61, the network of Fig. 7.63 will result. The resulting resistance R_s is important because it is often included in specification sheets since it affects the level of I_V.

The basic operation of the device can be reviewed through reference to Fig. 7.62. A device in the off state will not change state until the voltage V_P as defined by V_G and V_D is reached. The level of current until I_P is reached is very low, resulting in an open-circuit equivalent since $R = V$ (high)/I (low) will result in

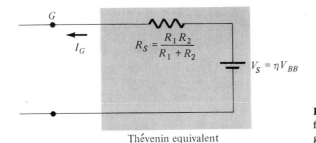

Thévenin equivalent

Figure 7.63 Thévenin equivalent for the network to the right of the gate terminal in Fig. 7.61.

a high resistance level. When V_P is reached, the device will switch through the unstable region to the on state, where the voltage is lower but the current higher, resulting in a terminal resistance $R = V(\text{low})/I(\text{high})$ which is quite small, representing short-circuit equivalent on an approximate basis. The device has therefore switched from essentially an open-circuit to a short-circuit state at a point determined by the choice of R_{B_1}, R_{B_2}, and V_{BB}. Once the device is in the on state, the removal of V_G will not turn the device off. The level of voltage V_{AK} must be dropped sufficiently to reduce the current below a holding level.

EXAMPLE 7.1 Determine R_{B_1} and V_{BB} for a silicon PUT if it is determined that η should be 0.8, $V_P = 10.3$ V, and $R_{B_2} = 5$ K.

Solution: From Eq. (7.10):

$$\eta = \frac{R_{B_1}}{R_{B_1} + R_{B_2}} = 0.8$$

$$R_{B_1} = 0.8(R_{B_1} + R_{B_2})$$

$$0.2 R_{B_1} = 0.8 R_{B_2}$$

$$R_{B_1} = 4 R_{B_2}$$

$$R_{B_1} = 4(5\text{ K}) = \mathbf{20\ K}$$

From Eq. (7.11):

$$V_P = \eta V_{BB} + V_D$$

$$10.3 = (0.8)(V_{BB}) + 0.7$$

$$9.6 = 0.8 V_{BB}$$

$$V_{BB} = \mathbf{12\ V}$$

One popular application of the PUT is in the relaxation oscillator of Fig. 7.64. The instant the supply is connected, the capacitor will begin to charge toward 10 V, since there is no anode current at this point. The charging curve appears in Fig. 7.65.

The period T required to reach the firing potential V_P is given approximately by

$$T \cong RC \log_e\left(\frac{V_{BB}}{V_{BB} - V_P}\right) \tag{7.14}$$

SEC. 7.17 PROGRAMMABLE UNIJUNCTION TRANSISTOR

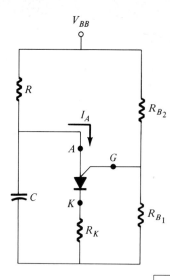

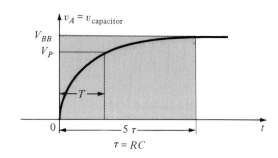

Figure 7.64 PUT relaxation oscillator.

Figure 7.65 Changing wave for the capacitor C of Fig. 7.64.

or

$$T \cong RC \log_e\left(1 + \frac{R_{B_1}}{R_{B_2}}\right) \quad (7.15)$$

The instant the voltage across the capacitor equals V_P, the device will fire and a current I_P will flow through the PUT. If R is too large, the current I_P cannot be established and the device will not fire. At the point of transition,

$$I_P R = V_{BB} - V_P$$

and

$$R_{\max} = \frac{V_{BB} - V_P}{I_P} \quad (7.16)$$

The subscript is included to indicate that any R greater than $R_{\max}$ will result in a current less than I_P. The level of R must also be such to ensure it is less than I_V if oscillations are to occur. In other words, we want the device to enter the unstable region and then return to the off state. From reasoning similar to that above:

$$R_{\min} = \frac{V_{BB} - V_V}{I_V} \quad (7.17)$$

The discussion above requires that R be limited to the following for an oscillatory system:

$$R_{\min} < R < R_{\max}$$

The waveforms of v_A, v_G, and v_K appear in Fig. 7.66. Note that T determines the maximum voltage v_A can charge to. Once the device fires, the capacitor will rapidly discharge through the PUT and R_K, producing the drop shown. Of course, v_K will peak at the same time due to the brief but heavy current flow. The voltage v_G will rapidly drop down from V_G to a level just greater than 0 volts. When the capacitor voltage drops to a low level, the PUT will once again turn off and the charging cycle repeated. The effect on V_G and V_K is shown in Fig. 7.66.

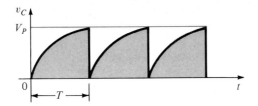
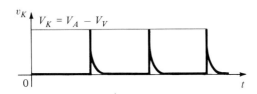

Figure 7.66 Waveforms for PUT oscillator of Fig. 7.64.

EXAMPLE 7.2 If $V_{BB} = 12$ V, $R = 20$ K, $C = 1$ μF, $R_K = 100$ Ω, $R_{B_1} = 10$ K, $R_{B_2} = 5$ K, $I_P = 100$μA, $V_V = 1$ V, and $I_V = 5.5$ mA, determine:
(a) V_P.
(b) R_{max} and R_{min}.
(c) T and frequency of oscillation.
(d) The waveforms of v_A, v_G, and v_K.

Solution:
(a) From Eq. (7.11):

$$V_P = \eta V_{BB} + V_D$$

$$= \left(\frac{R_{B_1}}{R_{B_1} + R_{B_2}}\right) V_{BB} + 0.7$$

$$= \left(\frac{10 \text{ K}}{10 \text{ K} + 5 \text{ K}}\right)(12) + 0.7$$

$$= (0.67)(12) + 0.7 = \mathbf{8.7 \text{ V}}$$

(b) From Eq. (7.16):

$$R_{max} = \frac{V_{BB} - V_P}{I_P}$$

$$= \frac{12 - 8.7}{100 \times 10^{-6}} = \mathbf{33 \text{ K}}$$

From Eq. (7.17):

$$R_{min} = \frac{V_{BB} - V_V}{I_V}$$

$$= \frac{12 - 1}{5.5 \times 10^{-3}} = \mathbf{2 \text{ K}}$$

$$R: \ 2\text{ K} < 20\text{ K} < 33\text{ K}$$

(c) Eq. (7.14):

$$T = RC \log_e \left(\frac{V_{BB}}{V_{BB} - V_P} \right)$$

$$= (20 \times 10^3)(1 \times 10^{-6}) \log_e \left(\frac{12}{12 - 8.7} \right)$$

$$= 20 \times 10^{-3} \log_e (3.64)$$
$$= 20 \times 10^{-3} (1.29)$$
$$= \mathbf{25.8 \text{ ms}}$$

$$f = \frac{1}{T} = \frac{1}{25.8 \times 10^{-3}} = \mathbf{38.8 \text{ Hz}}$$

(d) As indicated in Fig. 7.67.

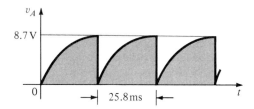

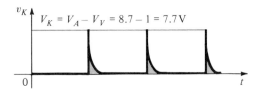

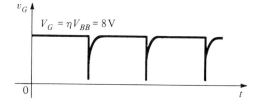

Figure 7.67 Waveforms for the oscillator of Example 7.1.

PROBLEMS

§ 7.3

1. Describe in your own words the basic behavior of the SCR using the two-transistor equivalent circuit.

2. Describe two techniques for turning an SCR off.

3. Consult a manufacturer's manual or specification sheet and obtain a turn-off network. If possible, describe the turn-off action of the design.

§ 7.4

4. (a) At high levels of gate current the characteristics of an SCR approach those of what two-terminal device?
   (b) At a fixed anode-to-cathode voltage less than $V_{(BR)F*}$, what is the effect on the firing of the SCR as the gate current is reduced from its maximum value to the zero level?
   (c) At a fixed gate current greater than $I_G = 0$, what is the effect on the firing of the SCR as the gate voltage is reduced from $V_{(BR)F*}$?
   (d) For increasing levels of I_G, what is the effect on the holding current?

5. (a) In Fig. 7.8, will a gate current of 50 mA fire the device at room temperature (25°C)?
   (b) Repeat part (a) for a gate current of 10 mA.
   (c) Will a gate voltage of 2.6 V trigger the device at room temperature?
   (d) Is $V_G = 6$ V, $I_G = 800$ mA a good choice for firing conditions? Would $V_G = 4$ V, $I_G = 1.6$ A be preferred? Explain.

§ 7.5

6. In Fig. 7.11b, why is there very little loss in potential across the SCR during conduction?

7. Fully explain why reduced values of R_1 in Fig. 7.12 will result in an increased angle of conduction.

8. Refer to the charging network of Fig. 7.13.
   (a) Determine the dc level of the full-wave rectified signal if a 1:1 transformer were employed.
   (b) If the battery in its uncharged state is sitting at 11 V, what is the anode-to-cathode voltage drop across SCR_1?
   (c) What is the maximum possible value of V_R ($V_{GK} \cong 0.7$ V)?
   (d) At the maximum value of part (c), what is the gate potential of SCR_2?
   (e) Once SCR_2 has entered the short-circuit state, what is the level of V_2?

§ 7.7

9. Fully describe in your own words the behavior of the networks of Fig. 7.17.

§ 7.8

10. (a) In Fig. 7.23, if $V_Z = 50$ V, determine the maximum possible value the capacitor C_1 can charge to ($V_{GK} \cong 0.7$ V).
    (b) Determine the approximate discharge time (5τ) for $R_3 = 20$ K.
    (c) Determine the internal resistance of the GTO if the rise time is one-half the decay period determined in part (b).

§ 7.9

11. (a) Using Fig. 7.25b, determine the minimum irradiance required to fire the device at room temperature (20°C).

(b) What percent reduction in irradiance is allowable if the junction temperature is increased from 0°C (32°F) to 100°C (212°F)?

§ 7.10

12. For the network of Fig. 7.29, if $V_{(BR)} = 6$ V, $V = 40$ V, $R = 10$ K, $C = 0.2$ μF, and V_{GK} (firing potential) $= 3$ V, determine the time period between energizing the network and the turning on of the SCR.

§ 7.11

13. Using whatever reference you require, find an application of a diac and explain the network behavior.

14. If V_{BR_2} is 6.4 V, determine the range for V_{BR_1} using Eq. (7.1).

§ 7.12

15. Repeat Problem 13 for the triac.

16. Sketch the characteristics of the CSB20 triac using the data provided in the electrical specifications of Fig. 7.35.

17. For the typical and maximum curves of Fig. 7.36, determine the appropriate value of V_o through a linearization of the curve.

18. What is the dynamic resistance of the typical curve at $I_T = 40$ A of Fig. 7.36?

19. Determine the cycles at 60 Hz from Fig. 7.37 that will reduce the nonrepetitive peak on-state current to 60% of its maximum value.

20. (a) Using the curve of Fig. 7.38, determine the peak value of the current that will result in a power dissipation of 25 W.
    (b) Linearize the curve of Fig. 7.38 and write an equation relating P_D to $I_{T_{rms}}$. Test the accuracy of your graph by finding P_D from the graph at 20 A and compare to the results obtained with the equation derived above.

21. (a) Resketch Fig. 7.39 with the current the vertical axis and the temperature the horizontal axis.
    (b) Calculate the derating factor for the current from the graph (after the curve is linearized). Assume 25 A to be the maximum.
    (c) Compare the results obtained with the derating factor to that obtained from the graph at a temperature of 80°C.

§ 7.13

22. For the network of Fig. 7.46, in which $V = 40$ V, $\eta = 0.6$, $V_v = 1$ V, $I_v = 8$ mA, and $I_p = 10$ μA, determine the range of R_1 for the triggering network.

23. For a unijunction transistor with $V_{BB} = 20$ V, $\eta = 0.65$, $R_{B_1} = 2$ K ($I_E = 0$), and $V_D = 0.7$ V, determine the following:
    (a) R_{B_2}.
    (b) R_{BB}.
    (c) V_{RB_1}.
    (d) V_P.

§ 7.14

24. Describe the difference in construction between the standard FET and the V-FET.

25. Request information on a commercially available V-FET and discuss its characteristics.

§ 7.15

26. For a phototransistor having the characteristics of Fig. 7.50, determine the photoinduced base current for a radian flux density of 5 mW/cm². If $h_{fe} = 40$, find I_C.

27. Design a high-isolation OR-gate employing phototransistors and LEDs.

§ 7.16

28. (a) Determine an average derating factor from the curve of Fig. 7.58 for the region defined by temperatures less than 25°C.
    (b) Is it fair to say that for temperature greater than room temperature (up to 100°C), the output current is somewhat unaffected by temperature?

29. (a) Determine, from Fig. 7.54, the change in I_{CEO} per degree change in temperature for the range 25 to 50°C.
    (b) Using the results of part (a), determine the level of I_{CEO} at 35°C.

30. Determine, from Fig. 7.55, the ratio of LED input current to detector output current for the linear region. Would you consider the device to be relatively efficient in its purpose?

31. (a) Sketch the maximum power curve of $P_D = 200$ mW on the graph of Fig. 7.56. List any noteworthy conclusions.
    (b) Determine β_{dc} (defined by I_C/I_F) for the system at $V_{CE} = 15$ V, $I_F = 10$ mA.
    (c) Compare the results of part (b) with those obtained from Fig. 7.55 at $I_F = 10$ mA. Do they compare? Should they? Why?

32. (a) Referring to Fig. 7.57, determine the collector current above which the switching time does not change appreciably for $R_L = 1$ K and $R_L = 100$ Ω.
    (b) At $I_C = 6$ mA, how does the ratio of switching times for $R_L = 1$ K and $R_L = 100$ Ω compare to the ratio of resistance levels?

§ 7.17

33. Determine η and V_G for a PUT with $V_{BB} = 20$ V and $R_{B_1} = 3R_{B_2}$.

34. Using the data provided in Example 7.2, determine the impedance of the PUT at the firing and valley points. Are the approximate open- and short-circuit states verified?

35. Can Eq. (7.15) be derived exactly as shown from Eq. (7.14)? If not, what element is missing in Eq. (7.15)?

36. (a) Will the network of Example 7.2 oscillate if V_{BB} is changed to 10 V? What minimum value of V_{BB} is required (V_V a constant)?
    (b) Referring to the same example, what value of R would place the network in the stable on state and remove the oscillatory response of the system?
    (c) What value of R would make the network a 2 ms time-delay network? That is, provide a pulse v_k 2 ms after the supply is turned on and then stay in the *on* state.

GLOSSARY

Silicon-Controlled Rectifier (SCR) As the terminology implies, a rectifier constructed of silicon material that has a third terminal for control purposes.

Anode Current Interruption A method of turning off an SCR that requires that the anode current be brought to the zero level.

Forced Commutation Technique A method of turning off an SCR that requires that a short-circuit path be placed across the SCR, thereby reducing the anode current to zero.

Forward Breakover Voltage ($V_{BR}*$) The voltage above which the SCR enters the conduction state.

Holding Current (I_H) The level of current below which the SCR switches from the conduction state to the forward blocking region.

Silicon Control Switch (SCS) A four-layer semiconductor device in which all layers are externally available—a switch whose behavior can be controlled by either the cathode or anode gate.

Gate Turn-off Switch (GTO) A three-terminal four-layer semiconductor device that can be turned on or off by applying the proper pulse to the cathode gate.

Light Activated SCR (LASCR) An SCR whose state is controlled by the light falling upon a silicon semiconductor layer of the device.

Shockley Diode A four-layer two-terminal semiconductor device having characteristics similar to an SCR with zero gate current.

Diac A two-terminal parallel-inverse combination of semiconductor layers that permits conduction in either direction.

Triac Fundamentally, a diac with a gate terminal for controlling the turn-on condition of the bilateral device in either direction.

Unijunction Transistor A three-terminal device having a single (hence the term uni-) metal-semiconductor junction that will enter a heavy conduction state when properly triggered.

Intrinsic Stand-off Ratio (η) A quantity determined by the ohmic base resistances of the UJT of importance in determining the firing potential of the device.

Negative Resistance Region A region on a set of characteristics in which an increase of voltage or current will result in a decrease in the other quantity.

V-FET A FET designed to have increased power dissipation levels.

Phototransistor A transistor whose on state is controlled by the level of base current established by incident light on the collector-base junction.

Opto-isolator A package that contains both an infrared LED and a photodetector such as a silicon diode, transistor Darlington pair, or SCR.

Programmable Unijunction Transistor (PUT) A four-layer semiconductor device having a control gate and the ability to control its stand-off ratio and firing potential.

PART II

Integrated

CHAPTER 8

Integrated Circuit Fabrication

8.1 INTRODUCTION

Integrated-circuit fabrication has advanced from individual transistors in the early 1950s to simple logic gates in the early 1960s to complex digital calculators, computers, and memories in the early 1970s. The basic planar process used to make individual transistors has grown and been refined into the present technology used to build large-scale integrated (LSI) circuits. This production technology is essentially the same whether used to build medium-scale integrated (MSI), or small-scale integrated (SSI), although the effort in quality control and manufacturing precision is far greater for LSI units. Extending the technology to pack even more components into a single IC as very-large-scale integrated (VLSI) units has required considerable skill, blending basic manufacturing techniques with new and refined techniques of production.

The basic process of manufacturing the various types of ICs uses highly purified silicon material, doped n-type or p-type. Starting with a *wafer* of a few inches diameter, many hundreds to thousands of circuits are formed at the same time, each integrated circuit containing hundreds to thousands of components. A process of photographically selecting sections of the wafer and diffusing n- or p-type dopants to produce the elements of a transistor or other electronic elements is then carried out so that the same operation takes place for many similar IC units. After the fabrication process is finished, the IC units are cut out of the wafer into separate dice or chips, each mounted into individual packages (after some initial testing). Figure 8.1 shows a processed silicon IC wafer with an individual die used to hold a complete integrated circuit. As shown, many hundreds of identical circuits

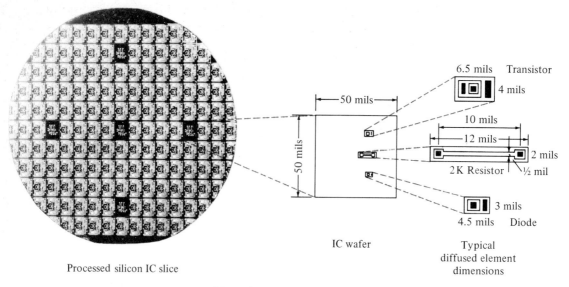

Figure 8.1 Processed monolithic IC wafer with the relative dimensions of the various elements. (Courtesy Robert Hibberd.)

are formed on a single wafer and then cut apart into individual chips for mounting in separate IC packages. Although many wafers containing large numbers of identical circuits are processed at the same time, description of the IC manufacturing process will generally focus on the information of a single circuit.

8.2 PLANAR PROCESS

The planar process is basic to the manufacture of individual semiconductor devices. The process includes a photographic masking to select a surface area of a die followed by a chemical etching to open up a *window* into the die interior and finally a diffusion process during which dopant gas is allowed to move into the die region to form part of a component. As example, during the *base diffusion* process, the bipolar transistor region is formed in the die. Figure 8.2a shows a

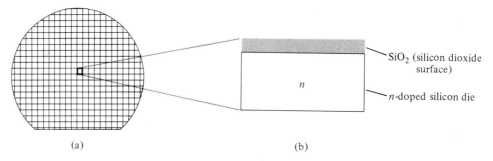

Figure 8.2 Wafer and individual die unit: (a) silicon wafer; (b) single die cross section.

wafer in which many chips are simultaneously formed. Figure 8.2b shows the cross section of one such die, starting as *n*-doped material with a protective surface covering of silicon dioxide (SiO_2). A basic step in IC fabrication is that of diffusing a selected region of semiconductor with either *n*- or *p*-type dopant.

Figure 8.3a shows the cross section of a die made of *n*-type semiconductor material with the protective SiO_2 covering. In order to diffuse a region with *p*-type dopant, it is first necessary to open a window in the silicon dioxide to the *n*-material as shown in Fig. 8.3b. The process involves exposing the SiO_2 surface, coated with a photoresist, to light, through a *mask*. The resist material that will remain is hardened, the rest of the surface to be subsequently removed by chemical etchant. This leaves an open window region (see Fig. 8.3b). A diffusion process then passes gas under high temperature and pressure over the surface for a period of time to change the *n*-type material into a *p*-type (see Fig. 8.3c). The *p*- and *n*-doped regions then form a *p-n* or diode junction. Steam is then passed over the wafer surface to form a new silicon dioxide layer which again closes off the surface of the semiconductor device. The multistep process of opening a window into a selected region of the wafer, diffusing in some dopant, and then closing the window is a basic process which is repeated a number of times to form various regions of the semiconductor devices in the wafer.

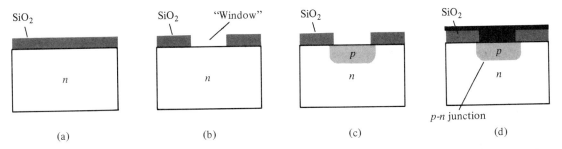

Figure 8.3 Steps in forming a *p-n* junction.

An overall view of the many steps required in fabricating an integrated circuit is shown in Fig. 8.4. The desired circuit is first designed to provide the electronic function using components of smallest area. This design is then made into a circuit layout and a sequence of photomasks prepared. The basic slice or wafer of silicon is first subjected to a polishing. Then a thin layer of epitaxial material is diffused onto the wafer and provides the basic substance into which the device will be formed, the starting silicon wafer providing the substrate of the IC device.

An isolation diffusion is one of the ways used to separate the many identical regions or dice of the wafer onto which identical circuits will be formed at the same time. Using different masks to open windows in various places on each die, a number of separate diffusion steps are performed to form the components, both active and passive. The process of opening a window is also followed, at the end, by a metallization deposition of aluminum to interconnect the circuit components and provide surface area to which the various device leads will be attached.

At this time hundreds of identical circuits exist on the single silicon wafer,

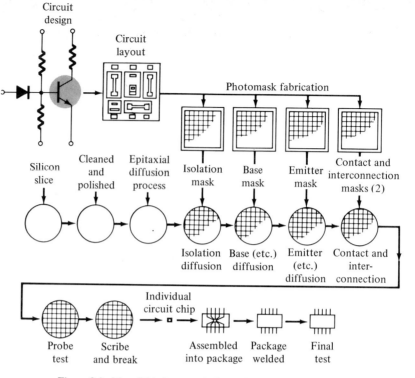

Figure 8.4 Monolithic integrated circuit fabrication. (Courtesy Robert Hibberd.)

and selective testing can be carried out to verify the manufacture of properly operating circuits. The wafer is then scribed and broken into individual dies, each of which is then packaged as an IC unit. Final testing of the circuit assures its proper functioning over the range of supply voltage, temperature, and environment conditions specified by the manufacturer.

8.3 MONOLITHIC CIRCUIT ELEMENTS

We shall now examine the basic construction of each in more detail.

Resistor

You will recall that the resistance of a material is determined by the resistivity, length, area, and temperature of the material. For the integrated circuit, each necessary element is present in the sheet of semiconductor material appearing in Fig. 8.5. As indicated in the figure, the semiconductor material can be either p- or n-type, although the p-type is most frequently employed.

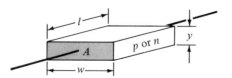

Figure 8.5 Parameters determining the resistance of a sheet of semiconductor material.

The resistance of any bulk material is determined by

$$R = \rho \frac{l}{A}$$

For $l = w$, resulting in a square sheet,

$$R = \frac{\rho l}{yw} = \frac{\rho l}{yl}$$

and

$$\boxed{R_s = \frac{\rho}{y} \quad \text{(ohms)}} \tag{8.1}$$

where ρ is in ohm-centimeters and y is in centimeters.

R_s is called the sheet resistance and has the units ohms per square. The equation clearly reveals that the sheet resistance is independent of the size of the square.

In general, where $l \neq w$,

$$\boxed{R = R_s \frac{l}{w} \quad \text{(ohms)}} \tag{8.2}$$

For the resistor appearing in Fig. 8.1, $w = \frac{1}{2}$ mil, $l = 10$ mils, and $R_s = 100\ \Omega/\text{square}$:

$$R = R_s \frac{l}{w} = 100 \times \frac{10}{1/2} = 2\ \text{K}$$

A cross-sectional view of a monolithic resistor appears in Fig. 8.6 along with the surface appearance of two monolithic resistors. In Fig. 8.6a the sheet resistive material (ρ) is indicated with its aluminum terminal connections. The n-isolation

Figure 8.6 Monolithic resistors: (a) cross section and determining dimensions; (b) surface view of two monolithic resistances in a single die. (Courtesy Motorola, Inc.)

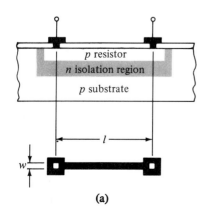
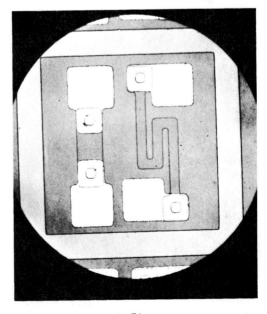

(a) (b)

region performs exactly that function indicated by its name; that is, it isolates the monolithic resistive elements from the other elements of the chip. Note in Fig. 8.6b the method employed to obtain a maximum l in a limited area. The resistors of Fig. 8.6 are called base-diffusion resistors, since the p-material is diffused into the p-type substrate during the base-diffusion process indicated in Fig. 8.4.

Capacitor

Monolithic capacitive elements are formed by making use of the transition capacitance of a reverse-biased p-n junction. At increasing reverse-bias potentials, there is an increasing distance at the junction between the p- and n-type impurities. The region between these oppositely doped layers is called the depletion region due to the absence of "free" carriers. The necessary elements of a capacitive element are therefore present—the depletion region has insulating characteristics that separate the two oppositely charged layers. The transition capacitance is related to the width (W) of the depletion region, the area (A) of the junction, and the permittivity (ϵ) of the material within the depletion region by

$$C_T = \frac{\epsilon A}{W} \tag{8.3}$$

The cross section and surface appearance of a monolithic capacitive element appear in Fig. 8.7. The reverse-biased junction of interest is J_2. The undesirable parasitic capacitance at junction J_1 is minimized through careful design. Owing to the fact that aluminum is a p-type impurity in silicon, a heavily doped n^+ region is diffused into the n-type region as shown to avoid the possibility of establish-

Figure 8.7 Monolithic capacitor: (a) cross section; (b) photograph. (Courtesy Motorola, Inc.)

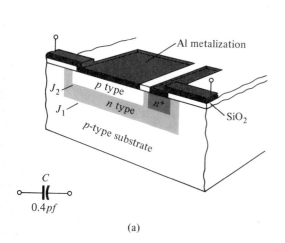

(a)

(b)

ing an undesired *p-n* junction at the boundary between the aluminum contact and the *n*-type impurity region.

Transistors

The cross section of a monolithic transistor appears in Fig. 8.8a. Note again the presense of the n^+ region in the *n*-type epitaxial collector region. The vast majority of monolithic IC transistors are *npn* rather than *pnp*, for reasons to be found in more advanced texts on the subject. Keep in mind when examining Fig. 8.8 that the *p*-substrate is only a supporting and isolating structure, forming no part of the active device itself. The base, emitter, and collector regions are formed during the corresponding diffusion processes of Fig. 8.4.

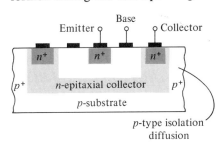

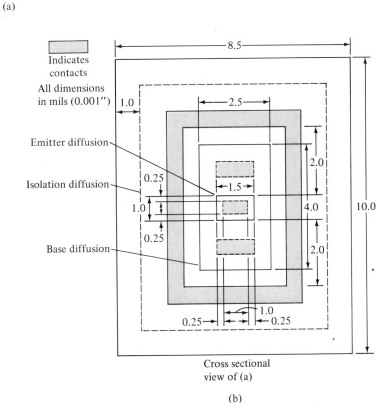

Figure 8.8 Monolithic transistor: (a) cross section; (b) surface appearance and dimension for a typical monolithic transistor. (Courtesy Motorola Monitor.)

SEC. 8.3 *MONOLITHIC CIRCUIT ELEMENTS*

The top view of a typical monolithic transistor appears in Fig. 8.8b. Note that two base terminals are provided while the collector has the outer rectangular aluminum contact surface.

Diodes

The diodes of a monolithic integrated circuit are formed by first diffusing the required regions of a transistor and then masking the diode rather than transistor terminal connections. There is, however, more than one way of hooking up a transistor to perform a basic diode action. The two most common methods applied to monolithic integrated circuits appear in Fig. 8.9. The structure of a *BC-E* diode appears in Fig. 8.10. Note that the only difference between the cross-sectional view of Fig. 8.10 and that of the transistor of Fig. 8.8a is the position of the ohmic aluminum contacts.

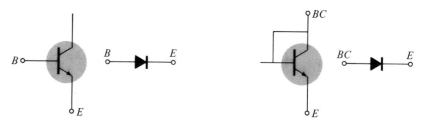

Figure 8.9 Transistor structure and connections employed in the formation of monolithic diodes.

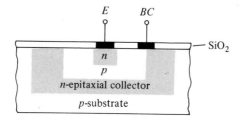

Figure 8.10 The cross-sectional view of a *BC-E* monolithic diode.

8.4 MASKS

The selective diffusion required in the formation of the various active and passive elements of an integrated circuit is accomplished through the use of masks such as those appearing in Fig. 8.11. This figure depicts and describes the major steps involved in the production of these masks. We shall find in the next section that the light areas are the only areas through which donor and acceptor impurities can pass. The dark areas will block the diffusion of impurities somewhat as a shade will prevent sunlight from changing the pigment of the skin. The next section will demonstrate the use of these masks in the formation of a computer logic circuit.

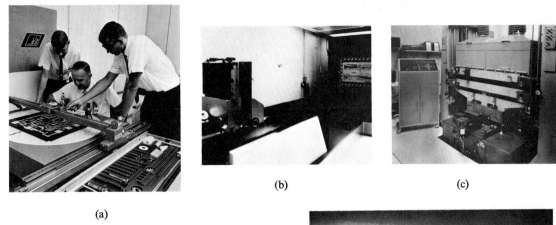

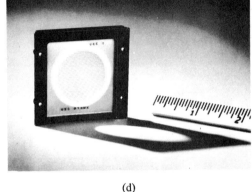

Figure 8.11 Mask preparation: (a) aristo handcutting of the mask pattern; (b) photo-reduction of the mask pattern; (c) step and repeat machine for the placement of a large number of the reduced mask pattern on a single production mask. (d) final mask. (*a* and *d,* courtesy Motorola, Inc.; *b* and *c,* courtesy Texas Instruments, Inc.)

8.5 MONOLITHIC INTEGRATED CIRCUIT—THE NAND GATE

This section is devoted to the sequence of production stages leading to a monolithic NAND-gate circuit (the operation of which is covered in Chapter 10). The circuit to be prepared appears in Fig. 8.12a. The criteria of space allocation, placement of pin connection, and so on require that the elements be situated in the relative positions indicated in Fig. 8.12b. The regions to be isolated from one another appear within the solid heavy lines. A set of masks for the various diffusion processes must then be made up for the circuit as it appears in Fig. 8.12b.

We shall now slowly proceed through the first diffusion process to demonstrate the natural sequence of steps that must be followed through each diffusion process indicated in Fig. 8.4

p-Type Silicon Wafer Preparation

After being sliced from the grown ingot, a *p*-type silicon wafer is polished and cleaned to produce the structure of Fig. 8.13.

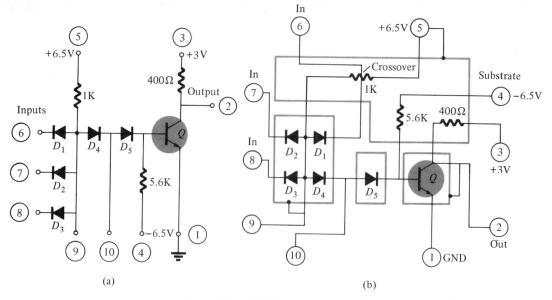

Figure 8.12 NAND gate: (a) circuit; (b) layout for monolithic fabrication.

Figure 8.13 (a) p-type silicon wafer; (b) polishing apparatus. (b, courtesy Texas Instruments, Inc.)

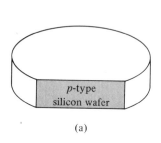

(a)

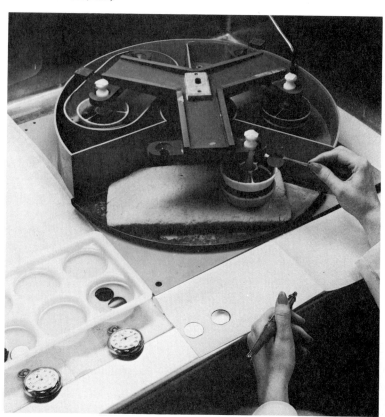

(b)

n-Type Epitaxial Region

An n-type epitaxial region is then diffused into the p-type substrate as shown in Fig. 8.14. It is *in* this thin epitaxial layer that the active and passive elements will be diffused. The p-type area remaining is simply adding some thickness to the structure to give it increase strength and permit easier handling.

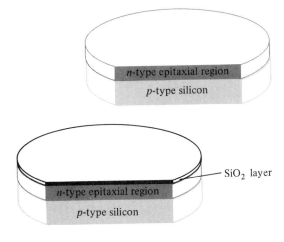

Figure 8.14 p-type silicon wafer after the n-type epitaxial diffusion process.

Figure 8.15 Wafer of Fig. 8.14 following the deposit of the SiO₂ layer.

Silicon Oxidation (SiO₂)

The resulting wafer is then subjected to an oxidation process resulting in a surface layer of SiO₂ (silicon dioxide) as shown in Fig. 8.15. This surface layer will prevent any impurities from entering the n-type epitaxial layer. However, selective etching of this layer will permit the diffusion of the proper impurity into designated areas of the n-type epitaxial region of the silicon wafer.

Photolithographic Process

The selective etching of the SiO₂ layer is accomplished through the use of a photolithographic process. A mask is first prepared on a glass plate as explained in Section 8.4. The first mask will determine those areas of the SiO₂ layer to be removed in preparation for the isolation diffusion process. The wafer is first coated with a thin layer of photosensitive material, commonly called *photoresist,* as demonstrated in Fig. 8.16a. This new layer is then covered by the mask and an ultraviolet light is applied that will expose those regions of the photosensitive material not covered by the masking pattern (Fig. 8.16b). The resulting wafer is then subjected to a chemical solution that will remove the unexposed photosensitive material. A cross section of a chip *(S-S)* of Fig. 8.16 will then appear as indicated in Fig. 8.17. A second solution will then etch away the SiO₂ layer from any region not covered by the photoresist material (Fig. 8.18). The final step before the diffusion process is the removal, by solution, of the remaining photosensitive material. The structure will then appear as shown in Fig. 8.19.

(a)

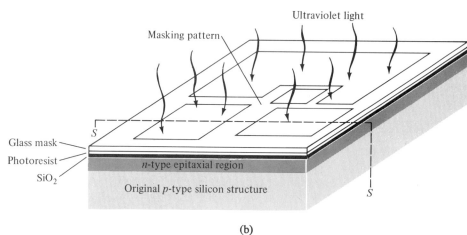

(b)

Figure 8.16 Photolithographic process: (a) applying the photoresist; the wafer is spun at a high speed to ensure an even distribution of the photoresist; (b) the application of ultraviolet light after the mask is properly set; the structure is only one of the 200, 400, or even 500 individual NAND gate circuits being formed on the wafer of Figs. 8.13 through 8.15. (*a*, courtesy Motorola, Inc.)

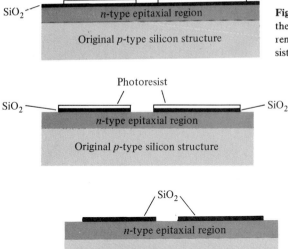

Figure 8.17 Cross section (s-s) of the chip of Fig. 8.16 following the removal of the unexposed photoresist.

Figure 8.18 Cross section of Fig. 8.17 following the removal of the uncovered SiO₂ regions.

Figure 8.19 Cross section of Fig. 8.18 following the removal of the remaining photoresist material.

Isolation Diffusion

The structure of Fig. 8.19 is then subjected to a p-type diffusion process, resulting in the islands of n-type regions indicated in Fig. 8.20. The diffusion process ensures a heavily doped p-type region (indicated by p^+) between the n-type islands. The p^+ regions will result in improved *isolation* properties between the active and passive components to be formed in the n-type islands. In preparation for the next masking and diffusion process, the entire surface of the wafer is coated with a SiO₂ layer, as indicated in Fig. 8.21.

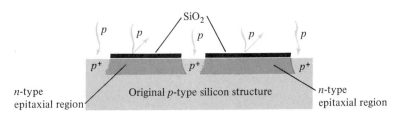

Figure 8.20 Cross section of Fig. 8.19 following the isolation diffusion process.

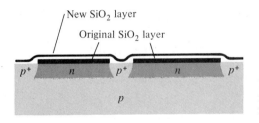

Figure 8.21 In preparation for the next diffusion process, the entire wafer is coated with a SiO₂ layer.

SEC. 8.5 MONOLITHIC INTEGRATED CIRCUIT—THE NAND GATE

Base and Emitter Diffusion Processes

The isolation diffusion process is followed by the base and emitter diffusion cycles. The sequence of steps in either case is the same as that encountered in the description of the isolation diffusion process. Although "base" and "emitter" refer specifically to the transistor structure, necessary parts (layers) of each element (resistor, capacitor, and diodes) will be formed during each diffusion process. The surface appearance of the NAND gate after the isolation base and emitter diffusion processes appears in Fig. 8.22. The mask employed in each process is also provided next to each photograph.

Isolation diffusion

Emitter diffusion

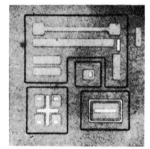

Emitter diffusion

Base diffusion

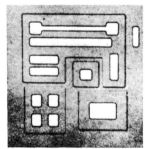

Base diffusion

Figure 8.22 The surface appearance of the monolithic NAND gate after the isolation, base, and emitter diffusion processes. The masks employed in each case are also included. (Courtesy Motorola Monitor.)

The cross section of the transistor of Fig. 8.12 will appear as shown in Fig. 8.23 after the base and emitter diffusion cycles.

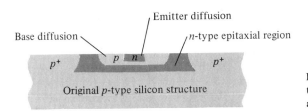

Figure 8.23 Cross section of the transistor of Fig. 8.12 after the base and emitter diffusion cycles.

Preohmic Etch

In preparation for a good ohmic contact, n^+ regions (see Section 8.3) are diffused into the structure, as clearly indicated by the light areas of Fig. 8.24. Note the correspondence between the light areas and the mask pattern.

Preohmic etch

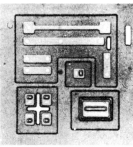

Preohmic etch

Figure 8.24 Surface appearance of the chip of Fig. 8.22 after the preohmic etch cycle. The mask employed is also included. (Courtesy Motorola Monitor.)

Metalization

A final masking pattern exposes those regions of each element to which a metallic contact must be made. The entire wafer is then coated with a thin layer of aluminum that after being properly etched will result in the desired interconnecting conduction pattern. A photograph of the completed metalization process appears in Fig. 8.25.

Metalization

Figure 8.25 Completed metalization process. (Courtesy Motorola Monitor.)

The complete structure with each element indicated appears in Fig. 8.26. Try to relate the interconnecting metallic pattern to the original circuit of Fig. 8.12a.

SEC. 8.5 MONOLITHIC INTEGRATED CIRCUIT—THE NAND GATE

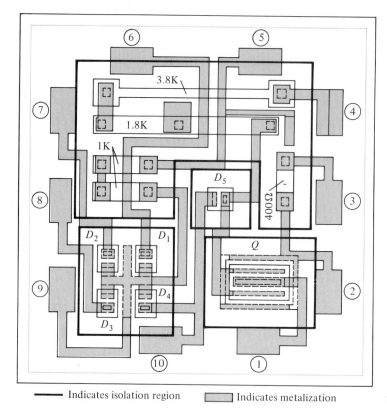

Figure 8.26 Monolithic structure for the NAND gate of Fig. 8.12.

Figure 8.27 (a) Scribing; (b) breaking of the monolithic wafer into individual chips. (*Left,* courtesy Autonetics, North American Rockwell Corporation; *middle,* courtesy Texas Instruments, Inc.; *right,* courtesy Motorola, Inc.)

(a)　　　　　　　　　　　　　　　　　　　　　　(b)

Packaging

Once the metalization process is complete, the wafer must be broken down into its individual chips. This is accomplished through the scribing and breaking processes depicted in Fig. 8.27. Each individual chip will then be packaged in one of the three forms indicated in Fig. 8.28. The name of each is provided in the figure.

Figure 8.28 Monolithic packaging techniques: (a) flat package; (b) TO (top-hat)-type package; (c) dual in-line plastic package. (Courtesy Texas Instruments, Inc.)

(a)

(b)

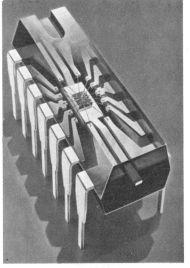

(c)

(a)

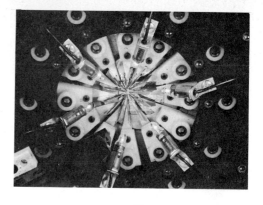

(b)

Figure 8.29 Production testing. (*a*, courtesy Autonetics, North American Rockwell Corporation; *b* and *c*, courtesy Texas Instruments, Inc.)

(c)

Testing

The final production stage, as with every commercial electronic package, is the testing of the system. This can demand a good percentage of the manufacturing costs. Photographs of various testing procedures appear in Fig. 8.29.

8.6 THIN AND THICK FILM INTEGRATED CIRCUITS

The general characteristics, properties, and appearance of thin and thick film integrated circuits are similar, although they both differ in many respects from the monolithic integrated circuit. They are not formed within a semiconductor wafer but *on* the surface of an insulating substrate such as glass or an appropriate ceramic material. In addition, *only* passive elements (resistors, capacitors) are formed through thin or thick film techniques on the insulating surface. The active elements (transistors, diodes) are added as *discrete* elements to the surface of the structure after the passive elements have been formed. The discrete active devices are frequently produced using the monolithic process.

Figure 8.30 Thin film integrated circuits. (Courtesy Autonetics, North American Rockwell Corporation.)

Two thin film integrated circuits appear in Fig. 8.30. Note the active elements added on the surface between the proper aluminum contacts. The interconnecting conduction pattern and the passive elements are prepared through masking techniques.

The primary difference between the thin and thick film techniques is the process employed for forming the passive components and the metallic conduction pattern. The thin film circuit employs an evaporation or cathode-sputtering technique; the thick film circuit employs silk-screen techniques.

In general, the passive components of film circuits can be formed with a broader range of values and reduced tolerances as compared to the monolithic IC. The use of discrete elements also increases the flexibility of design of film circuits, although, obviously, the resulting circuit will be that much larger. The cost of film circuits with a larger number of elements is also, in general, considerably higher than that of monolithic integrated circuits.

8.7 HYBRID INTEGRATED CIRCUITS

The term *hybrid integrated circuit* is applied to the wide variety of multichip integrated circuits and also those formed by a combination of the film and monolithic IC techniques. The multichip integrated circuit employs either the monolithic or film technique to form the various components, or set of individual circuits, which are then interconnected on an insulating substrate and packaged in the same container. Integrated circuits of this type appear in Fig. 8.31. In a more sophisticated type of hybrid integrated circuit the active devices are first formed within a semiconductor wafer which is subsequently covered with an insulating layer such as SiO_2. Film techniques are then employed to form the passive elements on the SiO_2 surface. Connections are made from the film to the monolithic structure through "windows" cut in the SiO_2 layer.

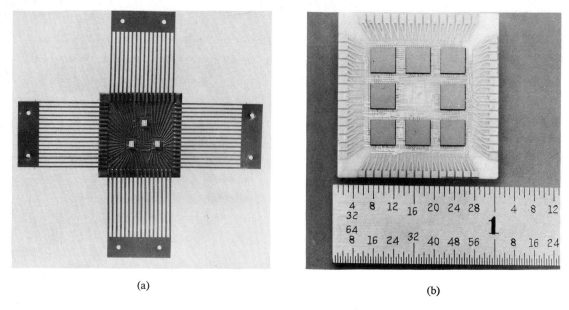

(a)   (b)

Figure 8.31 Hybrid integrated circuits. (Courtesy Texas Instruments, Inc.)

PROBLEMS

§ 8.1

1. Define SSI, MSI, and LSI.
2. What is the size of a typical IC wafer?
3. Which type of dopant results in excess holes?
4. What is the basic material used to manufacture ICs?

§ 8.2

5. What is the purpose of the silicon dioxide used in IC fabrication?
6. What is the purpose of windows in IC fabrication?
7. Describe the basic planar process.
8. What is the purpose of isolation diffusion?
9. What is the purpose of the epitaxial layer?

§ 8.6

10. Describe the difference between thin and thick film technology.

GLOSSARY

Base Diffusion A step in IC manufacture in which the base region of the bipolar transistor is diffused into semiconductor material.

Chip A single section of integrated circuit.

Die A single slice of semiconductor material containing a fully formed integrated circuit.

Dopant The chemical material diffused into semiconductor material to provide an excess of electrons (*n*-dopant) or holes (*p*-dopant).

Epitaxial The growth layer of semiconductor material of low resistivity into which the integrated circuit is formed.

Integrated Circuit (IC) A complete electronic circuit formed in semiconductor material.

Isolation Diffusion The step in IC manufacturing which diffuses an isolating region to separate components.

Large-Scale Integrated (LSI) Circuit An IC containing thousands of components in a single unit.

Mask The photographic plate used to define a part of the integrated circuit.

Medium-Scale Integrated (MSI) Circuit An IC containing hundreds of components in a single IC unit.

Metalization The step in IC fabrication in which aluminum is deposited to interconnect circuit components.

Monolithic An IC built in a single piece of semiconductor material.

Photoresist A layer of photosensitive material used to allow selective light exposure to define regions of semiconductor material.

Planar Process A popular manufacturing technique for constructing integrated circuits.

Small-Scale Integrated (SSI) Circuit An IC containing tens of components in a single IC unit.

Substrate The basic semiconductor material into which the IC components are formed.

Thick Film A process for forming passive components using silk-screening techniques.

Thin Film A process for forming passive components using evaporation techniques.

Very-Large-Scale Integrated (VLSI) Circuit An IC containing tens of thousands of components in a single IC unit.

Wafer The starting piece of semiconductor material (a few inches in diameter) into which many identical circuits are formed in the same process.

Window The opening of a section of semiconductor material into which dopant is diffused to form a part of an electronic device.

CHAPTER 9
Linear ICs

9.1 DIFFERENTIAL AMPLIFIER BASICS

A differential or difference amplifier provides an output that results from the difference of the signals applied to two separate input terminals. Figure 9.1a shows these two inputs, labeled noninverting (+) and inverting (−). The usual definition is that the output is in phase with the plus (+) input signal and opposite in phase with a signal applied to the minus (−) input.

A relatively simple bipolar circuit to demonstrate how a difference amplifier can be constructed is shown in Fig. 9.1b. A signal applied to the inverting input (transistor Q_2 base) results in an inverted (and amplified) signal at the collector of Q_2. A signal applied to Q_1 is coupled to the emitter of Q_2 and appears as an amplified, but not inverted, signal at the collector of Q_2.

Resistor R_C is a main factor in the amplifier gain, while resistor R_E has effect only on the common signals applied to both inputs. Analysis of the circuit ac operation will show a difference-mode gain, A_d,

$$A_d = \frac{h_{fe}R_C}{2h_{ie}} = \frac{R_C}{2r_e} \qquad (9.1)$$

and a common-mode gain, A_c,

$$A_c = \frac{R_C}{2R_E} \qquad (9.2)$$

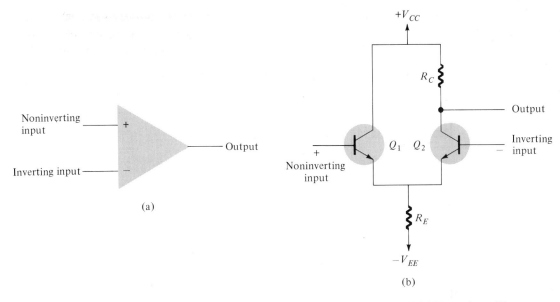

Figure 9.1 Basic differential (difference) amplifier.

The difference gain is generally referred to as the gain of the amplifier. The common-mode gain is used in specifying the amplifier rejection of undesired (noise) signals common to both input terminals. The relation that specifies how much the amplifier will provide signal gain while rejecting noise is the common-mode rejection, specified as

$$\text{CMR} = \frac{A_d}{A_c}$$

or in decibel units as

$$\boxed{\text{CMR} = 20 \log \frac{A_d}{A_c}} \tag{9.3}$$

EXAMPLE 9.1 The amplifier shown in Fig. 9.1b has $A_d = 150$, $A_c = 0.25$. Calculate the amplifier common-mode rejection.

Solution: Using Eq. (9.3) yields

$$\text{CMR} = 20 \log \frac{A_d}{A_c} = 20 \log \frac{150}{0.25} = 20 \log (600) = \mathbf{55.6 \text{ dB}}$$

Single-ended Operation

A signal may be applied to either input with the output being in phase with the noninverted input as shown in Fig. 9.2a or of opposite phase with the inverted input as in Fig. 9.2b. In either case the amplifier gain is ideally the same magnitude, A_d, for either input.

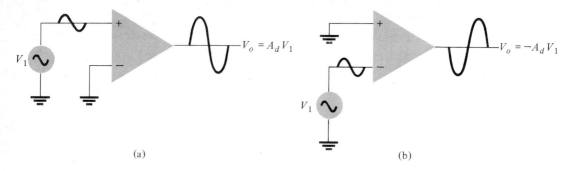

Figure 9.2 Single-ended operation of difference amplifier.

Differential Operation

When a signal is applied across the two input terminals, the resulting output is the amplified input, $A_d v_d$. If two separate input signals are applied, v_d is the difference of the two signals. The gain, A_d, is either specified by the manufacturer or can be determined from Eq. (9.1). Figure 9.3 summarizes these connections.

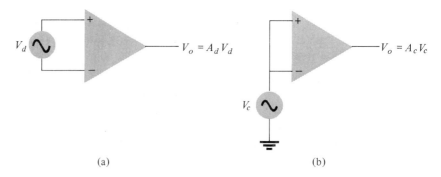

Figure 9.3 Difference and common-mode operation.

EXAMPLE 9.2 For two input signals applied to a differential amplifier ($A_d = 100$) calculate the output voltage for (a) $V_1 = 10$ mV, $V_2 = 0$ V; (b) $V_1 = +10$ mV, $V_2 = -10$ mV.

Solution:
(a) $V_o = A_d V_d = A_d(V_1 - V_2) = 100(10 \text{ mV} - 0) = \mathbf{1}$ **V**.
(b) $V_o = A_d V_d = A_d(V_1 - V_2) = 100[10 \text{ mV} - (-10 \text{ mV})]$
  $= 100 (20 \text{ mV}) = \mathbf{2}$ **V**.

Common-Mode Operation

When an input signal is applied in common to both inputs, the output is

$$V_o = A_c V_c$$

Ideally, the common-mode gain is zero, while practical common-mode gain is quite small.

EXAMPLE 9.3 Calculate V_o for common-mode inputs of $V_c = 10$ mV and $A_c = 10^{-3}$.

Solution: $V_o = A_c V_c = 10^{-3} (10 \text{ mV}) = \mathbf{10\ \mu V}$.

EXAMPLE 9.4 Calculate the common-mode rejection (CMR) for the values of Examples 9.2 and 9.3

Solution: $\text{CMR} = 20 \log \dfrac{A_d}{A_c} = 20 \log \left(\dfrac{100}{10^{-3}}\right) = 20 \log (10^5)$
$= \mathbf{100\ dB}$.

Input Resistance

The resistance seen looking into either input of a difference amplifier should be high to prevent it from loading the driving circuit. In the circuit of Fig. 9.1b, the input resistance is

$$R_i = 2 h_{ie} = 2\beta r_e \tag{9.4a}$$

typically a few kilohms. Higher input resistance can be obtained using bipolar transistors by the circuit shown in Fig. 9.4. The circuit of Fig. 9.4a has an input resistance of approximately

$$R_i = 4 h_{fe} h_{ie} = 4\beta^2 r_e \tag{9.4b}$$

while that of Fig. 9.4b is

$$R_i = 2(h_{ie} + h_{fe} R_e) = 2\beta(r_e + R_e) \tag{9.4c}$$

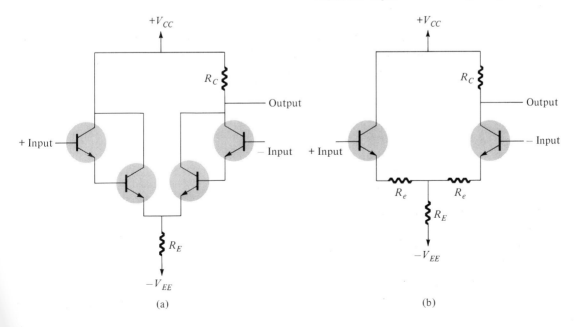

Figure 9.4 Bipolar circuits for higher input resistance.

EXAMPLE 9.5 Calculate the input resistance for the following circuit: (a) Fig. 9.1b for $h_{fe} = \beta = 80$, $h_{ie} = 1.8$ KΩ ($r_e = 22.5$ Ω); (b) Fig. 9.4a; (c) Fig. 9.4b ($R_e = 50$ Ω).

Solution:
(a) $R_i = 2\beta r_e = 2(80)(22.5) = 3600 \Omega =$ **3.6 K**.
(b) $R_i = 4\beta^2 r_e = 4(80)^2(22.5) =$ **576 K**.
(c) $R_i = 2\beta(r_e + R_e) = 2(80)(22.5 + 50) =$ **11.6 K**.

Output Resistance

To provide ideal output resistance the difference amplifier should have zero resistance. The circuits of Figs. 9.1b and 9.4 all have an output resistance of value R_C. Lower output resistance is provided from an emitter follower as that shown in Fig. 9.5. The output resistance for that circuit is

$$R_o = \frac{R_C}{\beta}$$

which is reduced from the previous circuit values by the output transistor current gain.

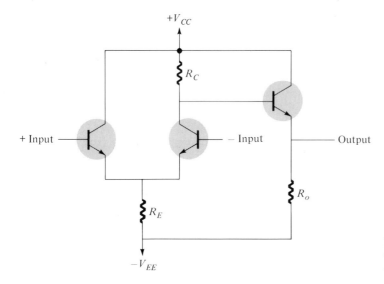

Figure 9.5 Difference amplifier with improved (lower) output resistance.

Common-Mode Rejection

The value of common-mode rejection can be increased by making A_d larger and also by making A_c smaller. The value of A_c can be reduced by increasing the effective value of R_E as in the circuit of Fig. 9.6. A constant-current source allows setting the dc bias value of I_E while providing a large value of resistance, R_E. Figure 9.6b shows a current source with I_E set by the Zener diode voltage, V_Z, and the effective value of R_E set by the transistor output resistance, h_{oe}.

$$R_E = h_{oe}$$

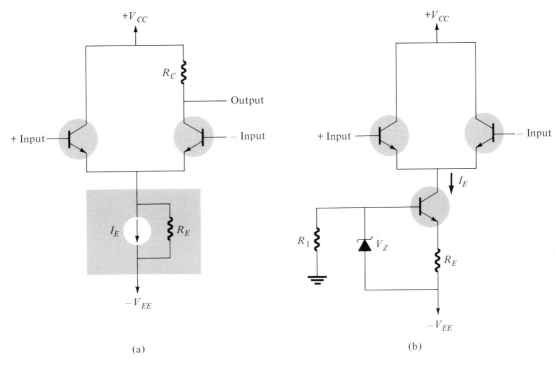

Figure 9.6 Constant current source used to provide larger CMR.

For a typical value of $h_{oe} = 1/100$ K, R_E is 100 K.

EXAMPLE 9.6 Calculate the value of CMR for a circuit as in Fig. 9.6 with $R_C = 10$ K, $r_e = 20$ Ω, and $R_E = 1/h_{oe} = 100$ K.

Solution:
$$A_d = \frac{R_C}{2r_e} = \frac{10{,}000}{2(20)} = 250$$
$$A_C = \frac{R_C}{2R_E} = \frac{10{,}000}{200{,}000} = 0.05$$
$$\text{CMR} = 20 \log \left(\frac{A_D}{A_c}\right) = 20 \log \left(\frac{250}{0.05}\right) = 20 \log 5000$$
$$= 74 \text{ dB.}$$

9.2 OP-AMP IC BASICS

Operational amplifiers (op-amps) are essentially difference amplifiers with a very large voltage gain. Signals are applied to either the noninverting input, the inverting input, or across both. The factor of common-mode rejection is present, typically, at a large value. A large variety of op-amps are available in IC form. A few of the more popular units will be discussed next. Present op-amps are built using bipolar, junction-field-effect transistors and MOSFET transistors to achieve nearly ideal operation in one or more device operation characteristics.

An ideal op-amp unit would have as device characteristics very high input resistance, low output resistance, low power consumption, high gain, large common-mode rejection, and insensitivity to temperature or power supply variations. Various combinations of bipolar and JFET circuits—biFETS, or bipolar and MOSFET circuits—biMOS provide nearly ideal operation in one or more device characteristics. Specifications for various op-amp units are covered after details of op-amp basics and op-amp connections are considered.

Constant-Gain Amplifier

The gain of op-amps vary from 1000 to over 100,000, with values typically in the tens of thousands. This large voltage gain is generally not the resulting gain desired when using the op-amp. Usual practice is to operate the op-amp in a circuit connection, as that of Fig. 9.7. The overall gain of the amplifier will be shown to be

$$A_v = \frac{V_o}{V_1} = -\frac{R_o}{R_1} \qquad (9.5)$$

Note from Eq. (9.5) that the overall gain is determined only by the resistors connected external to the op-amp and does not appear to depend at all on the op-amp gain. Equation (9.5) is, in fact, based on the assumption that A_d is very much larger than the resulting overall circuit gain, A_v; the larger A_d, the more closely A_v is to the ratio R_o/R_1.

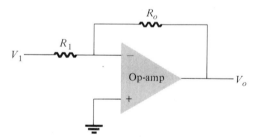

Figure 9.7 Basic op-amp constant-gain amplifier.

The main point of the discussion above is that the very large op-amp gain is usually not used directly but allows setting a circuit gain that can be precisely set by resistors or other passive components.

Virtual Ground

A useful way to analyze op-amp circuits such as that of Fig. 9.7 results using the concept of *virtual ground*. If the gain of the op-amp is large and the output is a finite voltage limited by the power supplies, the difference voltage across the inputs must be very small—virtually zero volts, or ground. As example, if V_o is ± 10 V and $A_d = 100,000$, then V_d is

$$V_d = \frac{V_o}{A_d} = \frac{10\text{ V}}{100{,}000} = 0.1\text{ mV}$$

which is nearly 0 V.

A simple example using the virtual ground approach is shown in Fig. 9.8.

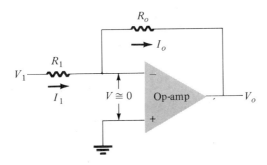

Figure 9.8 Op-amp gain using virtual ground.

The current through R_1 is

$$\frac{V_1 - 0}{R_1} = \frac{V_1}{R_1}$$

while the current through resistor R_o is

$$\frac{0 - V_o}{R_o} = -\frac{V_o}{R_o}$$

Since $I_1 = I_o$,

$$\frac{V_1}{R_1} = -\frac{V_o}{R_o}$$

from which

$$\frac{V_o}{V_1} = -\frac{R_o}{R_1}$$

as expected [see Eq. (9.5)]. This analysis is valid as long as the difference input voltage is small compared to V_1 or V_o, which is another way of saying that the op-amp gain, A_d, is very large (or at least large compared to V_o/V_1).

Summing Amplifier

A practical op-amp circuit connection is the *summing amplifier*, as shown in Fig. 9.9. Using the virtual ground analysis we get

$$\frac{V_1}{R_1} + \frac{V_2}{R_2} = -\frac{V_o}{R_o}$$

so that

$$\boxed{V_o = -\left(\frac{R_o}{R_1} V_1 + \frac{R_o}{R_2} V_2\right)} \qquad (9.6)$$

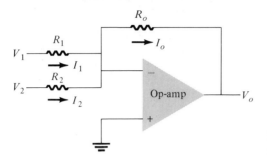

Figure 9.9 Op-amp summing amplifier.

The output is seen to be the sum of the input signals, each input multiplied by a separate gain factor depending on the feedback (R_o) and input resistor.

EXAMPLE 9.7 Calculate the output voltage for the circuit of Fig. 9.10.

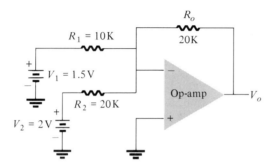

Figure 9.10 Circuit for Example 9.8.

Solution: Using Eq. (9.6), we obtain

$$V_o = -\left(\frac{R_o}{R_1}V_1 + \frac{R_o}{R_2}V_2\right) = -\left[\frac{20\text{ K}}{10\text{ K}}(1.5\text{ V}) + \frac{20\text{ K}}{20\text{ K}}(2\text{ V})\right]$$
$$= -5\text{ V}.$$

Other Op-Amp Configurations

Many other op-amp circuit connections can be made. Figure 9.11 shows a few of the more popular connections.

Unity Follower

The circuit of Fig. 9.11a is a unity follower for which

$$\boxed{\frac{V_o}{V_1} = 1} \qquad (9.7)$$

The unity follower is a useful buffer amplifier circuit used to avoid loading of a driver circuit while providing the signal from a low output resistance. Although the unity follower provides the same signal as output that it receives as input, it provides impedance matching between driver and load.

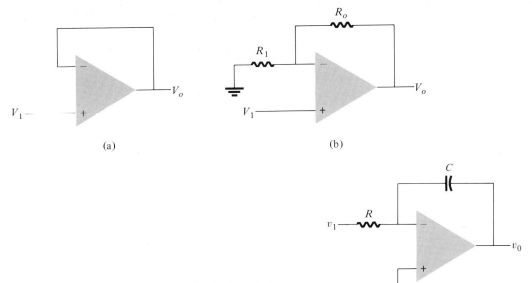

Figure 9.11 Op-amp circuits: (a) unity follower; (b) noninverting multiplier; (c) inverting multiplier.

Noninverting Constant-Gain Amplifier

The circuit of Fig. 9.11b is a noninverting constant-gain amplifier with voltage gain given by

$$\boxed{\frac{V_o}{V_1} = 1 + \frac{R_o}{R_1}} \tag{9.8}$$

The gain is about the same as that of the inverting amplifier of Fig. 9.7 without the inversion between input and output.

Integrator

The circuit of Fig. 9.11c, a very useful circuit in analog computers among other applications, is an integrator circuit for which

$$\boxed{v_o(t) = -\frac{1}{RC}\int v_i(t)\, dt} \tag{9.9}$$

The output is the time integral of the input signal multiplied by a gain factor of $1/RC$.

One popular application of the integrator circuit involves generation of a linear ramp voltage. A linear ramp is a voltage that rises (or falls) at a constant rate. Applying a constant dc voltage as input to the integrator circuit of Fig. 9.11c results in the desired ramp voltage for use in timing or analog-to-digital conversion circuits.

EXAMPLE 9.8 Sketch the output waveform for the circuit of Fig. 9.12a when the input switch is closed.

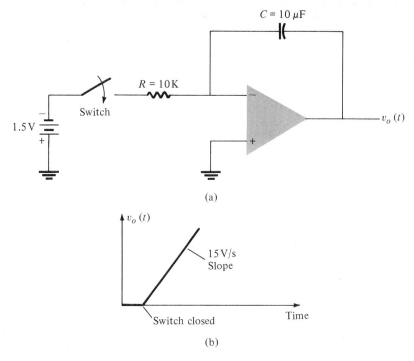

Figure 9.12 Circuit and waveforms for Example 9.8.

Solution: The voltage rises with a slope of

$$1.5 \text{ V}\left(\frac{1}{RC}\right) = 1.5 \text{ V}\left[\frac{1}{(10 \text{ k}\Omega)(10 \text{ }\mu\text{F})}\right] = 15 \text{ V/s}$$

(see Fig. 9.12b).

The output voltage will rise, starting when the input switch is closed, at a constant rate of 15 V/s until the upper saturation level of the op-amp, set by the supply voltage, is reached.

9.3 OP-AMP OPERATION AND APPLICATIONS

Having seen some basic op-amp circuit connections, we can now consider the specific features of a number of popular op-amp IC units. Op-amp IC units can be categorized in a number of ways. Categorizing by circuit components the types of op-amp IC units are:

1. Bipolar.
2. BiFET.
3. BiMOS.

Each of these construction methods provides nearly ideal op-amp performance, as described on the following pages.

Bipolar Op-Amps

One of the more popular ICs built using bipolar transistors is the 741 unit shown in Fig. 9.13. The IC can operate with supply voltage values up to ±22 V, with input voltage swings to ±15 V. The voltage gain of, typically, 200,000 is obtained from approximately 20 bipolar transistors connected in a few stages within the single IC package. Input resistance of the IC unit is typically 2 M and common-mode rejection is 90 dB.

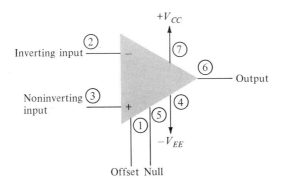

Figure 9.13 Pin connections of 741 op-amp (in 8-pin DIP).

Figure 9.14 shows the 741 connected as an inverting amplifier having an overall circuit gain of 100 as set by the resistor components R_o and R_1. The 10-K potentiometer provides a means of nulling the output voltage or setting it to 0 V when zero input signal is applied.

An input voltage, V_i, of 10 mV rms would appear amplified by a factor of 100, and opposite in phase to the input. As shown, the noninverting input is connected to ground through resistor R_2 of value 1 K.[1] The presence of R_2

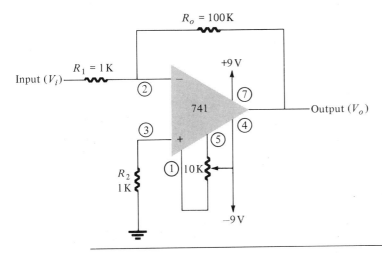

Figure 9.14 741 amplifier with gain of 100.

---

[1] The value of resistor R_2 should theoretically be

$$R_2 = R_o \| R_1 = \frac{R_o R_1}{R_o + R_1}$$

compensates for any offset current in both input terminals to help maintain the output at 0 V (dc).

A small signal, such as that from a phono cartridge or magnetic pickup, would be amplified by the circuit of Fig. 9.14 to a level suitable for driving a power amplifier unit, so that the present amplifier can be referred to as a preamplifier.

Another example of bipolar IC is the 124 quad op-amp unit in Fig. 9.15. As shown, the unit contains four separate op-amp stages in a single IC package.

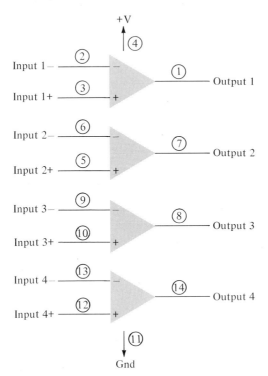

Figure 9.15 Quad op-amp (124) connection diagram.

Each of the four units is a separate op-amp capable of being used alone. Manufacturing technology has made it possible to construct four identical op-amp units in a single IC package, each made of many bipolar transistors to provide the gain and impedance levels desired for op-amp operation. These particular op-amp stages were designed to operate from a single power supply with a typical voltage gain of 100,000. In particular, the IC could easily operate from a single +5-V supply such as that found in digital applications. When using a 5-V supply, in fact, the current drain is about 800 μA in total, or 1 mW per op-amp stage, making this IC unit suitable even for battery operation. Figure 9.16 shows a few examples using the 124 unit to demonstrate how easily these stages are applied in building useful circuits.

Figure 9.16a is a noninverting amplifier with gain given by Eq. (9.8):

$$\frac{V_o}{V_i} = \left(1 + \frac{R_o}{R_1}\right) = 1 + \frac{1\text{ M}}{10\text{ K}} = 101$$

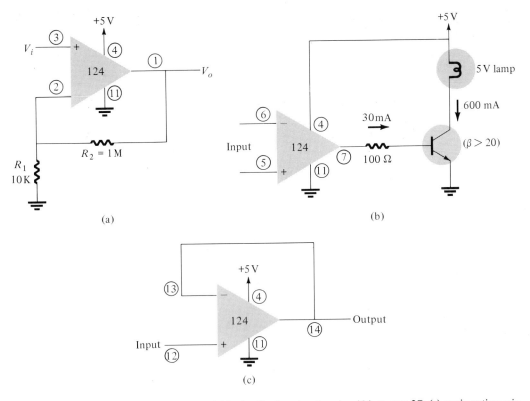

Figure 9.16 Application circuits using 124 op-amp IC: (a) noninverting gain amplifier ($A = +101$); (b) lamp driver; (c) voltage follower.

The circuit of Fig. 9.16b is an example of how the op-amp stage can be used to drive a lamp on or off. With input applied, an output current of at least 30 mA will result in sufficient base current to drive a collector current of 600 mA (transistor beta at least 20) to turn on the 5-V, 600-mA lamp.

The circuit of Fig. 9.16c is a unity follower providing gain of +1 for action as a buffer stage—driving the output load without excessively loading down the signal source.

BiFET Op-Amps

A second unit to consider is the 157 op-amp, which is a biFET unit containing bipolar and JFET transistors. Some of the advantages of this construction are low input bias and offset currents, low noise, and high input impedance at large dc voltage gain and large common-mode rejection. Typical specifications for this unit are:

dc voltage gain of 106 dB

common-mode rejection of 100 dB

input impedance of 10^{12} Ω

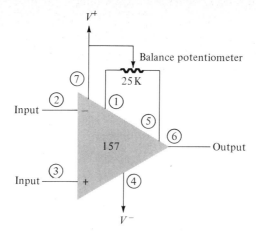

Figure 9.17 Connection diagram of 157 IC unit.

Figure 9.17 shows the connection diagram of an 8-pin DIP unit. Notice the usual inverting and noninverting pair of input terminals and the output terminal, typical of op-amp units. Dual voltage supply of $\pm V$ are provided to the IC. A balance potentiometer is shown connected to allow adjustment of the output offset voltage to obtain operation of 0 V output for 0 V input signal condition. If no offset bias adjustment is desired, pins 1 and 5 are left unconnected.

The 157 op-amp has very fast operation compared to those mentioned previously. A *slew rate* specified at, typically, 50 V/μs indicates op-amp operation up to high frequencies (20 MHz unity-gain bandwidth for the 157).

Figure 9.18 shows the connection diagram of the 157 unit operated as a large power-bandwidth (BW) amplifier providing a gain of 10 at a power bandwidth of 500 kHz.

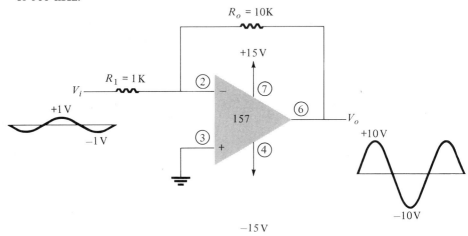

Figure 9.18 Large power-BW amplifier using 157 unit.

BiMOS Op-Amps

Figure 9.19 shows the connection diagram of a CA3140 BiMOS op-amp containing MOSFET and bipolar transistors within the single 8-pin IC package. The MOSFET transistor at the op-amp input provides extremely high input impedance—1.5 TΩ (1.5 × 10¹² Ω), typical, as well as very low input current—10 pA, typical. Otherwise, the biMOS op-amp can be applied in any of the various circuit applications described previously.

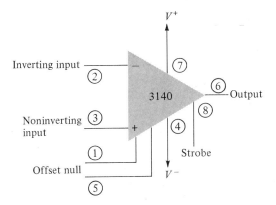

Figure 9.19 BiMOS op-amp (CA3140) in 8-pin DIP package.

9.4 VOLTAGE REGULATORS

Almost every electronic circuit needs a good dc voltage supply. The diode rectifier and capacitor filter convert available ac voltage into a dc voltage which is unregulated. Use of this unregulated dc voltage in an electronic circuit will provide a supply voltage that rises or drops as the load current drawn changes. Regulation of this dc voltage is provided by any one of a large number of voltage regulator ICs, as depicted in Fig. 9.20. Voltage regulation is defined by

$$\boxed{\%\text{V.R.} = \frac{V(\text{no-load}) - V(\text{load})}{V(\text{no-load})} \times 100} \qquad (9.10)$$

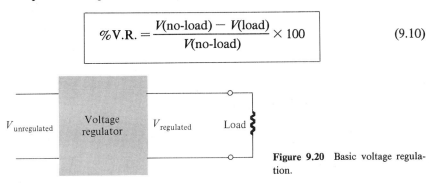

Figure 9.20 Basic voltage regulation.

EXAMPLE 9.9 Calculate the percent voltage regulation for a supply providing 9 V unloaded and 8.3 V when loaded.

Solution: $\%\text{V.R.} = \dfrac{V(\text{NL}) - V(\text{L})}{V(\text{NL})} \times 100 = \dfrac{9 - 8.3}{9} \times 100 = \mathbf{7.8\%}$

The best regulation is when little or no supply voltage change occurs as load changes or that the voltage regulation is smallest (ideally zero). Improved voltage regulation is obtained using a voltage-regulator circuit as described by the circuit of Fig. 9.21.

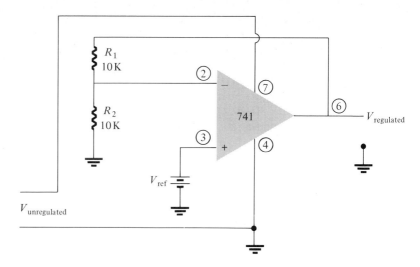

Figure 9.21 Operation of op-amp in voltage regulation.

Voltage Regulation Basics

The output voltage taken from a 741 op-amp can be regulated as described in the following discussion (refer to Fig. 9.21). A fixed reference voltage is applied to the noninverting input. The output voltage divided by a voltage divider of resistors R_1 and R_2 is then applied as the inverting input. Because of the high op-amp voltage gain, the output voltage adjusts until the resulting difference input voltage is near 0 V. The op-amp will now maintain this output voltage. If, for example, the output voltage drops due to increased load, the reduced value of the inverting input causes the output to get larger, which maintains it at the regulated value.

Such practical factors as the voltage stability of the reference voltage and the range of load current over which the output voltage is maintained regulated, among other factors, determine how well the output voltage is regulated.

Commercial voltage-regulator ICs are available to provide output-regulated voltages from a few volts to tens of volts and up to a few amperes of current. These IC units include op-amp circuit, voltage reference, and even adjustment resistors all in a single package. Many of these units are already set to operate at a fixed regulated voltage, typical values being 5, 6, 8, 12, 15, 18, and 24 V, either positive or negative polarity. Other ICs allow the regulated voltage to be adjusted by the user over a range of, typically, near 10 V to about 50 V. The IC packages usually provide connecting the body of the unit to the metal body of the board or case to promote dissipation of the heat developed in the regulator circuit.

Perhaps the most popular regulator ICs are the +5-V units used widely in digital circuitry. The 309 5-V regulator is shown in Fig. 9.22a. An input dc voltage in the range of 7 to 25 V will result in an output regulated at +5 V. The circuit inside the IC requires that the input be at least 2 V higher than the regulated 5-V output. Figure 9.22b shows that an unregulated supply voltage,

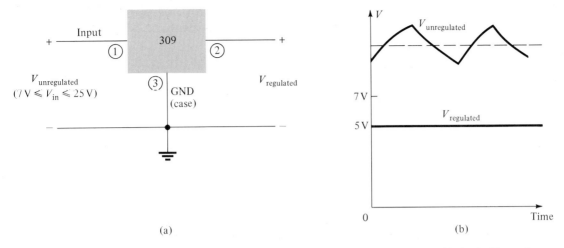

(a)   (b)

Figure 9.22 Operation of 309 voltage regulator IC: (a) circuit; (b) waveforms.

such as that obtained from the capacitor of a power supply, results in a steady dc output voltage. When the input voltage comes from a capacitor filter with some ripple voltage riding around a dc level, it is necessary that the ripple voltage not drop below the level to maintain a minimum 2-V differential across the voltage regulator, even though the average or dc level is higher.

EXAMPLE 9.10 A transformer rectifier-capacitor filter circuit, such as that shown in Fig. 9.23, provides 9 V dc at 10% ripple to a 309 IC regulator. Will the IC circuit operate properly with the output being maintained at +5 V?

Figure 9.23 +5 V power supply using 309 voltage regulator: (a) circuit; (b) waveform.

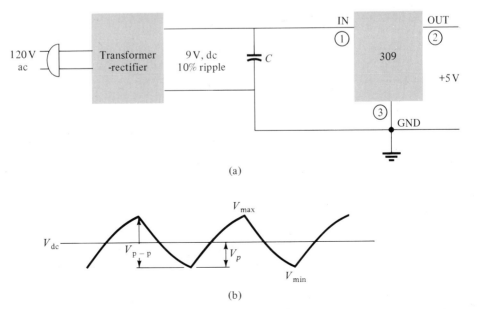

Solution: It is necessary to check that the input voltage never drops below 7 V at the 309 input. Referring to the input waveform of Fig. 9.23b, we obtain

$$r = \frac{V_{rms}}{V_{dc}}$$
$$V_{rms} = r \cdot V_{dc} = (0.1)(9 \text{ V}) = 0.9 \text{ V}$$
$$V_{p\text{-}p} = 2\sqrt{3} \, V_{rms} = 2(1.732)(0.9 \text{ V}) = 3.1 \text{ V}$$
$$V_p = \frac{V_{p\text{-}p}}{2} = 1.56 \text{ V}$$
$$V_{min} = V_{dc} - V_p = 9 \text{ V} - 1.56 \text{ V} = \mathbf{7.44 \text{ V}}$$

Since the voltage at the input remains above 7 V, even at the minimum voltage level of the input, the circuit inside the regulator IC will operate properly (at least 2 V differential voltage is maintained across the IC).

Adjustable Voltage Regulator

An example of a voltage regulator with adjustable regulated voltage setting is the 317 IC shown in Fig. 9.24a. The device's three terminals are input, output, and adjustment, the latter input allowing an input reference voltage to set the

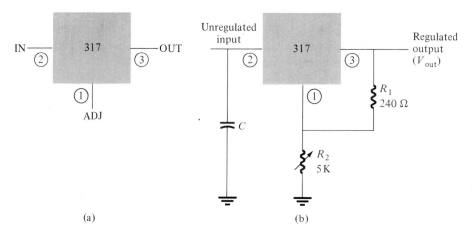

Figure 9.24 Variable voltage regulator: (a) IC; (b) adjustable voltage regulator circuit.

voltage value of the regulator output. Figure 9.24b shows the usual connection for adjustment of the regulated voltage. The regulated voltage can be calculated using the manufacturer's given relation

$$V_{out} \cong 1.25 \text{ V} \left(1 + \frac{R_2}{R_1}\right)$$

which would result in an adjustment range of 1.2 V to 25 V for the resistor values used. Using a fixed resistor for R_2 would provide an output regulated voltage set at a fixed value, somewhere in the range 1.2 V to 25 V.

EXAMPLE 9.11 Calculate the regulated output voltage of the regulator shown in Fig. 9.24b for resistor R_2 of fixed value 1.8 K.

Solution: The regulated voltage is

$$V_{out} = 1.25 \text{ V} \left(1 + \frac{1.8 \text{ K}}{240 \text{ }\Omega}\right) = \mathbf{10.6 \text{ V}}$$

Fixed Voltage Regulator

In addition to the popular +5-V regulator, a number of fixed voltage regulators, both positive and negative voltage, are available. As an example, the series of 340 IC units provide positive voltage regulators, while a similar series of 320 IC units provide negative voltage regulation. The IC unit number, in fact, provides the regulated output voltage. For example, a 320–12 unit provides a fixed −12-V regulated output voltage, while a 340–10 provides a +10-V regulated output. The series number indicates positive or negative polarity of the regulated voltage, the number after the hyphen indicating the value of the fixed output voltage. Many other IC regulator series exist, and reference to the various manufacturers' data handbooks will provide information not only on the IC unit but usually some application examples for using the regulator unit.

Figure 9.25 shows a circuit providing a regulated +6 V using the 340–6 IC regulator. The transformer steps down the 120-V rms line voltage through a center-tapped transformer to 8 V rms for each half. The ac voltage is full-wave-rectified and then filtered by the 1000-μF capacitor, providing a voltage of about 10 V with some ripple, as shown in the diagram. As long as the load connected to the output of the IC regulator is not too great, the output voltage will be maintained at a regulated or constant value of +6 V. Essentially, the load must not draw too much current, resulting in the input voltage across the filter capacitor dropping below about +8 V (maintaining at least 2 V across the regulator IC). Many other

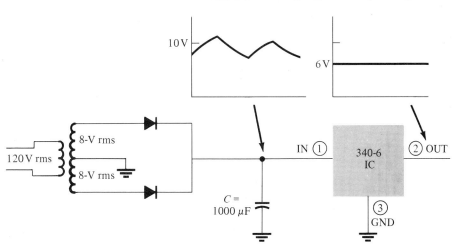

Figure 9.25 Voltage regulator using 6-V regulator IC.

regulated voltages can be achieved by using suitable transformer secondary voltages and any of the other 340 series ICs. For negative output voltage the rectifier diodes are reversed and any of the series 320 ICs then used.

PROBLEMS

§ 9.1

1. Calculate the common-mode rejection (in dB) for an amplifier as shown in Fig. 9.1b with a differential gain of 180 and a common-mode gain of 0.1.

2. Calculate the common-mode gain of an amplifier (as shown in Fig. 9.1b) having a common-mode rejection ratio of 65 dB and a difference gain of 135.

3. Calculate the output voltage of an amplifier such as that shown in Fig. 9.2b (with gain $A_d = 130$) for inputs of $V_1 = 25$ mV and $V_2 = -25$ mV.

4. Calculate the input resistance of an amplifier such as that shown in Fig. 9.1b for $h_{fe} = \beta = 120$, $h_{ie} = 2.2$ K ($r_e = 18.3$ Ω).

5. Calculate the input resistance of an amplifier such as that shown in Fig. 9.4b for $h_{fe} = \beta = 120$, $h_{ie} = 2.2$ K ($r_e = 18.3$ Ω), and $R_e = 51$ Ω.

6. Calculate the value of common-mode rejection (CMR) for a circuit such as that in Fig. 9.6 with $R_C = 15$ K, $r_e = 15$ Ω, and $R_E = 1/h_{oe} = 200$ K.

§ 9.2

7. Calculate the output voltage for the circuit as in Fig. 9.10 for resistor values of $R_o = 50$ K, $R_1 = 12$ K, $R_2 = 18$ K, and inputs of $V_1 = 1.2$ V and $V_2 = 1.5$ V.

8. Calculate the output voltage of a unity follower such as that shown in Fig. 9.11a for $V_1 = 2.5$ V.

9. Calculate the output voltage of a noninverting constant-gain amplifier such as that shown in Fig. 9.11b for resistor values of $R_o = 12$ K and $R_1 = 3$ K for inputs of $V_1 = 1.5$ V.

10. Sketch the output voltage waveform for the circuit of Fig. 9.12a when the input switch is closed (use component values of $R = 12$ K and $C = 15$ μF and $V_1 = -1.8$ V).

§ 9.3

11. Draw the functional diagram of the 124 quad op-amp, labeling all pin connections.

12. Draw the circuit diagram of a summing amplifier using a 124 quad op-amp for inputs with gains of 10 and 20 (using $R_o = 1$ M), including pin designations.

13. Draw the circuit diagram of a unity follower circuit driving an inverting gain of 10 circuits using a 124 quad op-amp IC, listing component values ($R_o = 1$ M) with all pin connections.

14. Draw the functional diagram of a 157 op-amp, including all pin connections.

§ 9.4

15. Calculate the percent voltage regulation for a supply providing 12 V when unloaded and 11.2 V when loaded.

16. A 317 IC voltage regulator is used as shown in Fig. 9.24b, with resistor values of $R_1 = 240\ \Omega$ and $R_2 = 2.4$ K. Calculate the regulated output voltage.

17. Draw the circuit diagram of a transformer full-wave rectifier-capacitor filter feeding a series 320 regulator IC to provide an output at -12 V. Sketch typical input and output voltage waveforms, including voltage levels.

GLOSSARY

BiFET Op-Amp An operational amplifier unit built using bipolar and JFET transistors.

BiMOS Op-Amp An operational amplifier circuit built using bipolar and MOSFET transistors.

Bipolar Op-Amp An operational amplifier built using bipolar transistors.

Common-Mode Gain The ratio of amplifier output to input when the same or common signal is applied to both inputs.

Common-Mode Rejection Ratio (CMR) The ratio between the amplifier difference gain and the common-mode gain in dB units.

Constant-Gain Amplifier An op-amp connection providing a fixed gain set by input and feedback resistor values.

Difference Amplifier An amplifier circuit with inputs resulting in output in phase and output out of phase with each respective input.

Difference Gain The gain of the amplifier calculated from the output to the difference of the inputs.

Integrator An op-amp circuit with capacitor feedback element providing output as the time integral of input.

Operational Amplifier (Op-Amp) An amplifier circuit having very high gain and differential inputs.

Single-ended Operation of the difference amplifier with one input grounded and the signal applied to the other input.

Slew Rate The op-amp characteristic providing an indication of the response time of the amplifier.

Unity Follower An op-amp circuit connected to provide a gain of +1 or unity.

Virtual Ground The concept that the input difference voltage is so small that it is virtually zero or ground, allowing a simplified technique for analyzing the op-amp circuit connection.

Voltage Regulator An IC circuit used to maintain or regulate a dc voltage, typically used in power or voltage supplies.

CHAPTER 10

Digital IC Units

10.1 DIGITAL FUNDAMENTALS

Digital circuits are built using bipolar transistors, or MOSFETs, in integrated circuits (ICs). Basically, digital circuits are simple since they are operated either fully saturated (ON) or in cutoff (OFF), usually represented by two distinct voltage levels. Our concern, then, is with the two states of a digital circuit. These can be at either one of two preselected voltage levels: +5 V and 0 V, for example. Logically, these levels may be related to the binary conditions of 1 and 0. For example, +5 V = 1 and 0 V = 0 or −10 V = 1 and 0 V = 0 are possible relations that may be assigned to voltage levels and logic conditions. Digital circuits provide manipulation of the logical conditions representing the two different voltage levels of the circuit. A number of logic gates or circuits are covered in this chapter. A logical AND gate, for example, provides an output only when *all* inputs are present, whereas a logical OR gate provides an output if *any* one input is present. An inverter circuit provides logical inversion—a 1 output for 0 input, or vice versa. The most popular configurations for a logic gate include an AND or OR gate, followed by an inverter, an AND-gate-inverter combination called a not-AND or NAND gate, and an OR inverter called a not-OR or NOR gate.

Also important in digital circuits is a class of multivibrator circuits. These circuits have two opposite output terminals (if one output is logical-1, the other is logical-0, or vice versa). The most useful of these is the bistable multivibrator or *flip-flop*, which can remain in either stable condition of output. The flip-flop can be used as a memory device—holding the state it was placed in after the initial pulse operating the circuit has passed. It can also be used to build a binary counter or a shift register for use in digital computers.

A second circuit, the monostable multivibrator, provides opposite voltage outputs but, as the name implies, it can remain stable in only one state. If the circuit is triggered to operate, it can go into the opposite state of output voltages but can remain in this state only for a fixed time interval. The monostable multivibrator, or one-shot, is used for pulse shaping, time delay, and other timing actions.

An astable multivibrator or clock circuit has no stable operating state and provides a changing output voltage (pulse train). The circuit is therefore a square-wave oscillator providing pulses to activate various computer operations.

A circuit that appears similar to the multivibrator class is the Schmitt trigger. This circuit operates from a slowly varying input signal and switches output voltage state when the input voltage goes above a preset voltage level or below a second voltage level.

Diode Logic Gates—AND, OR

A logic gate provides an output signal for a desired logical combination of input signals. An AND gate, for example, provides an output of "logical-1" only if all the inputs are present, that is, if each is at 1 (logical-1). This circuit can be compared to one using switches connected in series. Only if, as in Fig. 10.1

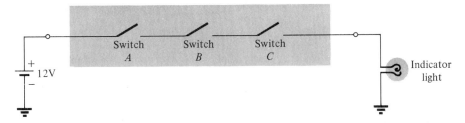

Figure 10.1 AND circuit using switches.

all switches are closed (at 1) does an output voltage appear (and drive the light indicator ON). Thus, a light ON (logical-1) appears if switch A, AND switch B, and AND switch C are closed. This can also be written as the Boolean or logical expression

$$\text{Light ON} = A \cdot B \cdot C$$

(The dot indicates AND; read as A AND B AND C.)

DIODE AND GATE

A more practical form of AND gate using diodes is shown in Fig. 10.2. Using positive-voltage levels of $+V$ for 1 and 0 V as 0, we see that the circuit provides a 1 output of $+V$ only if *all* inputs are $+V$ or 1. More specifically, any input at 0 V will hold the output at about 0 V, as shown on the next page.

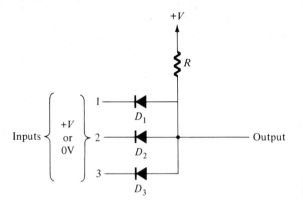

Figure 10.2 Diode AND-gate circuit.

1. A zero-volt input at any (one or more) input terminal(s) will cause that diode to short the output to ground. In logic form this means that 0 at any input produces 0 output.
2. A 1 input at inputs 1 and 2, but a 0 input at 3, will produce a 0 output because diode 3 shorts the output to ground.
3. Only when a 1 input is provided at inputs 1 AND 2 AND 3, none of the diodes is conducting and the output is 1.

Figure 10.3 shows a typical diode gate circuit and voltage truth table. Figure 10.4 shows the logic symbol and logic truth table for the circuit of Fig. 10.3 for positive logic operation.

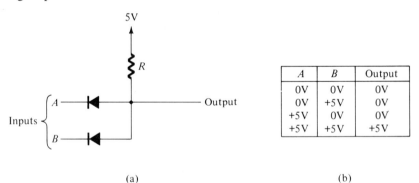

(a) (b)

Figure 10.3 (a) Diode logic circuit; (b) voltage truth table.

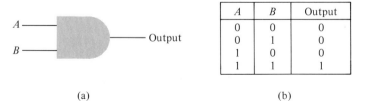

(a) (b)

Figure 10.4 (a) Diode AND-gate symbol; (b) logic truth table, for positive logic ($+5$ V = 1, 0 V = 0).

DIODE OR GATE

The circuit of Fig. 10.5 shows an OR-gate circuit connection using three switches connected in parallel. If either switch *A* OR *B* OR *C* (or any combination of these) is closed (logical-1), the indicator light will be turned ON. A practical version of such a circuit uses diodes as shown in Fig. 10.6. The circuit is a positive-logic diode OR gate, and for this example it uses the same $+V$ and 0-V levels as the AND-gate circuit previously considered.

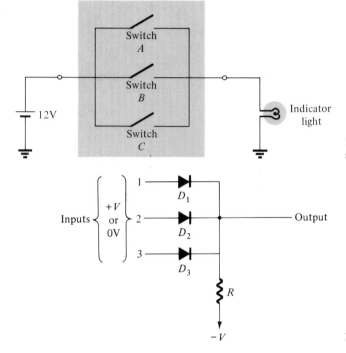

Figure 10.5 Logical OR gate using switches (Output = $A + B + C$).

Figure 10.6 Diode OR-gate circuit.

The operation of the circuit of Fig. 10.6 is the following:

1. A 1 $(+V)$ input at any (one or more) input terminal(s) will cause that diode to conduct, placing the output at the 1 level.
2. A 0 input at inputs 1 and 2, but a 1 input at 3, will produce a 1 output because diode 3 conducts, placing the output at $+V$, and thereby holding diodes 1 and 2 in cutoff.
3. Only when a 0 input is provided at all three inputs will the output be 0.

Figures 10.7 and 10.8 show a typical OR-gate circuit and logic symbol and the respective voltage and logic truth tables.

POSITIVE LOGIC—NEGATIVE LOGIC

In the two circuits just considered the voltage operation of these circuits is described by the voltage truth tables of Figs. 10.3 and 10.7. The logic operation

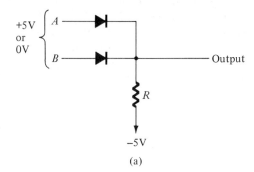

Figure 10.7 (a) OR-gate circuit; (b) voltage truth table.

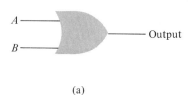

Figure 10.8 (a) OR-gate logic symbol; (b) logic truth table.

of these two circuits, however, is dependent on the definitions of logical-1 and logical-0. Positive-logic definitions were used so far, where **positive logic** meant that the *more positive* voltage was assigned as the logical-1 state. It is possible to use other logic definitions. Using the same two circuits and voltage levels of $-V$ and 0 V provides **negative-logic** operation—with the definition of the *more negative* voltage $(-V)$ as the logical-1 level (and 0 V as logical-0).

Transistor Inverter

A simple but important digital circuit is the inverter. Using voltage levels of 0 and +5 V, as an example, the inverter circuit will provide an output of 0 V

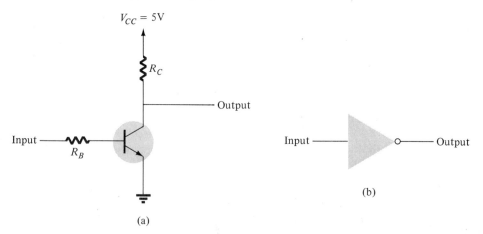

Figure 10.9 Inverter circuit and characteristics: (a) transistor inverter circuit; (b) inverter logic symbol.

for an input of +5 V and an output of +5 V for an input of 0 V. In logical terms the inverter, having single-input and single-output terminals, provides the opposite output—a 1 output for 0 input, or vice versa.

Figure 10.9 shows the circuit diagram of an inverter. The transistor in this circuit is operated either in saturation (ON) or in cutoff (OFF). To operate the transistor in saturation there must be sufficient base current drive corresponding to the current drawn by the collector. This requires that for a particular amount of collector current the base current must be larger than the amount specified by the transistor current gain (h_{fe} or β). To hold the transistor in cutoff the base-emitter must be reverse-biased or not forward-biased.

10.2 INTEGRATED-CIRCUIT (IC) LOGIC DEVICES

Digital integrated circuits (ICs) of various types find widespread use. It is economical and attractive to purchase a complete circuit of small size so that users of digital logic circuits depend on the IC units provided by the numerous manufacturers. At present there are a number of different types of logic circuits popular, each having some advantages and disadvantages. Since no one circuit type has been universally accepted, it seems reasonable to consider some of the more popular circuit types to understand their operation and their relative advantages and disadvantages. It should be clear that each type provides the same basic logical function and that other more practical factors about the overall system generally dictate which particular logic circuit type is chosen.

As a partial summary of the important factors used in selecting a circuit type we have (not necessarily in order of importance)

1. Cost.
2. Power dissipation.
3. Speed of operation.
4. Noise immunity.

Transistor-Transistor Logic (TTL or T²L) Circuit

Transistor-transistor logic (T²L) is one circuit form of logic gate that, although possible as discrete components, is appropriate for manufacture in integrated form. Figure 10.10 shows a basic form of the logic circuit. Notice that each input is made using an emitter, a single multiple emitter transistor providing the inputs.

Figure 10.11 shows the circuit operation for output logical-0 (0 V). All inputs must either have high inputs (+5 V) or not be connected, so that transistor Q_1 is off. Transistor Q_2 is then driven by a base current from $+V_{cc}$ through R_1, the base collector of Q_1 and forward-biased base emitter or Q_2. When Q_2 is on, the output voltage taken from the collector is $V_{CE_{sat}}$, approximately 0 V.

Figure 10.11 shows one (or more) input of 0 V allowing Q_1 to be biased *on*, resulting in the collector voltage of Q_1 to be near 0 V, thereby holding Q_2 *off*. When Q_2 is *off*, the collector voltage provides an output of +5 V. The circuit operates as a positive logic NAND gate.

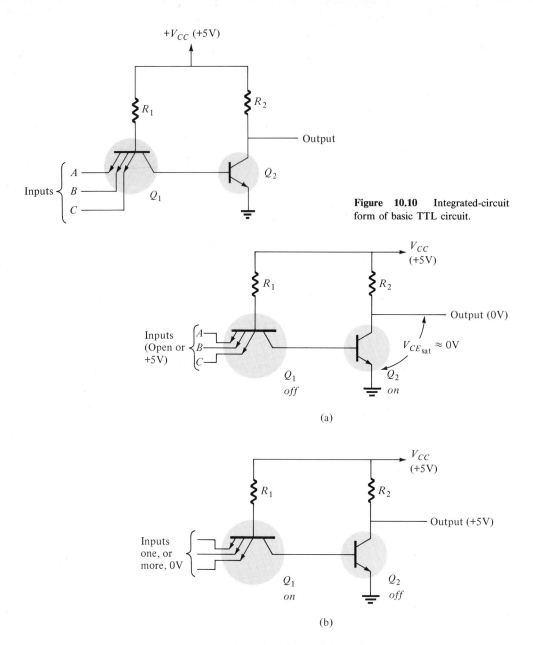

Figure 10.10 Integrated-circuit form of basic TTL circuit.

Figure 10.11 Operation of TTL (T²L) logic gate: (a) output transistor *on;* (b) output transistor *off.*

The T²L unit is now very popular because it has fast speed and good noise immunity and because of its low cost. A practical T²L logic unit is shown in Fig. 10.12a, the NAND logic symbol in Fig. 10.12b, and the package pin connections in Fig. 10.12c. Four T²L NAND gates are contained in the single SN7400 IC package, the entire unit being a quad, 2-input positive NAND gate. The basic T²L gate can switch state in typically 10 ns at power dissipation of 10 mW.

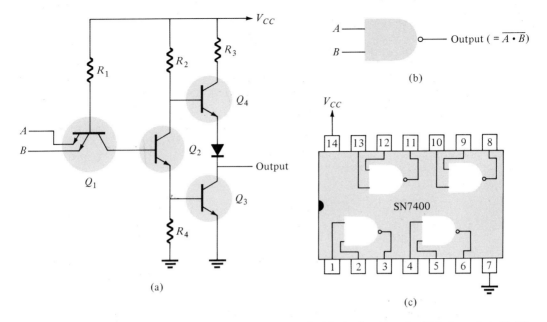

Figure 10.12 T²L logic unit: (a) circuit schematic; (b) logic symbol; (c) IC package of SN7400.

CMOS Logic Circuits

Another popular IC logic circuit is made using MOSFET devices of both *p*-channel and *n*-channel transistors, the complementary symmetry circuit being COS/MOS, COSMOS, or more simply CMOS logic. A basic CMOS inverter is shown in Fig. 10.13. A positive input voltage drives the *n*-channel FET *on* with output 0 V. An input of 0 V results in the *p*-channel FET *on* with the output a positive voltage.

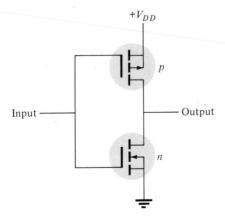

Figure 10.13 CMOS inverter circuit.

A positive logic NOR gate is made as shown in Fig. 10.14. When either input is positive, an *n*-channel FET is driven *on* with output then 0 V. When both

SEC. 10.2 INTEGRATED-CIRCUIT (IC) LOGIC DEVICES

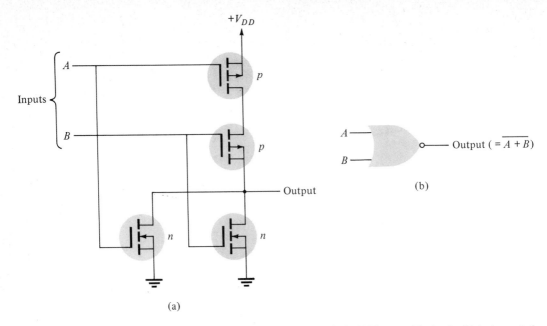

Figure 10.14 CMOS positive logic NOR gate: (a) circuit; (b) logic symbol.

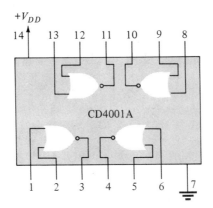

Figure 10.15 CMOS quad two-input positive NOR gate IC: CD4001A.

inputs are 0 V, the two *p*-channel FETs are turned *on* with output then a positive voltage. The CD4001A, for example, is an IC package of four 2-input NOR gates as shown in Fig. 10.15.

CMOS logic units offer very low-power dissipation at relatively slow speed. Typical propagation delay is 100 ns, with 5-V supply at only microwatts of power dissipation, while providing noise immunity of over 2 V. A CMOS logic gate is at least five times smaller than the comparable T²L gate.

Integrated Injection Logic (I²L or IIL)

A relatively new IC logic circuit uses bipolar transistors in a circuit and physical arrangement called *integrated injection logic* (I²L) and has the better characteristics of T²L and CMOS units: The I²L unit has low-power dissipation, fast speed, and small size.

Complementary bipolar transistors are used, the *pnp* injector transistor serving as a current source, with a multiple collector *npn* transistor forming the basic gate. Figure 10.16 shows a dual gate unit. I²L gates use microwatts of power at speeds in the tens of nanoseconds, and they require the least IC area of the gates considered.

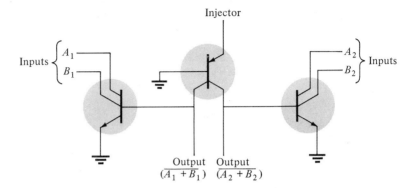

Figure 10.16 I²L dual NOR gate.

Emitter-coupled Logic (ECL) Circuit

The technique of emitter-coupled logic (ECL) or current-mode logic (CML) circuits differs from those covered previously in one main respect. All previous circuits allowed the transistors to saturate. This means that an amount of charge is stored in the transistor base and collector regions, resulting in a time delay in turning off the transistor. Current-mode logic operates the transistor in a nonsaturated condition, thereby providing shorter propagation delays through the circuits.

An emitter-coupled logic circuit, such as that of Fig. 10.17, uses a transistor for each input of the circuit, which is desirable for IC manufacture. Whereas a direct-coupled logic circuit would connect the transistor emitters to ground (thereby allowing the base drive to be dependent on the input voltage), the present circuit

Figure 10.17 Emitter-coupled logic (ECL) circuit.

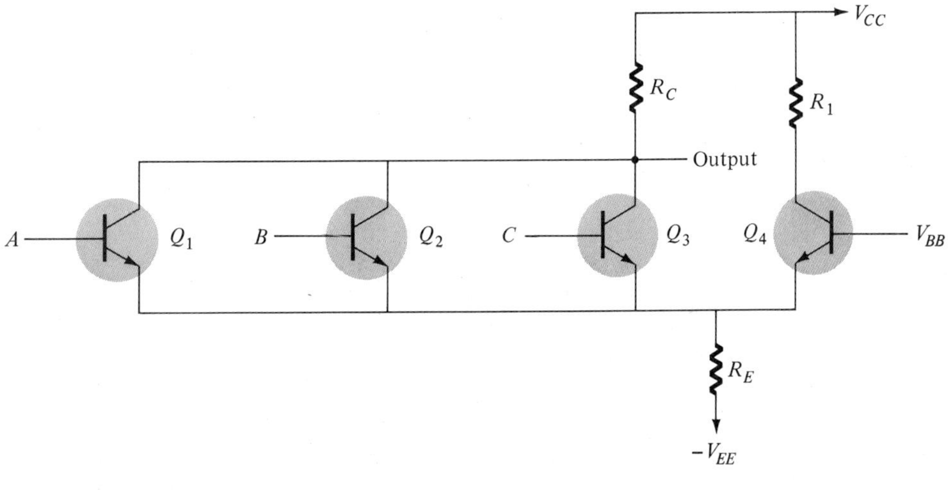

uses an additional transistor stage providing a reference point for the common-emitter terminals. If more than the needed turn-on drive is applied, this will cause the emitter point to rise in voltage, maintaining the logic transistor in a nonsaturated mode of operation.

In summary, then, the nonsaturated operation of an emitter-coupled logic (ECL) circuit provides very fast switching speeds, simultaneous OR and NOR outputs from each circuit, high fan-in and fan-out capacity, and constant noise immunity of the power supply due to the relative constant-current demand of the logic circuit whether *off* or *on*. Disadvantages are the need for a bias driver to provide the reference supply voltage, the higher cost of ECL circuits, and the larger size of a circuit.

10.3 BISTABLE MULTIVIBRATOR CIRCUITS

Of equal importance to logic circuits in digital circuitry is the class of multivibrator circuits. There are three basic forms of the multivibrator—bistable, monostable, and astable; the most important of these, by far, is the bistable multivibrator or flip-flop. As an indication of the applications of the multivibrator circuits, consider the following:

> Bistable (FLIP-FLOP)—storage stage, counter, shift register
> Monostable (ONE-SHOT)—delay circuit, waveshaping, timing circuit
> Astable (CLOCK)—timing oscillator (square-wave)

In a logic system there will typically be a large number of flip-flop stages used as counters, storage registers, shift-registers, a few one-shot circuits in special timing or pulse-shaping uses, and a limited number of CLOCK circuits (typically only one).

Characteristic of all three circuits is the availability of two outputs, where the outputs are logically inverse signals. One output is selected as the reference, this designation being indicated in a number of ways. The two outputs are sometimes marked as 0 and 1, FALSE and TRUE, or $\bar{A}$ and A, etc. The main point of the designation is to indicate that the outputs are logically opposite and to mark the output chosen as the reference output. Another means of indicating the state of the multivibrator circuit is the use of the designation of SET and RESET. When referring to the state of the circuit, the definitions of SET and RESET are the following:

> SET: Q output is logical-1
> $\bar{Q}$ output is logical-0
> RESET: Q output is logical-0
> $\bar{Q}$ output is logical-1

Bistable Multivibrator (Flip-Flop)

The flip-flop circuit, the most important of the multivibrator circuits, will be covered first. To provide some basic consideration of this circuit's operation a simple form of bistable circuit using two inverters is shown in Fig. 10.18. The

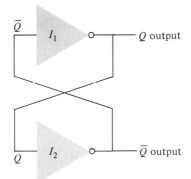

Figure 10.18 Two-inverter flip-flop circuit.

inverters are essentially connected in series, with two output points indicated. The two outputs are labelled Q and $\overline{Q}$, respectively. If the Q output is a logical-1, then inverter I_2 will provide $\overline{Q}$ as a logical-0. Since $\overline{Q}$ is connected as input to inverter I_1, it will cause the output of that stage to be a logical-1, as assumed. Thus, the state of logical conditions, or the voltage they represent, forms a stable situation with Q output logical-1 and $\overline{Q}$ output logical-0. If some external means is used to cause the Q signal to change to logical-0, then, through inverter I_2, $\overline{Q}$ would change to logical-1. The $\overline{Q}$ input of logical-1 would then result in Q being logical-0 as initially proposed. Thus, the circuit will also remain in a stable condition if the Q output is logical-0 and the $\overline{Q}$ output is logical-1. In effect, then, the circuit has two stable operating states that act as a memory of the last state it was placed into. Some external means is necessary, however, to cause the circuit to change state.

RS FLIP-FLOP

Figure 10.19 shows the use of NOR gates connected in series with additional inputs providing signals to cause the circuit to change state. The inputs are marked R for RESET and S for SET. Recall that a NOR gate provides logical-0 output if any of its inputs is logical-1. A logical-1 input to the S terminal will cause the output of N_2 to be logical-0. If it is assumed that no input is connected to the R terminal at this time, the inputs to N_1 are both logical-0 with output of logical-1. The result of a SET input signal then is to cause the circuit to become SET, where the SET state was previously defined as Q output = logical-1 and

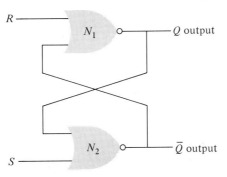

Figure 10.19 RESET and SET operation of flip-flop.

$\bar{Q}$ output = logical-0. Similarly, the application of only a logical-1 to the R input will cause the Q output to become logical-0 and $\bar{Q}$ output logical-1, which is the RESET state of the circuit. It should be obvious that simultaneous application of logical-1 signals to both S and R inputs is ambiguous, forcing both outputs to the logical-0 condition. This would not be an accepted operation of this circuit in which the two output signals should be always logically opposite. If the R and S inputs are both logical-0, the circuit remains in whatever state it was last placed into.

T-TYPE FLIP-FLOP

One of the more common type of flip-flop circuits is the T-type or triggered flip-flop. This circuit is also called a *complementing* flip-flop, or *toggle* flip-flop, since its action is to change state every time an input is applied to the single T-input terminal. Figure 10.20a shows a block diagram of a T-type flip-flop. The waveforms in Fig. 10.20b show that the output changes stage, or toggles, every negative edge of the input signal. Notice that the output goes through a full cycle every two input cycles so that the output is at one-half the frequency of the input or that every two input pulses result in one full cycle of the output. The T-type flip-flop is the basic circuit used as a counter stage in binary counters.

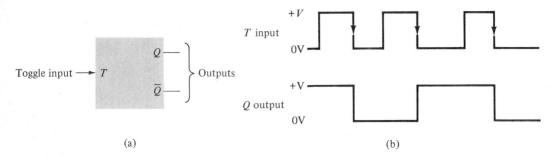

Figure 10.20 Toggle flip-flop (T-type): (a) block symbol; (b) operating waveforms.

JK FLIP-FLOP

The J and K input terminals (Fig. 10.21a) are used to provide information or data inputs. When a trigger pulse is then applied, the circuit changes state corresponding to the inputs to the J and K terminals. A JK circuit is built as an integrated circuit form using TTL or CMOS logic manufacture. It is not at all important here to consider the circuit details, and in fact one only purchases a complete unit in IC form and has little to do with the details of circuit operation. It is important to be aware of the differences in using these different logic types, in knowing whether they use current-sourcing, sinking, or emitter-coupled logic, in details of speed of operation, noise margin, power supply voltages, and so on. These factors, however, are descriptive of the overall circuit, and the details of what actually goes into building the actual circuit are often of little importance. For the *JK* flip-flop, then, we shall discuss its operation from the logic or block diagram point of view and consider some applications using the circuit. The *JK* circuit is versatile and is presently the most popular version of the flip-flop circuit. A logic symbol of a *JK* flip-flop is shown in Fig. 10.21a. The *J* and *K* terminals shown are the data input terminals receiving the information of logical-1 or logical-0. These information inputs do not, however, change the state of the flip-flop circuit, which will remain in its present state until a trigger pulse is applied. Thus, for example, if the circuit were presently RESET and the input data were such as to result in the SET condition, the circuit would still maintain the RESET condition, even with the *J* and *K* input data signals applied. Only when the trigger pulse occurs are the input data used to determine the new state of the circuit— the SET state for the present example.

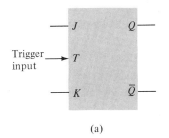

J	K	Circuit Action
0	0	Remains in same state
0	1	RESETS
1	0	SETS
1	1	Circuit toggles

(a)   (b)

Figure 10.21 Basics of flip-flop: (a) logic symbol; (b) truth table.

There are four possible combinations of the *J* and *K* inputs. To consider the complete operation of the circuit each of the possible conditions is listed in the truth table of Fig. 10.21b. If both *J* and *K* inputs are logical-0, the circuit remains in the same state (no change takes place). If the *J* input is logical-0 and *K* input logical-1, the circuit ends up in the RESET state. If the *K* input is logical-0 and the *J* input logical-1, the occurrence of the trigger pulse will cause the circuit to be set SET. Finally, if both *J* and *K* inputs are logical-1, the action of the trigger pulse's becoming logical-1 is to toggle or complement the circuit. Thus, the circuit would toggle (change state) from whichever condition it happened to be in on application of the trigger pulse. In this last case, with *J* and *K* inputs both logical-1, the circuit would operate as a *T* flip-flop and could be used as such. When opposite data input signals are applied as *J* and *K* inputs, the trigger pulse will shift the data into the present flip-flop state, the stage then acting as a shift-register stage. Thus, the *JK* flip-flop can be used as a shift-register stage, a toggle stage for counting operations, or generally as a control logic stage.

If positive logic is used (0 V = logical-0 and +V as logical-1, for example), the circuit triggers when the trigger pulse goes from the logical-0 to logical-1 condition—this being referred to as positive-edge triggering since the circuit is triggered at the time the voltage goes positive (from 0 to +V). When the circuit triggers on a voltage change from +V to 0 V, the triggering is trailing-edge triggering. The manufacturer's information sheets should indicate the type of triggering required to operate the particular circuit so that it may be properly used.

Monostable Multivibrator (One-Shot)

As a characteristic property of a multivibrator circuit the monostable provides two opposite-state output signals. As the name implies, the outputs are stable in only one of the two possible states (SET and RESET). Figure 10.22a shows a logic block symbol of a one-shot circuit in the stable RESET state (Q output = logical-0, and $\overline{Q}$ output = logical-1). The input trigger signal is a pulse that operates the circuit in an edge-triggered manner. Figure 10.22b shows a typical input trigger pulse and corresponding output waveform (assuming triggering on the trailing edge of the trigger pulse). The Q output is normally low (RESET state). When a negative-going voltage change triggers the circuit, the Q output goes high (SET state), which is the unstable circuit state. It will remain high only for a fixed time interval, T, which is determined basically by a timing capacitor whose value may be selected externally. Thus, the output state remains in the SET state only for a preselected time T, after which the output returns to the RESET state, where it remains until another trigger pulse is applied. In the waveform of Fig. 10.22b the output of the one-shot can be viewed as a delayed pulse whose negative-going edge occurs at some set time T after the trigger pulse.

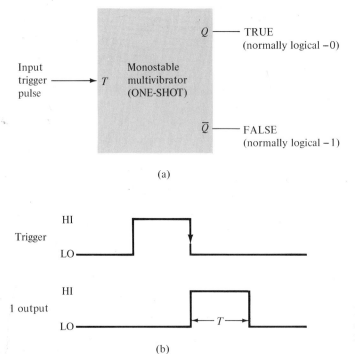

Figure 10.22 Operation of monostable multivibrator: (a) one-shot logic symbol; (b) input trigger and 1 output waveforms.

A popular TTL monostable multivibrator is the SN74121 shown in Fig. 10.23. The output provides a pulse of duration T,[1] the Q output going from its normally low state (0 V) to the high output state or SET state. Triggering can occur when

---

[1] For the SN74121 the pulse width of the output pulse is $T \cong 0.7 R_T C_T$ where R_T is the timing resistor, R_{ext} or R_{int} and C_T is the timing capacitor, C_{ext}.

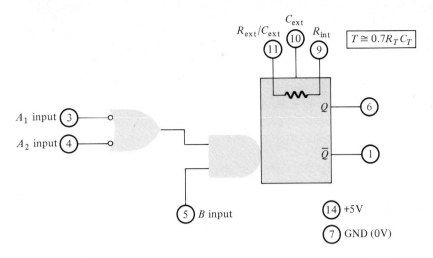

Figure 10.23 SN74121 monostable multivibrator.

either the A inputs are both grounded and the B input goes positive or when the B input is lifted high (or unconnected) and the A inputs go low. Example 10.1 will help show how the one-shot can be used.

EXAMPLE 10.1 Draw the Q output waveform for an SN74121 (see Fig. 10.24a) triggered by an input signal as shown in Fig. 10.24b.

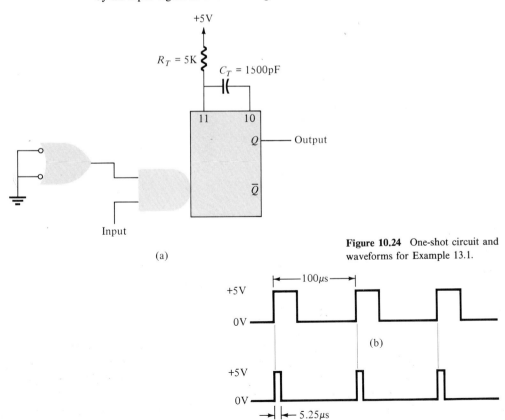

Figure 10.24 One-shot circuit and waveforms for Example 13.1.

Solution: The unit is triggered on a positive edge using the B input, and the time width of the output pulse from the normally low Q output is

$$T = 0.7\, R_T C_T = 0.7(5 \times 10^3)(1500 \times 10^{-12}) = 5.25\ \mu s$$

as shown in Fig. 10.24c.

Astable Multivibrator (Clock)

A third version of multivibrator has no stable operating state—it oscillates back and forth between RESET and SET states. The circuit provides a clock signal for use as a timing train of pulses to operate digital circuits. Figure 10.25

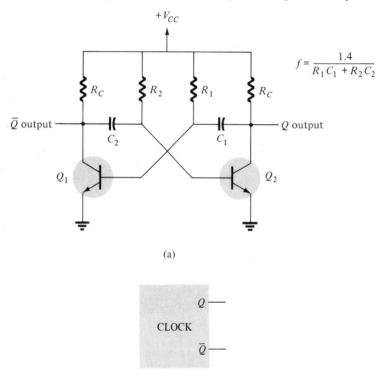

Figure 10.25 (a) Astable multivibrator circuit; (b) logic symbol.

shows a circuit diagram of an astable multivibrator. Notice that both cross-coupling components are capacitors, thereby allowing no stable operating state. If the resistors and capacitors used are of equal value, the frequency of the clock is

$$f = \frac{1}{2T} = \frac{1}{2(0.7RC)} = \frac{0.7}{RC} \tag{10.1}$$

Integrated circuit units may be used to build the clock circuit. Figure 10.26 shows a few examples including the circuit parameters affecting the clock frequency. A typical clock output waveform is shown in Fig. 10.27.

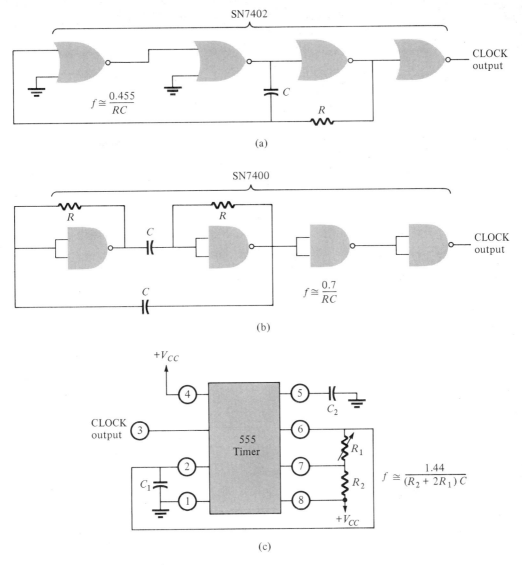

Figure 10.26 Clock circuits built with various IC units.

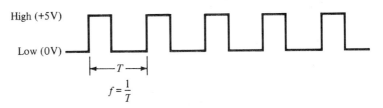

Figure 10.27 Clock output waveform.

Schmitt Trigger

A circuit that is somewhat like the multivibrator circuits considered is the Schmitt trigger circuit. Somewhat analogous to the ONE-SHOT, the Schmitt trigger is used for waveshaping purposes. Basically, the circuit has two opposite operating states as do all the multivibrator circuits. The trigger signal, however, is not typically a pulse waveform but a slowly varying ac voltage. The Schmitt trigger is level sensitive and switches the output state at two distinct triggering levels, one called a *lower-trigger level* (LTL) and the other an *upper-trigger level* (UTL). The circuit generally operates from a slowly varying input signal, such as a sinusoidal waveform, and provides a digital output—either the logical-0 or logical-1 voltage level.

The typical waveform of Fig. 10.28 shows a sinusoidal waveform input and squared waveform output. Note that the output signal frequency is exactly that of the input signal, except that the output has a sharply shaped slope and remains at the low or high voltage level until it switches. One example of a Schmitt trigger application is converting a sinusoidal signal to one that is useful with digital circuits. Signals such as a 60-Hz line voltage or a slowly varying voltage obtained from a magnetic pickup are squared up for digital use. Another possibility is using the Schmitt trigger to provide a logic signal that indicates whenever the input goes above a threshold level (UTL).

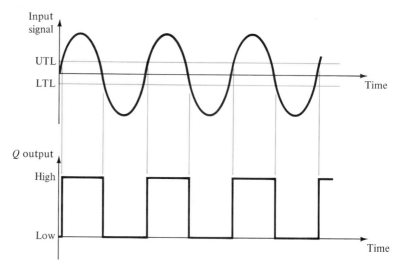

Figure 10.28 Typical waveforms for Schmitt trigger.

10.4 DIGITAL IC UNITS

The basic digital IC units are packaged as small-scale integrated (SSI) units—logic gates, flip-flops, one-shots, and clocks. More functional arrangements, such as medium-scale integrated (MSI) units, provide counters, data registers, and decoders among the list of functional units. While the range of MSI units, mostly built as TTL, is quite extensive, a description of some typical units will provide an introduction of what is available.

Serial Data Register

As an example, a 7491 (see Fig. 10.29) IC unit is an 8-bit serial input/serial output register containing eight flip-flop stages connected as a shift register plus some control gating. The data input is connected to the A, B terminals, which provide an ANDing of signals at those inputs. Usually, one of these inputs is not used, or used only for control, while the single serial data is connected as the other input.

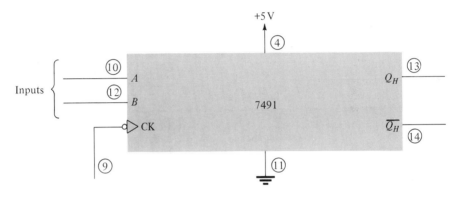

Figure 10.29 Serial data register, 7491.

The serial data coming from such devices as a teletype, or phone line, is clocked into the eight-stage register of the 7491 by shift pulses applied to the clock (CK) input. Data stored in the eight register stages of the 7491 may also be shifted out at the Q_H (or $\overline{Q_H}$) data output line when clocked by shift pulses provided at the clock input.

Parallel Data Register

An example of an 8-bit parallel input/output data register is the 74199 (see Fig. 10.30). Eight input bits are provided as input and transfer into eight flip-flop stages in the 74199 when the SHIFT/LOAD line goes low. These data remain stored in the IC unit with 8 bits appearing at the output terminals for transfer to other IC units. The register could be used, for example, to input an 8-bit character from a keyboard, the data being held in the 74199 IC until another keyboard character is entered, as indicated by a momentary low signal on the SHIFT/LOAD line. The IC can also be used as a teletype computer interface since it also operates as a shift register. Serial data from a teletype are applied at the JK input terminals and transferred into the eight register stages by shift pulses at the clock input (shifting can be prevented, if desired, by a high-level

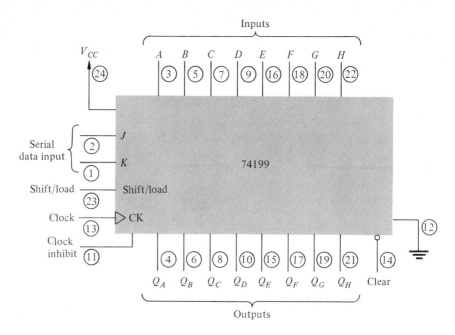

Figure 10.30 Parallel data register, 74199.

signal at the clock inhibit input). The data are then available in parallel as an 8-bit character to the computer. Another 74199 IC can be equally used to interface from the computer out to the teletype, with parallel transfer from the computer to the IC unit, followed by shifting out of the data using clock pulses (with data taken from the Q_H output).

Counters

Another important application of digital MSI units is in digital counters of various count steps. The 7490 IC, for example, a popular decade counter is shown in Fig. 10.31a. The IC unit contains four flip-flop stages which can easily be connected to operate as a decade counter. The four output bits (Q_A, Q_B, Q_C, Q_D) advance from zero to nine as indicated in Fig. 10.31b. The 7490 counter can be part of a voltage or frequency converter, providing binary output to represent a decimal count from 0 to 9. [For decade-count operation, the reset inputs—$R_0(1)$, $R_0(2)$, $R_9(1)$, $R_9(2)$—must all be grounded and the Q_A output connected as the B input.] The counter advances from 0 to 9 and back to 0 on the following clock pulse (at input A). If two 7490 units are used, the count sequence obtained can go from 00 to 99, using each IC as a decade counter. In that case the Q_D output of the first decade counter is used as input A of the second-decade-counter stage.

Other count sequences are possible using any of a large variety of IC units.

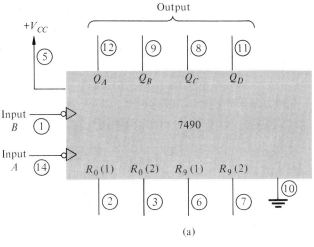

Count	Output			
	Q_D	Q_C	Q_B	Q_A
0	0	0	0	0
1	0	0	0	1
2	0	0	1	0
3	0	0	1	1
4	0	1	0	0
5	0	1	0	1
6	0	1	1	0
7	0	1	1	1
8	1	0	0	0
9	1	0	0	1

(b)

Figure 10.31 Decade counter, 7490: (a) IC unit; (b) BCD count sequence.

Decoders

A most useful operation obtained from MSI units is decoding from one binary code into another. A popular decoder unit is the 7447 IC, which accepts a 4-bit binary coded decimal (see Fig. 10.31b) character as input and then outputs the signals to directly operate a seven-segment display as used in meters and clocks. Figure 10.32a shows the 7447 IC, while Fig. 10.32b shows the BCD-to-seven-segment code conversion. A logic-0 signal to a common-cathode LED will result in the LED segment going on. The 7447 provides logic-0 (0 V) outputs, which will result in the segment connected to that signal going on. Figure 10.33 shows a 7490 connected as a decade counter driving a 7447 connected to a seven-segment LED display. The clock input advances the 7490 through its count sequence from 0 to 9 (and back to 0). For each count step resulting as output from the 7490, the 7447 decodes into the seven-segment codes to display the character on an LED. Operating the unit of Fig. 10.33 results in the LED displaying 0 to 9 in decimal as the count procedes in binary. This display could be one digit of a clock, a frequency counter, or a voltmeter.

The range and number of MSI units, although quite large, can be seen by the examples above to provide the small unit of a larger operating system. The particular units used depends on various features of the application.

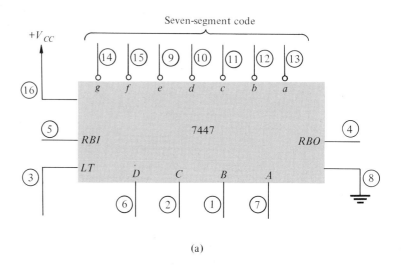

D	C	B	A	a	b	c	d	e	f	g
0	0	0	0	0	0	0	0	0	0	1
0	0	0	1	1	0	0	1	1	1	1
0	0	1	0	0	0	1	0	0	1	0
0	0	1	1	0	0	0	0	1	1	0
0	1	0	0	1	0	0	1	1	0	0
0	1	0	1	0	1	0	0	1	0	0
0	1	1	0	1	1	0	0	0	0	0
0	1	1	1	0	0	0	1	1	1	1
1	0	0	0	0	0	0	0	0	0	0
1	0	0	1	0	0	0	1	1	0	0

1 = open circuit; 0 = low level (0 V)

(b)

Figure 10.32 Decoder IC, 7447.

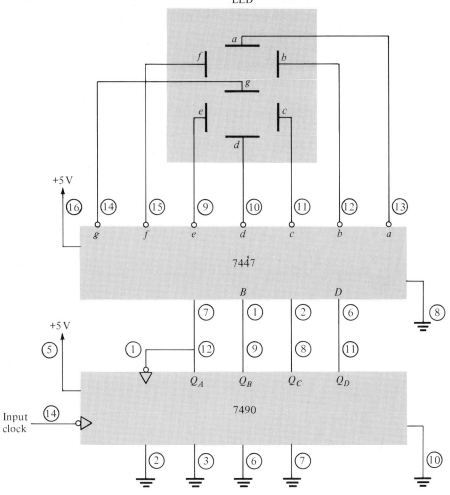

Figure 10.33 One-digit counter with display.

10.5 DIGITAL MEMORY UNITS

An important area of digital IC application is in semiconductor memory. Large-scale integrated (LSI) circuits provide memory cells arranged in various size groups to store information for use with computers, CRT terminals, and other digital devices. A typical IC unit contains many thousands of memory cells arranged in various ways. A 1K-bit (1024 memory cells) memory IC, for example, could be arranged as 1K memory groups of 1 bit each, or 256 groups (words) of 4 bits each, the first being referred to as a 1K × 1 (1K by 1), the second as 256 × 4. The designation of memory cell organization provides information of the necessary

IC signals to operate the unit. The number of words is a multiple of 2^n, $256 = 2^8$, $1K = 2^{10}$, $4K = 2^{12}$, as examples. The value n is the number of address bits needed to select each separate word. For 256 words it is necessary to use 8 bits to address each word of storage, the word size then being 1 bit or 4 bits or 8 bits. For a 1K × 4 memory the IC would require 10 address bits to select a word of memory, which is then 4 bits in size. Figure 10.34 shows a typical memory IC of 4K-bit capacity, organized as 1K × 4. The IC can be housed in an 18-pin package with 4 pins providing 4 bits of data as input or output, 10 bits for selecting among the 4096 (2^{10}) memory addresses, 2 pins providing control signals, and 2 pins to supply the IC power (typically 5 V). The READ/WRITE control signal is used to specify whether the 4 data bits are input to the memory (READ/WRITE signal low) or output taken from the IC (READ/WRITE signal high). The chip enable (CE) is a control signal that can be used to select operation of each separate IC when a group of ICs are used to make a larger memory than is provided by only one IC.

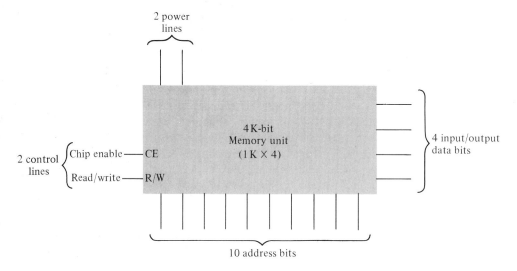

Figure 10.34 Typical 4K-bit memory IC.

Figure 10.35 shows an arrangement of two 4K × 1 ICs into an 8K × 1 memory. There are 11 bits ($2^{11} = 8192$) necessary to select a location in an 8K memory. In this example the 11th bit is used to select among 4K located in one IC or 4K located in a second IC. The address bits are labeled to represent the positional value in address selection (i.e., A_0 is the 2^0 bit, A_9 is the 2^9 selection bit). In this example the chip-enable control line allows for the additional bit needed to extend the IC for use in an 8K memory unit. This chip-enable input can also be used with additional logic gating or decoding to further extend the memory size above 8K using the basic 4K x 1 ICs. The data lines for both IC are connected in common with the address lines selecting which IC the data operate with for a specific address.

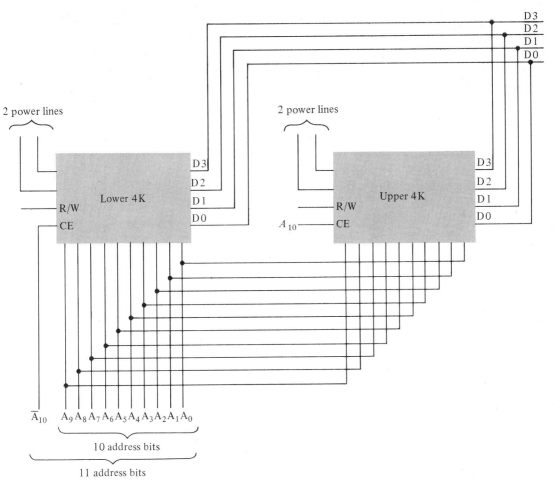

Figure 10.35 Organization of 8K × 1 memory.

Memory Types

Semiconductor memory cells are built using bipolar or field-effect transistors, the latter being most popular at present. In either case the memory cell can be built to operate as static or dynamic. A static memory cell will hold a stored bit as long as power is maintained, while a dynamic memory cell would only hold the data bit a short time and must be *refreshed* or rewritten to maintain the data. The dynamic cell requires fewer transistor components than the static cell so that the greatest-capacity ICs use that cell type. The static ICs, however, handle data at a faster rate.

Memories that provide both reading and writing of information are RAM (random-access memory). IC capacities of 64K bits are presently popular using dynamic MOSFET RAM cells. Eight such ICs could then be used, to result in a 64K-byte memory.

When fixed data, as in programmed calculators, fixed conversion routines, or basic control or operating systems are used, ROM (read-only memory) ICs can be used. These ICs are organized similar to that of Fig. 10.34 except ROMs do not need a READ/WRITE control signal. If the data are programmed at the time of manufacture, the IC is called a ROM. If data are written into the IC after the IC is manufactured by the user, it is called a pROM (programmable ROM) and the stored data are also permanent. Of great popularity is the eraseable programmable ROM (EPROM) made of MOSFET storage cells. Data are written into a cell as stored charge, trapped in the gate region of a MOSFET transistor.

While individual data bits cannot be altered, the total IC data can be erased using ultraviolet light, after which new information can be programmed into the IC memory cells. The erasure is a slow, offline process, which is useful for occasional rewriting of "permanently" stored data.

PROBLEMS

§ 10.1

1. A positive-logic diode AND gate has voltage levels of 0 V and +5 V. Draw a circuit diagram and prepare voltage and logic truth tables for a two-input gate. (Assume ideal diodes.)

2. Draw the circuit diagram of a negative-logic AND gate for voltage levels of −5 V and 0 V. Assume ideal diodes and prepare voltage and logic truth tables for a two-input gate.

§ 10.2

3. Draw the circuit diagram of a four-input TTL NAND gate indicating voltage levels and logic definitions.

4. State two differences between CMOS and TTL operation.

5. Draw the circuit diagram of a three-input CMOS NOR gate.

6. How does an ECL-type gate differ from a TTL?

§ 10.3

7. Describe the operation of an RS flip-flop.

8. Draw the output waveform of a T flip-flop triggered by a 10-kHz clock.

9. Describe the operation of a JK flip-flop.

10. Draw the Q-output waveform of an SN74121 circuit driven by an input clock signal of 100 kHz applied to the B input. The timing component values are $R = 10$ K and $C = 100$ pF.

11. What is the frequency of an astable multivibrator circuit (as in Fig. 10.26a) having timing component values of $R = 2.7$ K and $C = 750$ pF?

12. Draw the output voltage signal of a Schmitt trigger circuit for the input trigger signal of 5 V rms at 60 Hz. Circuit trigger levels are UTL = +5 V and LTL = 0 V.

§ 10.4

13. Describe the difference between a serial data register and a parallel data register.
14. What maximum count is possible using six count stages?
15. What maximum count is possible using two 7490 counter ICs?

§ 10.5

16. How many 1K x 4 ICs are needed to build a 4K x 8 memory?
17. What is the difference between a ROM and a RAM?

GLOSSARY

AND Gate   A logic circuit with output logic-1 when all inputs are logic-1.

Astable Multivibrator   A digital circuit providing constantly alternating logic state or clock output.

Bistable Multivibrator   A logic circuit that remains in either logic-1 or logic-0 output (either SET or RESET).

Clock   A digital circuit providing constantly alternating logic output (astable multivibrator operation).

CMOS Logic Circuit   A digital logic gate built using complementary MOSFET transistors for low power dissipation.

Counter   A digital circuit made of toggle flip-flops to count pulses in binary.

Data Register   A logic circuit made of flip-flops to store a group of bits.

Decoder   A logic circuit that accepts inputs in one code and provides outputs in another code.

Digital Logic   The two signal levels of low (0 V, false) and high (+5 V, true).

Digital Memory Unit   A computer circuit made of flip-flops allowing read and write of stored binary data.

Emitter-coupled Logic (ECL)   A digital circuit built using nonsaturated operation for fastest speed.

Eraseable Programmable Read-Only Memory (EPROM)   A digital storage device for holding fixed data that may be read by computer. Stored data may be erased by long exposure to intense UV radiation and then reprogrammed.

Flip-Flop   A digital circuit having two stable output states, SET and RESET, used in storage and counting operations.

Integrated Injection Logic (IIL or I^2L)   A digital circuit built using complementary bipolar transistors in LSI units.

Inverter A logic circuit accepting input of logic-1 or logic-0, providing the opposite output of logic-0 or logic-1, respectively.

JK Flip-Flop A flip-flop circuit with *J* and *K* inputs which determine action of flip-flop when trigger pulse is applied.

Logic Truth Table An exhaustive method of describing a logic function or operation for all combinations of logic input.

Monostable Multivibrator A circuit having only one stable output state, remaining in the opposite state for a time fixed by external resistor/capacitor elements.

NAND Gate A logic circuit providing a low (0) output when all inputs are high (1).

Negative Logic A logic definition with the lower voltage level defined as logic-1.

NOR Gate A logic circuit providing a low (0) output when any input is high (1).

One-Shot A monostable multivibrator providing an output that remains high for a time interval set by external resistor/capacitor elements, remaining low until triggered.

OR Gate A logic circuit with output high (1) when any input is high (1).

Parallel Data Register A group of flip-flops used to transfer all bits at the same time.

Positive Logic A logic definition with the more positive voltage level defined as logic-1.

Programmable Read-Only Memory (PROM) An IC having permanently stored data that may be read by the other computer circuits, these data being programmed once by the user.

Random-Access Memory (RAM) A computer unit made of flip-flops to allow read and write of binary data.

Read-Only Memory (ROM) A permanent memory fixed at the time the IC is manufactured.

RESET The logic state with reference or *Q*-output logic-0.

RS Flip-Flop A flip-flop circuit that may be SET and RESET by logic signals applied to *R* and *S* inputs.

Schmitt Trigger A digital circuit that provides logic levels which change state when the input signal goes above an upper trigger level (UTL) or below a lower trigger level (LTL).

Semiconductor Memory Cell A 1-bit storage circuit built with either bipolar or FET transistors.

Serial Data Register A group of flip-flops that transfer data bit by bit.

SET The logic state with reference *(Q)* output logic-1.

Toggle (T-Type) Flip-Flop A digital circuit that alternates state (complements or toggles) when a trigger input pulse is applied.

Transistor-Transistor Logic (TTL or T^2L) A digital logic circuit built using bipolar transistors as input elements and active output elements.

Voltage Truth Table An exhaustive method of describing the operation of a logic circuit for all conditions of input voltage.

CHAPTER 11

Linear/Digital IC Units

11.1 INTRODUCTION

A number of IC units containing both linear and digital circuits are used as basic devices in many applications. One such unit is the comparator IC unit, which allows comparing two analog voltages with digital-output indication of which input is larger. One application using a comparator is also a basic IC circuit used in a wide range of electronic units, this being the analog-to-digital (A/D) or digital-to-analog (D/A) converter unit. With a large variety of different types of circuits and a large number of different electronic units which must be operated together, one type of IC circuit of great use is the interface unit. Probably one of the best known IC units containing linear and digital electronic circuits is the 555 timer. This chapter introduces a number of IC devices which are basic to many areas of application.

11.2 COMPARATORS

A very useful IC unit is the voltage comparator, which compares linear input voltages and provides a digital output indicating when one input is less or greater than a reference voltage level. The comparator is essentially an op-amp circuit, as shown by the basic unit of Figure 11.1a. A 741 op-amp is used to demonstrate how its very high gain provides operation as a comparator. Because of the extremely high gain of the comparator, the output will rise to a positive saturation level,

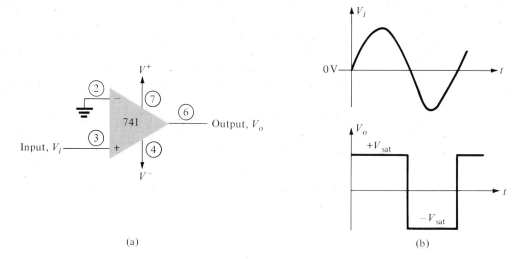

Figure 11.1 Operation of a 741 op-amp as a comparator.

$+V_{sat}$, as soon as the input, V_i, goes slightly positive. The output stays at this level until V_i drops slightly negative (below the 0-V reference), at which time the output switches to a negative saturation level, $-V_{sat}$. Figure 11.1b shows the output at one of two distinct levels, switching whenever the input crosses the 0-V reference level. Since the op-amp has a very high gain a small positive input voltage results in the output going to its largest positive value, $+V_{sat}$. As the input goes more positive, the output remains at its positive saturation level. The input then dropping below the reference level causes the output to quickly switch to its negative saturation level, remaining there as long as the input stays below the reference level.

In general, the reference level need not be 0 V, but any desired value, either positive or negative. Either op-amp (or comparator) input may be used as the reference input, the other then connected to the input signal.

Figure 11.2a shows a circuit operating with a positive voltage reference level, with output driving an indicator LED. The reference level is set at

$$V_{ref} = \frac{10 \text{ K}}{10 \text{ K} + 10 \text{ K}} (+12 \text{ V}) = +6 \text{ V}$$

Since the reference voltage is connected to the inverting input, the output will switch to its positive saturation level when the input, V_i, goes more positive than the +6-V reference voltage level. The output voltage, V_o, then drives the LED on, as an indication that the input is more positive than the reference level.

As an alternative connection, the reference voltage could be placed as the noninverting input (see Fig. 11.2b). With this connection the input signal going below the reference level would cause the output to drive the LED on. The LED can thus be made to go on when the input signal goes either above or below the reference level, depending on which input is connected to the input signal and which to the reference.

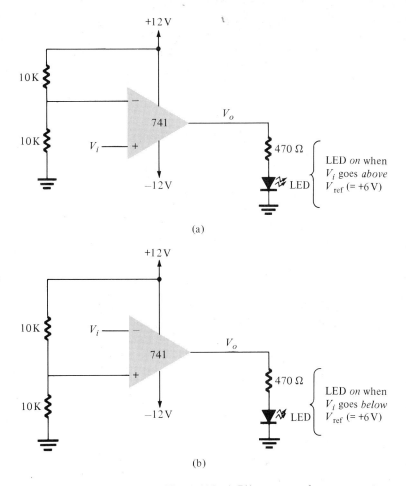

Figure 11.2 A 741 op-amp used as a comparator.

While op-amp ICs can be used as comparator circuits, separate comparator ICs are available for use in such applications. Some improvements built into comparator IC units are faster switching between the two output levels, have built-in noise immunity to prevent the output from oscillating when the input passes by the reference level, and have outputs capable of directly driving a variety of loads. A few popular IC comparators are covered next to show how they are defined and how they may be used.

The 311 voltage comparator shown in Fig. 11.3 contains a comparator circuit that can operate as well from dual power supplies of ±15 V as from a single +5-V supply (as used for digital logic circuits). The output can provide voltage at one of two distinct levels or can be used to drive lamps or relays. Notice that the output is taken from a bipolar transistor to allow driving a variety of loads. The unit also has balance and strobe inputs, the strobe input allowing gating of the output. A few examples will show how the comparator unit can be used for common applications.

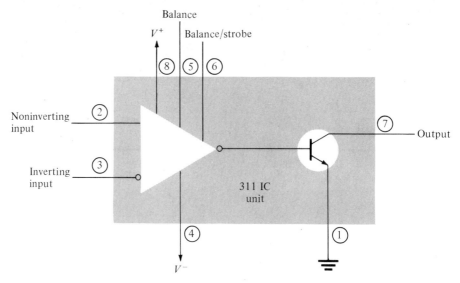

Figure 11.3 A 311 comparator (8-pin DIP unit).

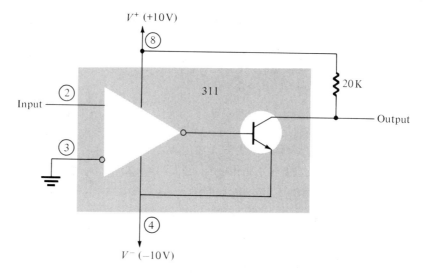

Figure 11.4 Zero-crossing detector using a 311 IC.

A zero-crossing detector can be built using the 311 as shown in Fig. 11.4. The input going positive (above 0 V) drives the output transistor on and the output goes low, to −10 V in this connection. The input going below 0 V will drive the output transistor off and the output goes to +10 V.

Figure 11.5 shows how a 311 comparator can be used with strobing. In this example the output will go high when the input goes above the reference level— but only if the TTL strobe input is off (or 0 V). If the TTL strobe input goes high, it drives the 311 strobe input at pin 6 low, causing the output to remain in

CH. 11 LINEAR/DIGITAL IC UNITS

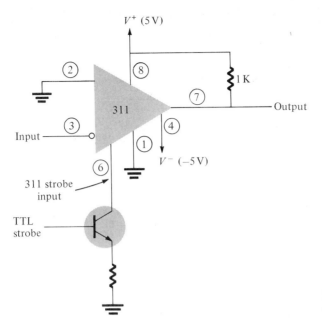

Figure 11.5 Operation of a 311 comparator with strobe input.

the off state (with output high) regardless of the input signal. In effect, the output remains high unless strobed. If strobed, the output then acts normally, switching from high to low depending on the input signal level. In operation, the comparator output will respond to the input signal only during the time the strobe signal allows such operation.

Figure 11.6 shows the comparator output driving a relay. When the input goes below 0 V, driving the output low, the relay is activated, closing the normally open (N.O.) contacts at that time. These contacts can then be connected to operate a large variety of devices.

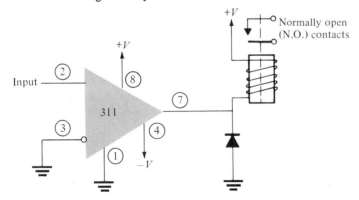

Figure 11.6 Operation of a 311 comparator with relay output.

Another popular comparator unit is packaged with four independent voltage comparator circuits in a single IC. The 339 is a quad comparator IC, all four comparator circuits connected to the external pins as shown in Fig. 11.7. Each comparator has inverting and noninverting input and an output. The supply

SEC. 11.2 COMPARATORS

voltage applied to a pair of pins connects to all four comparator circuits. Even if one wishes to use only some of the comparator circuits, all four are active, drawing power from the supply.

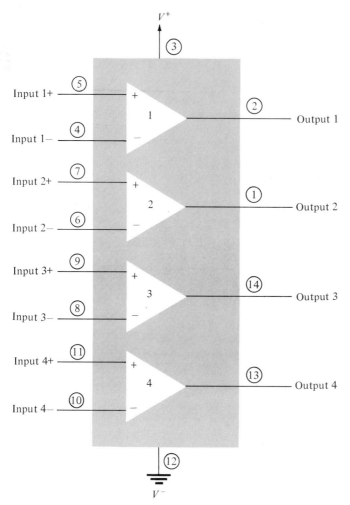

Figure 11.7 Quad comparator IC (339).

To see how these comparator circuits can be used, Fig. 11.8 shows one of the 339 comparator circuits connected as a zero crossing detector. Whenever the input signal goes above 0 V, the output switches to V^+. The output switches to V^- only when the input goes below 0 V.

A reference level of other than 0 V can also be used and either input terminal could be used as reference, the other input then being the signal input. The operation of one of the comparator circuits is described next.

The differential input (voltage difference across input terminals) going positive drives the output transistor off (open-circuit), whereas a negative differential input drives the output transistor on—the output then at the supply low level.

If the negative input is set at a reference level, V_{ref}, the positive input going

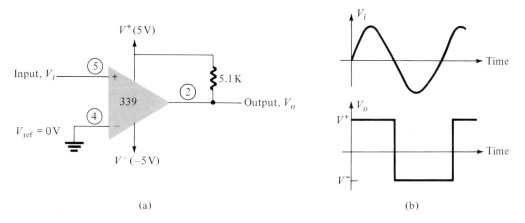

Figure 11.8 Operation of one 339 comparator circuit as a zero-crossing detector.

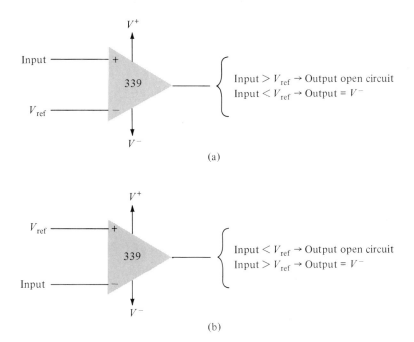

Figure 11.9 Operation of a 339 comparator circuit with reference input at: (a) minus input; (b) plus input.

above V_{ref} results in a positive differential input with output driven to the open-circuit state. When the noninverting input goes below V_{ref}, resulting in a negative differential input, the output will be driven to V^-.

If the positive input is set at the reference level, the inverting input going below V_{ref} results in the output open circuit, while the inverting input going above V_{ref} results in the output at V^-. This operation is summarized in Fig. 11.9.

Since the output of one of these comparator circuits is open-circuit collector,

SEC. 11.2 COMPARATORS   393 applications in which the outputs from more than one circuit can be wire-ORed are possible. Figure 11.10 shows two comparator circuits connected with common output, and also with common input. Comparator 1 has a +5-V reference voltage input connected to the noninverting input. The output will be driven low by

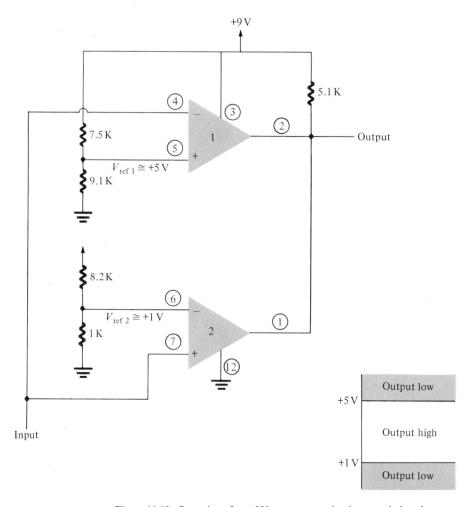

Figure 11.10 Operation of two 339 comparator circuits as a window detector.

comparator 1 when the input signal goes above +5 V. Comparator 2 has a reference voltage of +1 V connected to the inverting input. The output of comparator 2 will be driven low when the input signal goes below +1 V. In total the output will go low whenever the input is below +1 V or above +5 V, as shown in Fig. 11.10, the overall operation being that of a voltage window detector. The output high indicates that the input is within a voltage window of +1 V to +5 V (these values being set by the reference voltage levels used).

11.3 DIGITAL/ANALOG CONVERTERS

Many of the voltage and current signals occurring in electronics are linear, in that they vary continuously over some range of values. In digital devices and computers the signals are digital at one of two levels representing the binary values of 1 or zero.

If the signals to be used in some digital operations are linear (analog) voltages (e.g., dc voltages representing temperature or pressure, or position), a circuit must convert this analog voltage into a digital value—this conversion circuit being an analog-to-digital or A/D converter. When a computer has a digital value to be output as an analog voltage, a digital-to-analog or D/A converter circuit is used.

Digital-to-Analog (D/A) Conversion

Digital-to-analog conversion can be achieved using a number of different methods. One popular scheme uses a network of resistors, called a ladder network. A ladder network accepts inputs of binary values at, typically, 0 V or V_{ref}, and provides an output voltage proportional to the binary input value. Figure 11.11a shows a ladder network with four input voltages, representing 4 bits of digital data and a dc voltage output. The output voltage is proportional to the digital

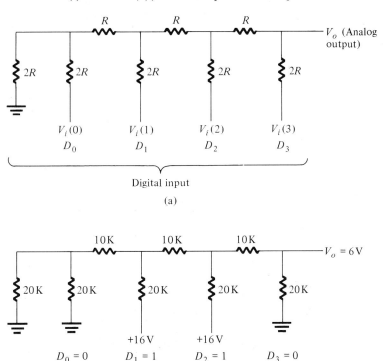

Figure 11.11 Four-stage ladder network used as a digital/analog (D/A) converter: (a) basic circuit; (b) circuit example with 0110 input.

input value as given by the relation

$$V_o = \frac{D_0 \times 2^0 + D_1 \times 2^1 + D_2 \times 2^2 + D_3 \times 2^3}{2^4 = 16} \times V_{\text{ref}}$$

In the example shown in Figure 11.11b, the output voltage resulting should be

$$V_o = \frac{0 \times 1 + 1 \times 2 + 1 \times 4 + 0 \times 8}{16} \times 16 \text{ V} = 6 \text{ V}$$

Therefore, 0110_2 converts to 6 V.

Use superposition to verify that the resulting value of V_o is indeed 6 V. The function of the ladder network is to convert the 16 possible binary values from 0000 to 1111 into one of 16 voltage levels in steps of $V_{\text{ref}}/16$. Using more sections of ladder allows having more binary inputs and greater quantization for each step. For example, a 10-stage ladder network could extend the number of voltage steps or voltage resolution to $V_{\text{ref}}/2^{10}$ or $V_{\text{ref}}/1024$. A reference voltage of 10 V would then provide output voltage steps of 10 V/1024 or approximately 10 mV. More ladder stages provide greater voltage resolution, in general the voltage resolution for n ladder stages being

$$V_{\text{ref}}/2^n$$

A block diagram of the main components of a typical IC D/A converter is shown in Fig. 11.12. The ladder network, referred to in the diagram as an R-$2R$ ladder, is sandwiched between the reference current supply and currents switches connected to each binary input with a resulting output current proportional to the input binary value. The binary inputs turn on selected legs of the ladder, the output current being a weighted summing of the reference current. Connecting the output across a resistor will produce an analog voltage, if desired.

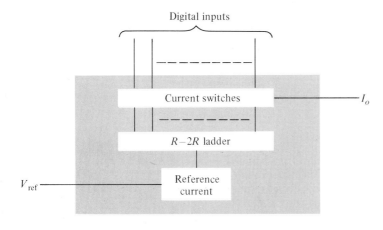

Figure 11.12 D/A converter IC using R–$2R$ ladder network.

Analog-to-Digital (A/D) Conversion

A popular method of converting analog voltage into a digital value is the dual-slope method. Figure 11.13a shows a block diagram of the basic dual-slope converter. The analog voltage to be converted is applied through an electronic

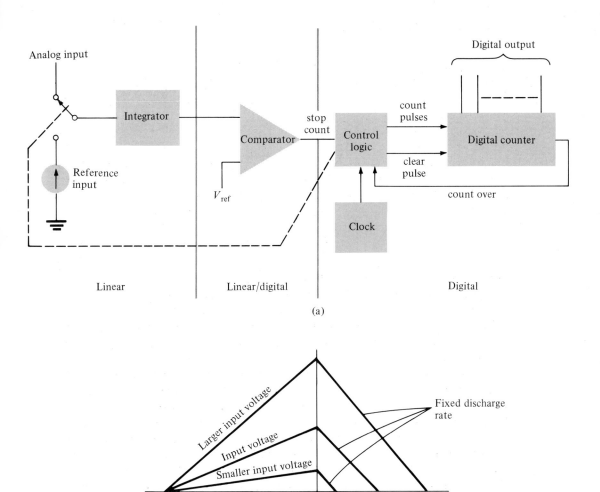

Figure 11.13 Analog/digital (A/D) conversion using dual-slope method: (a) logic diagram; (b) waveform.

switch to an integrator or ramp generator circuit (essentially a constant current charging a capacitor to produce a linear ramp voltage). The digital output is obtained from a counter operated during both positive and negative slope intervals of the integrator.

The conversion method proceeds as follows. For a fixed time interval (usually the full count range of the counter), the analog input voltage, connected to the integrator, raises the voltage to the comparator to some positive level. Figure 11.13b shows that at the end of the fixed time interval the voltage from the integrator is greater for larger input voltages. At the end of the fixed count interval, the count is set at zero and the electronic switch connects the integrator to a reference or fixed input. The integrator output (or capacitor input) then decreases at a fixed rate. The counter advances during this time. The integrator output voltage decreases at a fixed rate until it drops below the comparator reference voltage, at which time the control logic receives a signal (the comparator output) to stop the count. The digital value stored in the counter is then the digital output of the converter.

Using the same clock and integrator to perform the conversion during positive and negative slope intervals tends to compensate for clock frequency drift and integrator accuracy limitations. Setting the reference input value and clock rate

Figure 11.14 A/D conversion using ladder network: (a) logic diagram; (b) waveform.

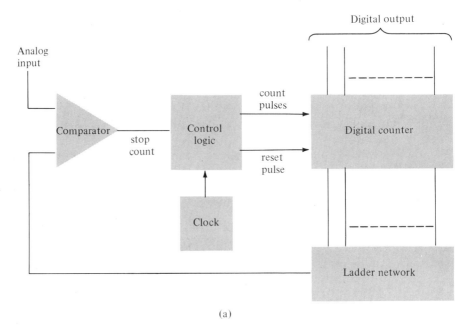

(a)

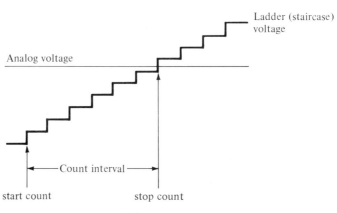

(b)

can scale the counter output as desired. The counter can be binary or BCD or other digital form, if desired.

Another popular method of analog-to-digital conversion uses the ladder network along with counter and comparator circuits (see Fig. 11.14). A digital counter advances from a zero count while a ladder network driven by the counter outputs a staircase voltage as shown in Fig. 11.14b, which increases one voltage increment for each count step. A comparator circuit, receiving both staircase voltage and analog input voltage, provides a signal to stop the count when the staircase voltage rises above the input voltage. The counter value at that time is the digital output.

The amount of voltage change stepped by the staircase signal depends on the reference voltage applied to the ladder network and on the number of count bits used. A 12-stage counter operating a 12-stage ladder network using a reference voltage of 10 V would step each count by a voltage of

$$V_{\text{ref}}/2^{12} = 10 \text{ V}/4096 = 2.5 \text{ mV}$$

This would result in a conversion resolution of 2.5 mV. The clock rate of the counter would affect the time required to carry out a conversion. A clock rate of 1 MHz operating a 12-stage counter would need a maximum conversion time of

$$4096 \times 1 \text{ } \mu\text{s} = 4096 \text{ } \mu\text{s} \cong 4.1 \text{ ms}$$

The minimum number of conversions that could be carried out each second would then be

$$\text{Number of conversions} = 1/4.1 \text{ ms} \cong 244 \text{ conversions/second}$$

Since on the average, with some conversions requiring little count time and others near maximum count time, a conversion time of 4.1 ms/2 = 2.05 ms would be needed, the average number of conversions would be $2 \times 244 = 488$ conversions/second. A slower clock rate would result in fewer conversions per second. A converter using fewer count stages (and less conversion resolution) would carry out more conversions per second.

The conversion accuracy would depend on the accuracy of the voltage reference used in the ladder network and on the accuracy of the comparator.

11.4 INTERFACING

Connecting different types of circuits, different analog or digital units, and inputs or loads to other electronics all require some sort of interfacing. Interface circuits may be categorized as either driver or receiver units. A receiver essentially accepts inputs, providing high input impedance to minimize loading of the input signal. A driver circuit provides the output signal at voltage or current levels suitable to operate a number of loads, or to operate such devices as relays, displays, and power units. Furthermore, these inputs or outputs may require strobing, which provides the interface signal connection during specific time intervals as established by the strobe.

Figure 11.15a shows a dual line driver, each driver having input operation capable of accepting TTL signals and output which can operate TTL, DTL, or

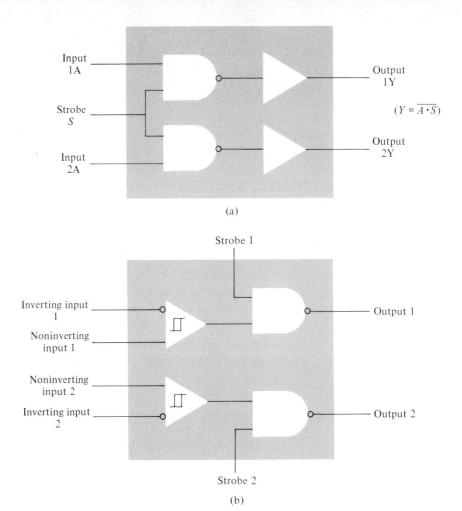

Figure 11.15 Interface units: (a) dual line drivers (SN75150); (b) dual line receivers (SN75152).

MOS devices. Often, the interface circuit is needed to receive signals from one type of circuit (TTL, DTL, ECL, MOS) and transmit the signal to another circuit type. The relation between the input and output signal may be such that the interface unit is a noninverting unit or an inverting unit. Interface circuits or all these various configurations are necessary and exist as IC units.

The circuit of Fig. 11.15b shows a dual line receiver having both inverting and noninverting inputs so that either operating condition can be selected. As example, connection of an input signal to the inverting input would result in an inverted output from the receiver unit. Connecting the input to the noninverting input would provide the same interfacing except that the output obtained would be in phase with the received signal.

In the case of both driver and receiver circuits shown in Fig. 11.15, the outputs will be present only when the strobe signal is present—high level in the present circuits.

Another type of interfacing that is quite important occurs when connecting signals between various terminals of a digital system. Signals from such devices as a teletype, video terminal, card reader, or line printer are usually one of a number of signal forms. One popular electronic industry standard is referred to as RS-232-C. Complete details of the expected signal conditions for this standard can be stated simply here as binary signals representing mark (logic-1) and space (logic-0) corresponding to the voltage levels of -12 V and $+12$ V, respectively. TTL circuits operate with signals defined as $+5$ V as mark and 0 V as space. Teletype units are sometimes wired to operate with current loop signals, for which 20 mA represents a mark with no current representing a space. These different types of signals may occur as either input or output of a particular terminal, so that a variety of interface circuitry is necessary to convert from one signal type to one of the other. Some popular examples of interfaces will be described next.

Figure 11.16a shows the defined mark and space conditions for current loop, RS-232-C, and TTL signals.

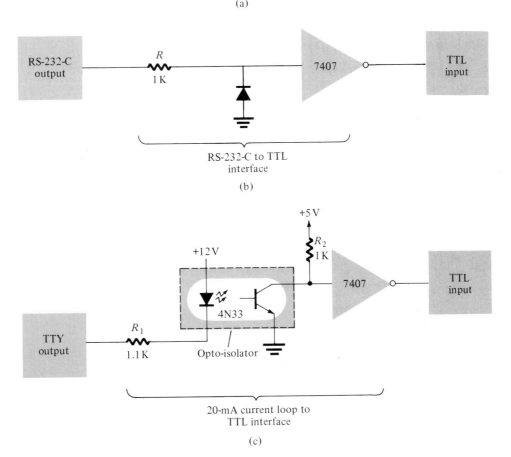

Figure 11.16 Interfacing signal standards and converter circuits.

RS-232-C-to-TTL Converter

If a unit having output defined by RS-232-C is to operate into another unit which operates with TTL signal levels, the interface circuit of Fig. 11.16b could be used. A mark output from the driver (−12 V) would get clipped by the diode so that the input to the inverter circuit is near 0 V, resulting in an output of +5 V or a TTL level mark. A space output at +12 V would drive the inverter output low for a 0-V space (TTL).

Another example of interface is that between a current loop input and TTL, as shown in Fig. 11.16c. An input mark results when 20 mA current is drawn through the output line of the Teletype (TTY). This current then flows through the diode element of an opto-isolator driving the output transistor on. The input to the inverter going low results in a +5-V signal to the TTL input, so that a mark from the Teletype results in a mark into the TTL input. A space from

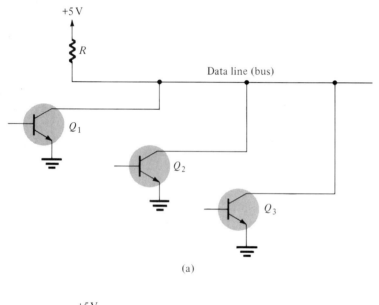

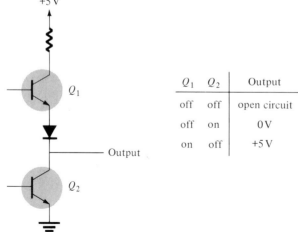

Q_1	Q_2	Output
off	off	open circuit
off	on	0 V
on	off	+5 V

Figure 11.17 Connections to data lines: (a) open collector output; (b) tri-state output.

the Teletype current loop provides no current with opto-isolator transistor remaining off and inverter output then 0 V, which is a TTL space signal.

Other types of interface circuits can be considered, those of Fig. 11.16 being representative. Another means of interfacing digital signals is made using open-collector output and using tri-state buffer outputs. When a signal is output from a transistor collector (see Fig. 11.17) which is not connected to any other electronic components, the output is open-collector. This allows connecting a number of signals to the same signal wire or signal bus. Then any transistor going on provides a low output condition, while all transistors off provides a high output. An equally popular connection for tieing a number of digital signals to a common bus uses tri-state output as shown in Fig. 11.18b. The output can be either high level (near +5 V), low level (0 V), or open-circuit. With this circuit connection the

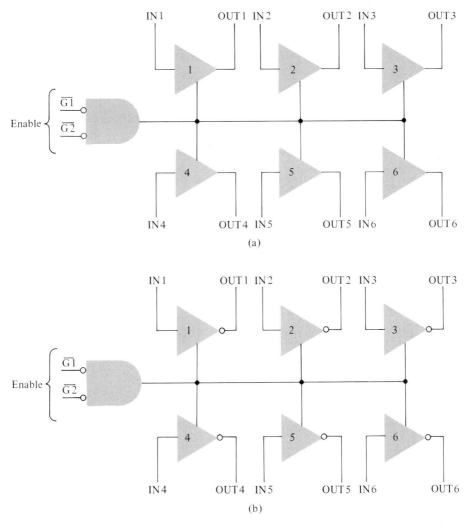

Figure 11.18 Tri-state buffer ICs: (a) noninverting (74365); (b) inverting (74366).

various logic circuits connected to the common line must be gated so that only one circuit can operate the bus, the other outputs being open-circuit at that time.

The tri-state buffer circuits could be either noninverting output or inverting output. Figure 11.18 shows two popular IC units, each packaged to hold six identical gates. The enable inputs provide the strobe signal to disable the set of six gates or cause the output to be open-circuit. When the buffer gates are enabled, the output can go to either +5 V or 0 V, depending on the input signal.

11.5 TIMERS

Another popular analog/digital integrated circuit is the versatile 555 timer unit. The IC is made of a combination of linear comparators and digital flip-flop as described in Fig. 11.19. The entire circuit is usually housed in an eight-pin DIP package with pin numbers as specified in Fig. 11.19. A series connection of three resistors set the reference level inputs to the two comparators at $\frac{2}{3}V_{CC}$ and $\frac{1}{3}V_{CC}$, the outputs of these comparators setting or resetting the flip-flop unit. The flip-flop circuit output is then brought out through an output amplifier stage. The flip-flop circuit also operates a transistor inside the IC, the transistor collector usually being driven low to discharge a timing capacitor.

One popular application of the 555 timer IC is as an astable multivibrator or clock circuit. The following analysis of the operation of the 555 as an astable circuit will include details of the different parts of the unit and how the various inputs and outputs are utilized. Figure 11.20 shows an astable circuit using external resistor and capacitor to set the timing interval of the output signal.

Figure 11.19 Details of a 555 timer IC.

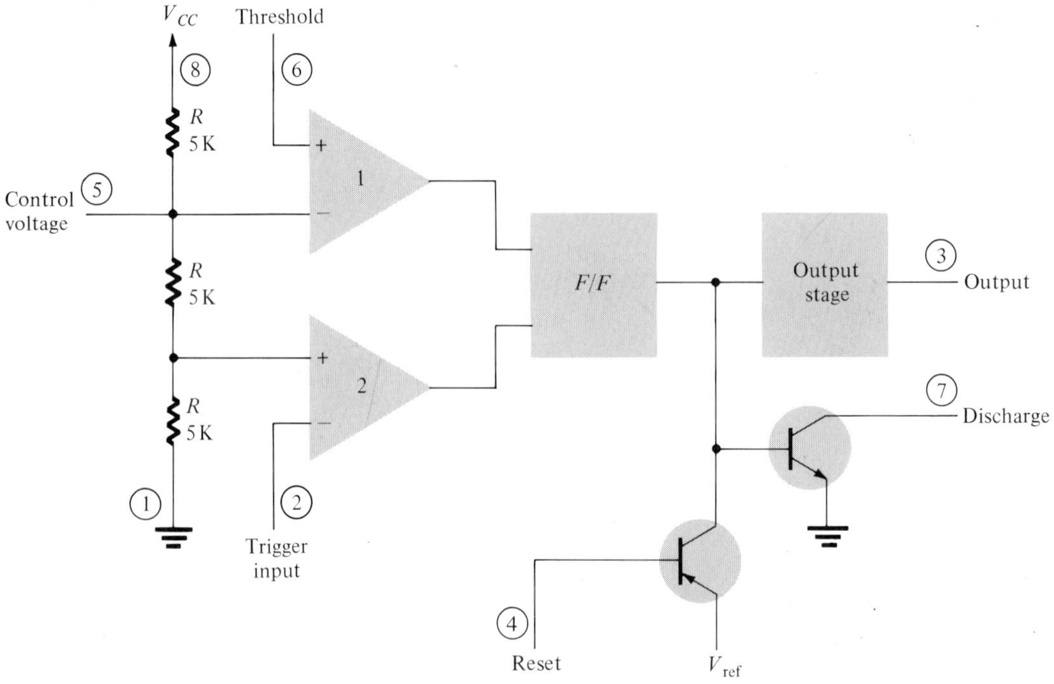

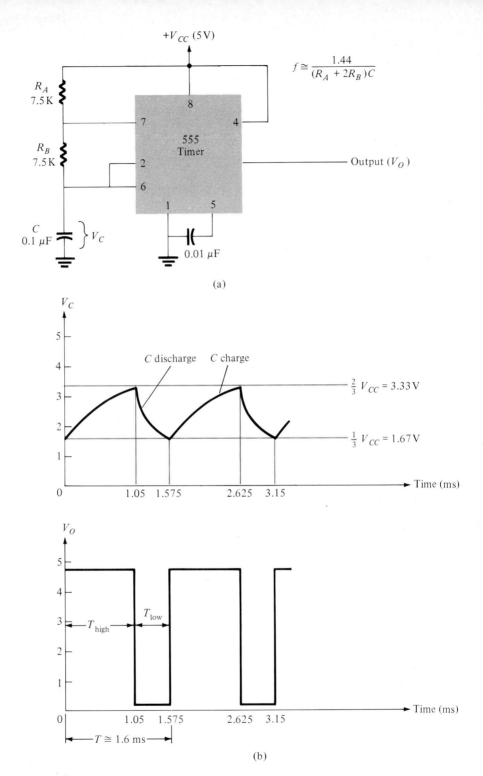

Figure 11.20 Astable multivibrator using a 555 timer IC: (a) circuit; (b) waveforms.

Capacitor C charges toward V_{CC} through external resistors R_A and R_B. Referring to Fig. 11.19, the capacitor voltage rises until it goes above $\frac{2}{3}V_{CC}$ [$=\frac{2}{3}(5\text{ V})=3.33$ V, in this example]. This voltage is the threshold voltage at pin 6, which drives comparator 1 to trigger the flip-flop so that the output at pin 3 goes low. In addition, the discharge transistor is driven on, causing the output at pin 7 to discharge the capacitor through resistor R_B. The capacitor voltage then decreases until it drops below the trigger level ($V_{CC}/3 = 5\text{ V}/3 = 1.67$ V). The flip-flop is triggered so that the output goes back high and the discharge transistor is turned off, so that the capacitor can again charge through resistors R_A and R_B toward V_{CC}.

Figure 11.20b shows the capacitor and output waveforms resulting from the astable circuit connection. Calculation of the time intervals during which the output is high and low can be made using the relations:

$$T_{high} \cong 0.7(R_A + R_B)C = 0.7(7.5\text{ K} + 7.5\text{ K})0.1\text{ }\mu\text{F}$$
$$= 1.05\text{ ms}$$
$$T_{low} \cong 0.7R_BC = 0.7(7.5\text{ K})0.1\text{ }\mu\text{F} = 0.525\text{ ms}$$

The total period is then

$$\text{period} = T = T_{high} + T_{low} = (1.05 + 0.525)\text{ ms} = 1.575\text{ ms}$$

The frequency of the astable circuit is then calculated using[1]

$$f = \frac{1}{T} = \frac{1}{1.575\text{ ms}} = 635\text{ Hz}$$

The duty cycle of the output waveform can also be determined from[2]

$$\text{duty cycle} = \frac{T_{low}}{T}$$
$$= \frac{0.525\text{ ms}}{1.575\text{ ms}} = 0.333\ (= 33.3\%)$$

The 555 timer can also be used as a one-shot or monostable multivibrator circuit. Figure 11.21 shows such connection. When the trigger input signal goes negative, it triggers the one-shot with output at pin 3 then going high for a time period

$$T_{high} = 1.1R_AC$$

---

[1] The period can be directly calculated from

$$T \cong 0.693\ (R_A + 2R_B)C$$

and frequency from

$$f \cong \frac{1.44}{(R_A + 2R_B)C}$$

[2] The duty cycle could also be calculated directly from

$$\text{duty cycle} = \frac{R_B}{R_A + 2R_B}\ 100\ \%$$

In the circuit of Fig. 11.21 this would be

$$T_{high} = 1.1(7.5 \text{ K})(0.1 \text{ }\mu\text{F}) \cong 0.8 \text{ ms}$$

Referring back to Fig. 11.19, the negative edge of the trigger input causes comparator 2 to trigger the flip-flop with output at pin 3 going high. Capacitor C charges toward V_{CC} through resistor R_A. During the charge interval the output remains high. When the voltage across the capacitor reaches the threshold level of $2V_{CC}/3$, comparator 1 then triggers the flip-flop with output going low. The discharge transistor also goes low, causing the capacitor to remain at near 0 V until triggered again.

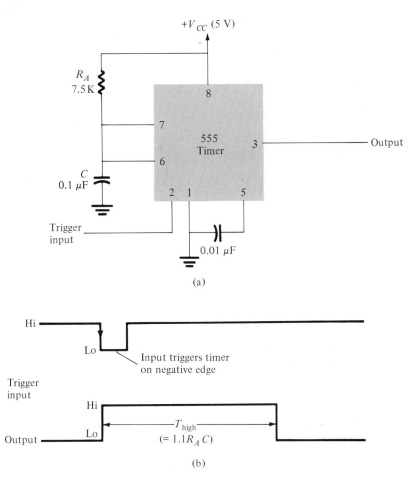

Figure 11.21 Operation of a 555 timer as a one-shot: (a) circuit; (b) waveforms.

Figure 11.21 shows the input trigger signal and the resulting output waveform for the 555 timer IC operated as a one-shot. Time periods for this circuit can range from microseconds to many seconds, making this IC useful for a large range of applications.

PROBLEMS

§ 11.2

1. Draw the diagram of a 741 op-amp operated from ±15-V supplies with input to minus input terminal and plus input terminal connected to a +5-V reference voltage. Include terminal pin connections.

2. Sketch the voltage waveforms for the circuit of Problem 1, with input of 10 V rms.

3. Draw the connection diagram of a 311 op-amp showing 10 V rms input applied to pin 3 and ground to pin 2.

4. Using ±12-V supply and output connected to positive supply through a 10-K resistor, sketch the input and resulting output waveforms for a 311 op-amp connected as in Problem 3.

5. Describe the operation of the circuit of Fig. 11.6 with input of 10 V rms applied.

6. Draw the circuit diagram of a zero-crossing detector using a 339 comparator stage with ±12-V supplies.

7. Sketch the voltage waveforms with 10-V rms input applied to the minus input and plus input grounded.

8. Describe the operation of a window detector circuit as in Fig. 11.10 for resistor values of 7.5 K and 8.2 K changed to 6.2 K.

§ 11.3

9. Sketch a three-input ladder network using 15-K and 30-K resistor values.

10. For a reference voltage of 16 V, calculate the output of the network in Problem 9 with an input of binary 110.

11. What voltage resolution is possible using a 12-stage ladder network with a 10-V reference voltage?

12. Describe what occurs during the fixed time interval and during the count interval of a dual-slope conversion.

13. How many count steps occur using a 12-stage digital counter as the output of an A/D converter?

14. What is the maximum count interval resulting using a 12-stage counter operated at a clock rate of 2 MHz?

§ 11.4

15. Describe the signal conditions for current loop and RS-232-C interfaces.

16. What is meant by a data bus?

17. What is the difference between open-collector and tri-state outputs?

§ 11.5

18. Sketch a 555 timer connected as an astable multivibrator for operation at 100 kHz. Determine the value of capacitor C needed if $R_A = R_B = 7.5$ K.

19. Draw a 555 timer used as a one-shot using $R_A = 7.5$ K for a time period of 25 μs. Determine the value of capacitor C needed.

20. Sketch the input and output waveforms for a one-shot as in Problem 19 for input triggered by a 10-kHz clock.

GLOSSARY

Analog-to-Digital (A/D) Converter A circuit that converts analog or linear signal inputs into an equivalent digital value.

Comparator A circuit that provides for comparing an input linear voltage to a reference level providing a digital output as an indication that the input is above or below the reference level.

Current Loop A serial data transmission path represented by current in a closed path (loop) or no current.

Digital-to-Analog (D/A) Converter A circuit used to convert a digital value into an analog or linear signal of proportional or equivalent value.

Driver Circuit A circuit used to feed a signal to a load.

Interface An electronic circuit used to connect various electronic units.

Ladder Network A circuit made of resistors to develop a voltage proportional to logic inputs.

Opto-Isolator An electronic device using a light path to couple a digital signal while providing direct connection isolation.

Quad Comparator A single IC unit containing four comparator circuits.

Receiver Circuit A circuit that accepts input of a signal.

RS-232-C The digital interface signal standard widely used by the electronics industry.

Saturation Level The maximum voltage level that the output can reach.

Slope Conversion A technique for A/D conversion using a linear ramp signal to compare to the input to be converted.

Staircase Signal A linear signal that rises an incremental amount or step as the digital count advances during A/D conversion.

Strobe A signal used to gate another signal at a desired time.

Timer An IC circuit used to build astable or monostable circuits.

Tri-State Buffer An interface logic circuit providing output either 0 V, +5 V, or open-circuit.

Zero-Crossing Detector A circuit providing an output that changes logic state whenever the input crosses zero volts.

APPENDIX A

Hybrid Parameters— Conversion Equations (Exact and Approximate)

A.1 EXACT

Common-Emitter Configuration

$$h_{ie} = \frac{h_{ib}}{(1+h_{fb})(1-h_{rb})+h_{ob}h_{ib}} = h_{ic}$$

$$h_{re} = \frac{h_{ib}h_{ob} - h_{rb}(1+h_{fb})}{(1+h_{fb})(1-h_{rb})+h_{ob}h_{ib}} = 1 - h_{rc}$$

$$h_{fe} = \frac{-h_{fb}(1-h_{rb}) - h_{ob}h_{ib}}{(1+h_{fb})(1-h_{rb})+h_{ob}h_{ib}} = -(1+h_{fc})$$

$$h_{oe} = \frac{h_{ob}}{(1+h_{fb})(1-h_{rb})+h_{ob}h_{ib}} = h_{oc}$$

Common-Base Configuration

$$h_{ib} = \frac{h_{ie}}{(1+h_{fe})(1-h_{re})+h_{ie}h_{oe}} = \frac{h_{ic}}{h_{ic}h_{oc} - h_{fc}h_{rc}}$$

$$h_{rb} = \frac{h_{ie}h_{oe} - h_{re}(1+h_{fe})}{(1+h_{fe})(1-h_{re})+h_{ie}h_{oe}} = \frac{h_{fc}(1-h_{rc})+h_{ic}h_{oc}}{h_{ic}h_{oc} - h_{fc}h_{rc}}$$

$$h_{fb} = \frac{-h_{fe}(1-h_{re}) - h_{ie}h_{oe}}{(1+h_{fe})(1-h_{re}) + h_{ie}h_{oe}} = \frac{h_{rc}(1+h_{fc}) - h_{ic}h_{oc}}{h_{ic}h_{oc} - h_{fc}h_{rc}}$$

$$h_{ob} = \frac{h_{oe}}{(1+h_{fe})(1-h_{re}) + h_{ie}h_{oe}} = \frac{h_{oc}}{h_{ic}h_{oc} - h_{fc}h_{rc}}$$

Common-Collector Configuration

$$h_{ic} = \frac{h_{ib}}{(1+h_{fb})(1-h_{rb}) + h_{ob}h_{ib}} = h_{ie}$$

$$h_{rc} = \frac{1+h_{fb}}{(1+h_{fb})(1-h_{rb}) + h_{ob}h_{ib}} = 1 - h_{re}$$

$$h_{fc} = \frac{h_{rb} - 1}{(1+h_{fb})(1-h_{rb}) + h_{ob}h_{ib}} = -(1+h_{fe})$$

$$h_{oc} = \frac{h_{ob}}{(1+h_{fb})(1-h_{rb}) + h_{ob}h_{ib}} = h_{oe}$$

A.2 APPROXIMATE

Common-Emitter Configuration

$$h_{ie} \cong \frac{h_{ib}}{1+h_{fb}} \cong \beta r_e$$

$$h_{re} \cong \frac{h_{ib}h_{ob}}{1+h_{fb}} - h_{rb}$$

$$h_{fe} \cong \frac{-h_{fb}}{1+h_{fb}} \cong \beta$$

$$h_{oe} \cong \frac{h_{ob}}{1+h_{fb}}$$

Common-Base Configuration

$$h_{ib} \cong \frac{h_{ie}}{1+h_{fe}} \cong \frac{-h_{ic}}{h_{fc}} \cong r_e$$

$$h_{rb} \cong \frac{h_{ie}h_{oe}}{1+h_{fe}} - h_{re} \cong h_{rc} - 1 - \frac{h_{ic}h_{oc}}{h_{fc}}$$

$$h_{fb} \cong \frac{-h_{fe}}{1+h_{fe}} \cong \frac{-(1+h_{fc})}{h_{fc}} \cong -\alpha$$

$$h_{ob} \cong \frac{h_{oe}}{1+h_{fe}} \qquad \cong \frac{-h_{oc}}{h_{fc}}$$

Common-Collector Configuration

$$h_{ic} \cong \frac{h_{ib}}{1+h_{fb}} \cong \beta r_e$$

$$h_{rc} \cong 1$$

$$h_{fc} \cong \frac{-1}{1+h_{fb}} \cong -\beta$$

$$h_{oc} \cong \frac{h_{ob}}{1+h_{fb}}$$

APPENDIX B

Charts and Tables

TABLE B.1 Greek Alphabet and Common Designations

Name	Capital	Lowercase	Used to Designate
alpha	A	α	Angles, area, coefficients
beta	B	β	Angles, flux density, coefficients
gamma	Γ	γ	Conductivity, specific gravity
delta	Δ	δ	Variation, density
epsilon	E	ϵ	Base of natural logarithms
zeta	Z	ζ	Impedance, coefficients, coordinates
eta	H	η	Hysteresis coefficient, efficiency
theta	Θ	θ	Temperature, phase angle
iota	I	ι	
kappa	K	κ	Dielectric constant, susceptibility
lambda	Λ	λ	Wave length
mu	M	μ	Micro, amplification factor, permeability
nu	N	ν	Reluctivity
xi	Ξ	ξ	
omicron	O	o	
pi	Π	π	Ratio of circumference to diameter = 3.1416
rho	P	ρ	Resistivity
sigma	Σ	σ	Sign of summation
tau	T	τ	Time constant, time phase displacement
upsilon	Υ	υ	
phi	Φ	ϕ	Magnetic flux, angles
chi	X	χ	
psi	Ψ	ψ	Dielectric flux, phase difference
omega	Ω	ω	Capital: ohms; lower case: angular velocity

Logarithms

Formulas: $\log ab = \log a + \log b$

$\log \dfrac{a}{b} = \log a - \log b$

$\log a^n = n \log a$

TABLE B.2 Common Logarithms

No.	0	1	2	3	4	5	6	7	8	9
0	....	0000	3010	4771	6021	6990	7782	8451	9031	9542
1	0000	0414	0792	1139	1461	1761	2041	2304	2553	2788
2	3010	3222	3424	3617	3802	3979	4150	4314	4472	4624
3	4771	4914	5051	5185	5315	5441	5563	5682	5798	5911
4	6021	6128	6232	6335	6435	6532	6628	6721	6812	6902
5	6990	7076	7160	7243	7324	7404	7482	7559	7634	7709
6	7782	7853	7924	7993	8062	8129	8195	8261	8325	8388
7	8451	8513	8573	8633	8692	8751	8808	8865	8921	8976
8	9031	9085	9138	9191	9243	9294	9345	9395	9445	9494
9	9542	9590	9638	9685	9731	9777	9823	9868	9912	9956
10	0000	0043	0086	0128	0170	0212	0253	0294	0334	0374
11	0414	0453	0492	0531	0569	0607	0645	0682	0719	0755
12	0792	0828	0864	0899	0934	0969	1004	1038	1072	1106
13	1139	1173	1206	1239	1271	1303	1335	1367	1399	1430
14	1461	1492	1523	1553	1584	1614	1644	1673	1703	1732
15	1761	1790	1818	1847	1875	1903	1931	1959	1987	2014
16	2041	2068	2095	2122	2148	2175	2201	2227	2253	2279
17	2304	2330	2355	2380	2405	2430	2455	2480	2504	2529
18	2553	2577	2601	2625	2648	2672	2695	2718	2742	2765
19	2788	2810	2833	2856	2878	2900	2923	2945	2967	2989
20	3010	3032	3054	3075	3096	3118	3139	3160	3181	3201
21	3222	3243	3263	3284	3304	3324	3345	3365	3385	3404
22	3424	3444	3464	3483	3502	3522	3541	3560	3579	3598
23	3617	3636	3655	3674	3692	3711	3729	3747	3766	3784
24	3802	3820	3838	3856	3874	3892	3909	3927	3945	3962
25	3979	3997	4014	4031	4048	4065	4082	4099	4416	4133
26	4150	4166	4183	4200	4216	4232	4249	4265	4281	4298
27	4314	4330	4346	4362	4378	4393	4409	4425	4440	4456
28	4472	4487	4502	4518	4533	4548	4564	4579	4594	4609
29	4624	4639	4654	4669	4683	4698	4713	4728	4742	4757
30	4771	4786	4800	4814	4829	4843	4857	4871	4886	4900
31	4914	4928	4942	4955	4969	4983	4997	5011	5024	5038
32	5051	5065	5079	5092	5105	5119	5132	5145	5159	5172
33	5185	5198	5211	5224	5237	5250	5263	5276	5289	5302
34	5315	5328	5340	5353	5366	5378	5391	5403	5416	5428
35	5441	5453	5465	5478	5490	5502	5514	5527	5539	5551
36	5563	5575	5587	5599	5611	5623	5635	5647	5658	5670
37	5682	5694	5705	5717	5729	5740	5752	5763	5775	5786
38	5798	5809	5821	5832	5843	5855	5866	5877	5888	5899
39	5911	5922	5933	5944	5955	5966	5977	5988	5999	6010
No.	0	1	2	3	4	5	6	7	8	9

TABLE B.2 (Continued)

No.	0	1	2	3	4	5	6	7	8	9
40	6021	6031	6042	6053	6064	6075	6085	6096	6107	6117
41	6128	6138	6149	6160	6170	6180	6191	6201	6212	6222
42	6232	6243	6253	6263	6274	6284	6294	6304	6314	6325
43	6335	6345	6355	6365	6375	6385	6395	6405	6415	6425
44	6435	6444	6454	6464	6474	6494	6493	6503	6513	6522
45	6532	6542	6551	6561	6571	6580	6590	6599	6609	6618
46	6628	6637	6646	6656	6665	6675	6684	6693	6702	6712
47	6721	6730	6739	6749	6758	6767	6776	6785	6794	6803
48	6812	6821	6830	6839	6848	6857	6866	6875	6884	6893
49	6902	6911	6920	6928	6937	6946	6955	6964	6972	6981
50	6990	6998	7007	7016	7024	7033	7042	7050	7059	7067
51	7076	7084	7093	7101	7110	7118	7126	7135	7143	7152
52	7160	7168	7177	7185	7193	7202	7210	7218	7226	7235
53	7243	7251	7259	7267	7275	7284	7292	7300	7308	7316
54	7324	7332	7340	7348	7356	7364	7372	7380	7388	7396
55	7404	7412	7419	7427	7435	7443	7451	7459	7466	7474
56	7482	7490	7497	7505	7513	7520	7528	7536	7543	7551
57	7559	7566	7574	7582	7589	7597	7604	7612	7619	7627
58	7634	7642	7649	7657	7664	7672	7679	7686	7694	7701
59	7709	7716	7723	7731	7738	7745	7752	7760	7767	7774
60	7782	7789	7796	7803	7810	7818	7825	7832	7839	7846
61	7853	7860	7868	7875	7882	7889	7895	7903	7910	7917
62	7924	7931	7938	7945	7952	7959	7966	7973	7980	7987
63	7993	8000	8007	8014	8021	8028	8035	8041	8048	8055
64	8062	8069	8075	8082	8089	8096	8102	8109	8116	8122
65	8129	8136	8142	8149	8156	8162	8169	8176	8182	8189
66	8195	8202	8209	8215	8222	8228	8235	8241	8248	8254
67	8261	8267	8274	8280	8287	8293	8299	8306	8312	8319
68	8325	8331	8338	8344	8351	8357	8363	8370	8376	8382
69	8388	8395	8401	8407	8414	8420	8426	8432	8439	8445
70	8451	8457	8463	8470	8476	8482	8488	8494	8500	8506
71	8513	8519	8525	8531	8537	8543	8549	8555	8561	8567
72	8573	8579	8585	8591	8597	8603	8609	8615	8621	8627
73	8633	8639	8645	8651	8657	8663	8669	8675	8681	8686
74	8692	8698	8704	8710	8716	8722	8727	8733	8739	8745
75	8751	8756	8762	8768	8774	8779	8785	8791	8797	8802
76	8808	8814	8820	8825	8831	8837	8842	8848	8854	8859
77	8865	8871	8876	8882	8887	8893	8899	8904	8910	8915
78	8921	8927	8932	8938	8943	8949	8954	8960	8965	8971
79	8976	8982	8987	8993	8998	9004	9009	9015	9020	9025
80	9031	9036	9042	9047	9053	9058	9063	9069	9074	9079
81	9085	9090	9096	9101	9106	9112	9117	9122	9128	9133
82	9138	9143	9149	9154	9159	9165	9170	9175	9180	9186
83	9191	9196	9201	9206	9212	9217	9222	9227	9232	9238
84	9243	9248	9253	9258	9263	9269	9274	9279	9284	9289
No.	0	1	2	3	4	5	6	7	8	9

TABLE B.2 (Continued)

No.	0	1	2	3	4	5	6	7	8	9
85	9294	9299	9304	9309	9315	9320	9235	9330	9335	9340
86	9345	9350	9355	9360	9365	9370	9375	9380	9385	9390
87	9395	9400	9405	9410	9415	9420	9425	9430	9435	9440
88	9445	9450	9455	9460	9465	9469	9474	9479	9484	9489
89	9494	9499	9504	9509	9513	9518	9523	9528	9533	9538
90	9542	9547	9552	9557	9562	9566	9571	9576	9581	9586
91	9590	9595	9600	9605	9609	9614	9619	9624	9628	9633
92	9638	9643	9647	9652	9657	9661	9666	9671	9675	9680
93	9685	9689	9694	9699	9703	9708	9713	9717	9722	9727
94	9731	9736	9741	9745	9750	9754	9759	9763	9768	9773
95	9777	9782	9786	9791	9795	9800	9805	9809	9814	9818
96	9823	9827	9832	9836	9841	9845	9850	9854	9859	9863
97	9868	9872	9877	9881	9886	9890	9894	9899	9903	9908
98	9912	9917	9921	9926	9930	9934	9939	9943	9948	9952
99	9956	9961	9965	9969	9974	9978	9983	9987	9991	9996
100	0000	0004	0009	0013	0017	0022	0026	0030	0035	0039
No.	0	1	2	3	4	5	6	7	8	9

Answers To Selected Odd-Numbered Problems

CHAPTER 1

7. 6.40×10^{-19} C.

CHAPTER 2

5. No positive excursions, negative sine wave section having a peak value of -10 V; no positive excursions, negative square wave section of -15 V. **7.** Same results as Problem 5. **11.** 13.73 mA. **13.** 12 μA. **15.** (a) At -10 V, $C_T \cong 1.0$ pF; at -25 V, $C_T \cong 0.05$ pF; 0.033 pF/V; (b) at -10 V, $C_T \cong 1.0$ pF; at -1 V, $C_T \cong 2.25$ pF; 1.39 pF/V. **21.** 22.22 W. **23.** (a) $\theta_{JC} = 5.33°$C/W; (b) $\theta_{JA} = 8.33°$C/W. **25.** (a) 1.667 W; (b) $D_F = 0.067$ W/°C, $\theta_{JC} = 15.0°$C/W; (c) 115.0°C; (d) $\theta_{CS} = 2.0°$C/W, $\theta_{SA} = 8°$C/W, $\theta_{CA} = 10°$C/W. **27.** 100 mW. **31.** 26.54 K. **33.** $I_{peak} = 628.93$ mA. **35.** (a) 2.27 mV/°C; (b) 45.4 mV. **37.** $\cong 450$ Ω (log scale). **39.** 240 mW. **41.** 320 mA, 1 V. **47.** $\cong 101$ Ω. **49.** $R_{dc} = 170$ Ω, $r_d = 30$ Ω. **51.** $r_d = 5$ Ω (graphically), $r_d = 4.6$ Ω.

CHAPTER 3

3. 0.053%/°C. **5.** 13.0 Ω. **7.** graph: 175°C, calculated; 195.12°C. **11.** $R_S = 100$ Ω. **15.** $\cong 45°$C, low levels. **17.** (0 $\rightarrow$ 2 V) 35%, (8 $\rightarrow$ 16 V) 6%, $\cong$ 6 : 1. **19.** (a) $\cong 27$ pF; (b) at -8 V, $\cong 2.14$ pF/V; at -2 V,

417

$\cong 10.33$ pF/V; $\cong 5:1$. **21.** 6.59. **27.** 1 MHz, 31.85 K; 100 MHz, 318 Ω; 100 MHz has most effect; 1 MHz, $X_L = 0.0377$ Ω; 100 MHz, $X_L = 3.77$ Ω,; neither of sufficient magnitude. **29.** 3.97×10^{-19} J. **31.** 430 μA. **33.** 4.1 V. **39.** green to yellow. **41.** (a) 0.77. **45.** 2.25 V. **47.** (a) 40 mA; (b) 60 mA. **59.** 20 K. **61.** 90 Ω.

CHAPTER 4

9. 7.92 mA. **11.** 25. **13.** (a) $\cong 4.95$ mA; (b) $\cong 4.0$ mA; (c) 740 mA. **17.** (a) $\cong 118$; (b) 0.992; (c) 250 μA; (d) 2.12 μA. **23.** 0.972. **27.** (a) 3.4 mA; (b) 28 V; (c) 25 μA. **29.** (a) Free air: -4.57 mW/°C, Core temp.: -17.14 mW/°C; (b) 1.715 W; (c) 1.7 W. **31.** (a) 8 nA; (b) 1.6 μA; (c) 0.53 nA/°C.

CHAPTER 5

1. $I_E = 0.8$ mA, $I_C = .788$ mA, $V_{CB} = 8.927$ V. **3.** $I_B = 55.30$ μA, $I_C = 2.49$ mA, $V_{CE} = 3.77$ V. **5.** $I_B = 194$ μA, $I_C = 10.691$ mA, $V_{CE} = 4.506$ V. **7.** $I_E = 1.544$ mA $= I_C$, $V_{CE} = 12.340$ V. **9.** $I_B = 21.023$ μA, $I_C = 1.787$ mA, $V_{CE} = 5.783$ V. **11.** $V_E = 0.934$ V. **13.** $V_{CE} \cong 8.3$ V, $I_C \cong 5.7$ mA. **15.** (a) $h_{11} = 2$ Ω, $h_{12} = \frac{2}{3}$, $h_{21} = -\frac{2}{3}$, $h_{22} = 4.9$ S. **17.** greatest: h_{oe}, least: h_{fe}. **19.** greatest: h_{ie}, least: h_{oe}. **21.** (a) 80; (b) -180; (c) 0.8 K; (d) 1.8 K; (e) 14.4×10^3. **23.** (a) -0.988; (b) 270; (c) 13 Ω; (d) 3.6 K; (e) 266.76. **25.** $Z_i = 122.45$ K, $Z_o \cong 5.6$ K, $A_v = -4.67$. **27.** $Z_i = 62.86$ K, $Z_o \cong 18.8$ Ω, $A_v = 0.994$, $A_i = 19$. **29.** $r_e \cong 7$ Ω, $A_i = 60$, $A_v = -571.43$, $Z_i = 0.42$ K, $Z_o = 4$ K. **31.** $r_e = 15.04$ Ω, $Z_i = 122.45$ K, $Z_o = 5.6$ K, $A_v = -4.67$. **33.** $r_e = 19.01$ Ω, $Z_i = 62.86$ K, $Z_o = 18.8$ Ω, $A_v = 0.994$, $A_i = 19$. **35.** $r_e = 26$ Ω, $A_v = 14.43$, $A_i = -0.25$, $Z_i = 207.88$ Ω, $Z_o = 12$ K. **39.** $A_{i_T} = 1783.38$, $|A_v| = 4.4$, $Z_o = 2.2$ K, $Z_i = 0.89$ M. **41.** $Z_i = 0.44$ M, $A_{i_T} = 2952$, $Z_o = 2.2$ K, $|A_{v_T}| = 14.76$. **43.** (a) 42 W; (b) 0.57 W/°C. **45.** (a) through 0.1 MHz; (b) 140 kHz.

CHAPTER 6

3. $V_{GS} = -1.5$ V, $I_D = 4.7$ mA, $V_{DS} = 3.5$ V. **5.** $V_{DS_Q} = -9.6$ V, $I_{D_Q} = 1.6$ mA. **7.** $V_{GS} \cong +0.94$ V, $I_D \cong 3.9$ mA, $V_{DS_Q} = -4.2$ V. **9.** $V_{GS} = +0.88$ V, $I_D = 7.3$ mA, $V_{DS} = -2.65$ V. **11.** $V_{GS} = +1$ V, $I_D = 6.75$ ma, $V_{DS} = -4.2$ V. **13.** $V_{GS} = +1.8$ V, $I_D = 3$ mA, $V_{DS} = -7.9$ V. **15.** $A_v = -13.95$. **17.** $A_v = -5.4$. **19.** $A_v = -7.2$. **21.** $|V_o| = 127.5$ mV. **23.** $V_{DS} = V_{GS} = 6.4$ V, $I_D = 3.5$ mA. **25.** $|V_o| = 257$ mV.

CHAPTER 7

5. (a) yes; (b) no; (c) no; (d) $V_G = 6$ V, $I_G = 0.8$ A preferred. **11.** (a) 0.75 mW/cm². (b) 82.35%. **17.** 1.1 V, 1.8 V. **19.** 3.4 cycles. **21.** (b) 0.67 A/°C. **23.** (a) 1.077 K; (b) 3.077 K; (c) 13.0 V; (d) 13.70 V. **29.** 2 nA/°C, 6 nA. **31.** (b) 0.41. **33.** 0.75, 15 V.

CHAPTER 9

1. $CMR = 65.1$ dB. **3.** $V_o = 6.5$ V. **5.** $R_i = 16.6$ K. **7.** $V_o = -9.17$ V. **9.** $V_o = +7.5$ V. **15.** %V.R. $= 6.7$%.

CHAPTER 10

11. $f = 224.7$ kHz. **15.** Maximum count $= 99$.

CHAPTER 11

11. 2.4 mV. **13.** 4096 count steps.

Index

A

Ac analysis, transistor, 179–227
Acceptor atom, 11–12
Acceptors, 44
Alarm circuit, 277–78
Alloy-junction transistor, 151, 160
Alloy process, 43
Alpha (*see* transistors)
Ambient temperature, 32–37
AND gate, 358
Angstrom, 98, 126
Anode current interruption, 267, 312
Antimony, 9–10
Arrays, diode, 46–48
Arsenic, 9
Astable multivibrator, 374
Avalanche breakdown, 29
Average ac resistance, 55–61
Average rectified current, 39–41
Axial luminous intensity, 105, 126

B

Base diffusion, 316, 328
Battery charging regulator, 273–74
Beta (*see* transistors)
BiFET, 342, 349
BiMOS, 342, 350
Bipolar junction transistor, 159
Bipolar O-amps, 347
Bistable multivibrator circuit, 368
Bohr model, 6
Boron, 10–12
Bulk resistance, 4, 18, 27, 54–55

C

Capacitance:
   diffusion, 30–31
   transition, 30–31

Capacitance temperature coefficient, 91–93
Channel, 238
Chip, 316
Clampers, 63, 65–67
Clippers, 62–64
CMOS logic, 365
Common-base configuration (see transistors)
Common-collector configuration (see transistors)
Common-emitter configuration (see transistors)
Common-mode gain, 336
Common-mode operation, 338
Common-mode rejection, 337, 340
Comparator, 387
Conductor, 4–5, 7–8
Constant gain amplifier, 345
Contact resistance, 27, 54–55
Conventional flow, 21, 22
Counter, 378
Covalent bonding, 6, 18
Crystal lattice, 5, 13, 18
Current:
  average rectified, 39–41
  diffusion, 12, 13, 19
  drift, 12, 13, 19
  peak forward surge current, 39–41
  peak repetitive forward current, 39–41
Current loop, 402
Czochralski technique, 13–15, 19, 152

D

Dark current, 99–100
Darlington compound configuration, 217–27
Dc analysis transistor, 161–78, 204–16
Dc bias JFET, 245–52
Decoder, 379
Depletion MOSFET, 252–54, 256–58
Depletion region, 23–25, 240–42
Diac:
  applications, 285–88
  characteristics, 285
  operation, 284–85
Difference-mode gain, 336

Differential amplifier, 336
Differential operation, 338
Diffusion capacitance, 30–31
Diffusion current, 12, 13, 19
Diffusion process, 43–44, 152–53, 160, 320–22
Diffusion transistor, 151–53, 160
Digital circuits, 358
Digital memory unit, 381
Digital-to-analog conversion, 395
Diodes, 20–67
  alloy, 43
  array, 46–48
  average ac resistance, 55–61
  average rectified current, 39–41
  clampers, 62, 65–67
  clippers, 62–64
  dc conditions, 49–52, 58–67
  depletion region, 23–25
  diffusion, 43–44
  diffusion capacitance, 30–31
  dynamic resistance, 52–55
  epitaxial growth, 44–45
  equivalent circuit, 56–61
  fabrication, 42–48
  forward-bias, 25–32
  grown junction, 42–43
  heat sinks, 32–37
  ideal, 20–23
  light emitting, 105–10
  load line, 50–52
  minority carriers, 32
  notation, 41–42
  ohmmeter check, 42
  peak forward surge current, 39–41
  peak inverse voltage (PIV), 30
  peak repetitive forward current, 39–41
  photo, 97–101
  point contact, 45
  power, 93–94, 126
  power derating curve, 36–37
  reverse-bias, 25, 28–32
  reverse recovery time, 31–32
  reverse saturation current, 25–30
  Schottky barrier, 84–88, 126
  specification sheet, 37–41
  static resistance, 51–52
  temperature effects, 32–40
  transition capacitance, 30–31

Diodes (*cont.*)
   tunnel, 94–97, 126
   varicaps, 88–93, 126
   Zener, 29, 76–84, 126
   Zener region, 28–30
Discrete elements, 332
Dopants, 315
Doping, 5, 9–12
Drain, 237
Drain characteristic, 242–44
Drift current, 12, 13, 19
Duty cycle, 406
Dynamic memory, 383
Dynamic resistance, 52–55

E

Electroluminescence, 105
Electron, 5, 9–12, 18
Electron flow, 11, 22
Electron volt, 8, 18
Emergency lighting system, 275
Emitter-coupled logic (ECL), 367
Emitter follower, 173–76, 199–203
Energy levels, 7–8
Enhancement MOSFET, 252, 254, 258–60
Epitaxial mesa transistor, 152–53, 160
Epitaxial region, 325
EPROM, 384
Equivalent circuits, 56–61
   diodes, 56–61
Extensive materials, 9–12, 18

F

Fabrication transistors, 151–53
Field effect transistor (FET), 237–63
Flip-flop, 358
Floating zone technique, 13–15, 19
Foot-candles, 98–101
Forbidden region, 8, 18
Forced commutation technique, 267, 312
Forward-bias, 25–32
Forward current amplification factor (*see* transistors)

Forward transfer current ratio parameter, 184–227

G

Gallium, 10
Gate, 237
Germanium, 4–16, 18, 32
   acceptor atom, 11, 12
   Bohr model, 6
   covalent bonding, 6, 18
   Czochralski technique, 13–15, 19
   donor atoms, 9–12
   floating zone technique, 13–15, 19
   hole, 10–12
   intrinsic state, 6, 18
   majority carrier, 11, 12, 19
   manufacturing techniques, 13–16
   resistivity, 4, 18
   zone refining, 13–14
Graphical analysis transistor, 176–78
Grown-junction transistor, 151–52, 160
GTO:
   application, 280
   construction, 279
   operation, 279–80
   terminal identification, 279

H

Half-wave variable resistance phase control, 272–73
Heat sinks, 32–37
Hole, 10–12
Hole flow, 11, 19
Hot-carrier diode (*see* Schottky barrier diode)
Hybrid equivalent circuit, 182–227
Hybrid integrated circuit, 333

I

IC fabrication, 315–35
Ideal diode, 20–23

I_{DSS}, 240-52
Impedance matching, 142-43
Indium, 10
Input impedance parameter, 183-227
Insulator, 4-5, 7-8, 35
Integrated circuits, 46-48
Integrated injection logic (IIL), 366
Integrator, 345
Interfacing, 399
Intrinsic materials, 6, 18
Intrinsic standoff ratio, 292-94, 303-5, 312
Inverter, 362
Ionization potential, 6, 18
IR emitters, 103-5, 126
Isolation diffusion, 317, 327

J

JFET, 237-52
JK flip-flop, 370
Junction temperature, 35-39

L

Ladder network, 395-96
LASCR:
   applications, 283
   construction, 280-81
   operation, 280-83
   terminal identification, 281
Latching relay, 283
Lattice structure, 5, 18
LCDs (liquid crystal displays), 110-13
   dynamic scattering, 110-13
   field-effect, 112-13, 127
   nematic, 110-13, 126
   operation, 110-13
   transmissive, 112-13
   twisted nematic, 112-13
Leakage current, 131-32, 134-50, 159
LEDs (light emitting diodes), 105-10, 126, 299, 379
   characteristics, 105-10

LEDs (*cont.*)
   operation, 105
Linear IC, 336
Load line diodes, 50-52
Logic circuitry, 283
Lumen, 98, 126
Luminous efficacy, 105, 126
Luminous flux, 98-101

M

Majority carriers, 11-12, 19, 24-26, 130-50
Mash, 317, 322
Metal gate, 254
Metalization, 329
Minority carriers, 11, 12, 19, 24-25, 32, 130-50
Monolithic circuit elements, 318-22
Monolithic integrated circuit, 323
Monostable multivibrator, 359, 372
MOSFET, 237, 252-60

N

NAND gate, 358, 363
Negative logic, 361
Negative resistance, 293, 312
Negative temperature coefficient, 7, 18
Neutron, 5, 18
n-type materials, 9-12, 19, 23-26, 42-45, 129-253

O

Ohmic contact resistance, 4, 18
Op-amp, 341-51
Opto-electronics, 97-118
Opto-isolators, 403
   characteristics, 301-2
   operation, 300-2

Opto-isolators (*cont.*)
   schematic representation, 302
OR gate, 358
Output conductance parameter, 184-227
Oxide layer, 254

P

Packaging, 331
Parallel data register, 377
Peak forward surge-current, 39-41
Peak inverse voltage (PIV), 30
Peak repetitive forward current, 39-41
Phase (power) control, 288
Phosphorus, 9
Photo conductive cell, 101-3, 126
   application, 102-3
   characteristics, 101-2
   symbol, 101
Photo diodes, 97-101, 126
   characteristics, 99-100
   construction, 98-100
   spectral response, 98-101
Photolithographic process, 325
Photon, 98, 126
Photoresist, 325
Phototransistors:
   application, 298-99
   characteristics, 298-99
   operation, 297-98
   terminal identification, 299
Pinch off, 240, 244
Pinch-off voltage, V_P, 242
Planar process, 316
Planar transistor, 152-53, 160
Planck's constant, 98
PNPN devices:
   diac, 284-86
   GTO, 279-80, 312
   LASCR, 280-83, 312
   Shockley diode, 284, 312
   SCS, 275-78, 312
   SCR, 264-75, 312
   triac, 287-90
Point-contact transistor, 151, 160
Positive logic, 361
Positive temperature coefficient, 7, 18

Power derating curve, 36-37
Power diodes, 93-94, 126
Preohmic etch, 329
Prom, 384
Protone, 9-10, 18
Proton, 5, 18
Proximity detector, 285-87
PUT:
   applications, 285-86, 305-8
   characteristics, 303-4
   construction, 303
   operation, 303-5

Q

Quad comparator, 391
Quiescent point transistors, 161-78

R

Rectification, 22, 23
Reference voltage, 81-84
Relaxation oscillator, 296, 305-8
RESET, 368
Resistance:
   average ac, 55-61
   bulk, 4, 18, 27, 54-55
   contact, 27, 54-55
   dynamic, 52-55
   insulator, 35
   ohmic contact, 4, 18
   resistivity, 4, 18
   static resistance, 51-52
Resistivity, 4, 18
Reverse-bias, 25, 28-32
Reverse recovery time, 31-32
   Schottky barrier diode, 87
Reverse saturation current, 25-30, 98
Reverse transfer voltage ratio parameter, 183-227
ROM, 384
RS flip-flop, 369
RS-232-C, 402

S

Saturation region (*see* transistors)
Sawtooth generator, 280
Schmitt trigger, 359, 376
Schottky barrier diode, 84–88, 126
   construction, 84–86
   equivalent circuit, 87–88
SCR:
   anode current interruption, 267
   applications, 272–75, 277–78, 284, 295
   construction, 265–71, 275–76
   forced commutation technique, 267
   operation, 264–70, 275–77
   symbol, 265, 276
   terminal identification, 271, 278
Self-bias JFET, 248–52
Semiconductor materials, 3–16, 18, 129–33
   acceptor atom, 11, 12
   Czochralski technique, 13–15, 19
   donor atoms, 9–12
   energy levels, 7–8
   extrinsic materials, 9–12, 18
   floating zone technique, 13–15, 19
   germanium, 4–16, 18
   hole, 10–12
   majority carrier, 11, 12, 19, 24–26
   manufacturing techniques, 13–16
   minority carrier, 11, 12, 19
   negative temperature coefficient, 7, 18
   n-type, 9–12, 19, 23–26, 42–45
   p-type, 9–12, 19, 23–26, 42–45
   silicon, 4–16, 18
   temperature effects, 28, 30, 32–40
   wafer preparation, 15, 16
   zone refining, 13–14
Serial data register, 377
Series static switch, 272
SET, 368
Shockley diode:
   application, 284
   characteristics, 284
   construction, 284
   operation, 284
Silicon, 32
   acceptor atom, 11, 12
   Bohr model, 6
   Czochralski technique, 13–15, 19
   donor atoms, 9–12

Silicon (*cont.*)
   floating zone technique, 13–15, 19
   hole, 10–12
   majority carrier, 11, 12, 19
   manufacturing techniques, 13–16
   resistivity, 4, 18
   zone refining, 13–14
Silicon oxidation, 325
Single-ended operation, 337
Slew rate, 350
Solar cells, 114–18, 127
   characteristics, 115–18
   construction, 114–15
   efficiency, 117–18
Source, 237
Specification sheet diodes, 37–41
Square wave generator, 82–84
Static memory, 383
Static resistance, 51–52
Strobe, 390
Summing amplifier, 343
Superposition theorem, 60–61
Surface barrier, 84–85

T

Teletype, 402
Temperature coefficient, 78–80
Temperature controller, 274–75
Temperature effects:
   ambient temperature, 32–37
   diodes, 28, 30, 32–40
   junction temperature, 35–39
   power derating curve, 36–37
   thermal conductivity, 34
   thermal resistance, 34
Terminal identification transistors, 153–55
Tetravalent, 6, 18
Theorems superposition, 60–61
Thermal conductivity, 34
Thermal resistance, 34
Thermistors, 119–21, 127
   application, 120–21
   characteristics, 119
   operation, 119

Thick film, 332-33
Thin film, 332-33
Time constant, 65-67
Timer, 404
Timer (555), 406
Transconductance, 246
Transfer characteristic, 242-43
Transistors, 128-56, 159-60
  ac analysis, 179-227
  amplifying action, 132-33
  casing, 153-55
  common-base configuration, 132-36, 159, 163-66, 178, 194-96
  common-collector configuration, 142-44, 159, 173-76, 178
  common-emitter configuration, 136-42, 159, 166-73, 176-78, 189-93, 196-98
  construction, 129-30
  cut-off region, 136-39, 144, 159
  Darlington compound configuration, 217-27
  dc analysis, 161-78, 204-16
  fabrication, 151-53
  forward current amplification factor (beta), 140-42, 159
  graphical analysis, 176-78
  hybrid equivalent circuit, 182-227
  hybrid parameter variations, 187-89
  leakage current, 131-32, 134-50, 159
  maximum ratings, 144-50, 159
  operation, 130-50
  quiescent point, 161-78
  saturation region, 137-38, 144
  short-circuit amplification factor (alpha), 132-50, 159
  specification sheet, 145-50
  terminal identification, 153-55
  testing, 155-56
Transistor-transistor logic (TTL), 363
Transition capacitance, 30-31
Triac application, 288
  characteristics, 287-90
  operation, 287-88
  symbol, 287-88
  terminal identification, 287
Tri-state buffer, 404
Truth table, 360
T-type flip-flop, 370
Tubes, vacuum, 128

Tunnel diodes, 94-97, 126
  construction, 94-97
  equivalent circuit, 95-96
  symbols, 95
Two-port theory, 182-86

U

UJT (unijunction transistor):
  applications, 295-96
  characteristics, 293-94
  construction, 291
  operation, 291-96
  triggering, 295-96
Unity follower, 344

V

Vacuum tubes, 128
Valence electrons, 6-12, 18
Varicap diodes, 88-93, 126
  capacitance temperature coefficient, 91-93
  characteristics, 88-93
  equivalent circuit, 91
  symbols, 91
Velocity of light, 98
V-FET, 296-97, 312
Virtual ground, 342
Voltage regulator, 102-3, 351, 356

W

Wafer, 315-17, 325
Wavelength, 98, 126
Window, 316-17, 333

Z

Zener diodes, 29, 76–84, 126, 280
   applications, 81–84
   characteristics, 76–78
   power derating curve, 80

Zener diodes (*cont.*)
   symbol, 77, 80
   temperature coefficient, 78–80
Zener region, 28–30
Zero-crossing detector, 390
Zone refining, 13–14